Differential-Algebraic Equations Forum

Editors-in-Chief

Achim Ilchmann (TU Ilmenau, Ilmenau, Germany)
Timo Reis (University of Hamburg, Hamburg, Germany)

Editorial Board

Larry Biegler (Carnegie Mellon University, Pittsburgh, USA)
Steve Campbell (North Carolina State University, Raleigh, USA)
Claus Führer (Lunds Universitet, Lund, Sweden)
Roswitha März (Humboldt Universität zu Berlin, Berlin, Germany)
Stephan Trenn (TU Kaiserslautern, Kaiserslautern, Germany)
Peter Kunkel (Universität Leipzig, Leipzig, Germany)
Ricardo Riaza (Universidad Politécnica de Madrid, Madrid, Spain)
Vu Hoang Linh (Vietnam National University, Hanoi, Vietnam)
Matthias Gerdts (Universität der Bundeswehr München, Munich, Germany)
Sebastian Sager (Otto-von-Guericke-Universität Magdeburg, Magdeburg, Germany)
Sebastian Schöps (TU Darmstadt, Darmstadt, Germany)
Bernd Simeon (TU Kaiserslautern, Kaiserslautern, Germany)
Eva Zerz (RWTH Aachen, Aachen, Germany)

Differential-Algebraic Equations Forum

The series "Differential-Algebraic Equations Forum" is concerned with analytical, algebraic, control theoretic and numerical aspects of differential algebraic equations (DAEs) as well as their applications in science and engineering. It is aimed to contain survey and mathematically rigorous articles, research monographs and textbooks. Proposals are assigned to an Associate Editor, who recommends publication on the basis of a detailed and careful evaluation by at least two referees. The appraisals will be based on the substance and quality of the exposition.

For further volumes:
www.springer.com/series/11221

Bernd Simeon

Computational Flexible Multibody Dynamics

A Differential-Algebraic Approach

Bernd Simeon
Felix-Klein-Zentrum für Mathematik
Technische Universität Kaiserslautern
Kaiserslautern, Germany

ISBN 978-3-642-35157-0   ISBN 978-3-642-35158-7 (eBook)
DOI 10.1007/978-3-642-35158-7
Springer Heidelberg New York Dordrecht London

Library of Congress Control Number: 2013942394

Mathematics Subject Classification (2010): 74S05, 65M12, 65L80

© Springer-Verlag Berlin Heidelberg 2013
This work is subject to copyright. All rights are reserved by the Publisher, whether the whole or part of the material is concerned, specifically the rights of translation, reprinting, reuse of illustrations, recitation, broadcasting, reproduction on microfilms or in any other physical way, and transmission or information storage and retrieval, electronic adaptation, computer software, or by similar or dissimilar methodology now known or hereafter developed. Exempted from this legal reservation are brief excerpts in connection with reviews or scholarly analysis or material supplied specifically for the purpose of being entered and executed on a computer system, for exclusive use by the purchaser of the work. Duplication of this publication or parts thereof is permitted only under the provisions of the Copyright Law of the Publisher's location, in its current version, and permission for use must always be obtained from Springer. Permissions for use may be obtained through RightsLink at the Copyright Clearance Center. Violations are liable to prosecution under the respective Copyright Law.
The use of general descriptive names, registered names, trademarks, service marks, etc. in this publication does not imply, even in the absence of a specific statement, that such names are exempt from the relevant protective laws and regulations and therefore free for general use.
While the advice and information in this book are believed to be true and accurate at the date of publication, neither the authors nor the editors nor the publisher can accept any legal responsibility for any errors or omissions that may be made. The publisher makes no warranty, express or implied, with respect to the material contained herein.

Printed on acid-free paper

Springer is part of Springer Science+Business Media (www.springer.com)

To Shery.

توانا بُوَد، هَرکه دانا بُوَد

Adorn thy mind with knowledge, for knowledge maketh thy worth.
Firdausi, Persian poet (940–1020 CE)

Preface

Flexible multibody dynamics is a rapidly growing field with various applications in vehicle analysis, aerospace engineering, robotics, and biomechanics. Overall, the mathematical models in this field seem reasonably settled and, in computational terms, great strides have been made over the last two decades as sophisticated software packages are nowadays capable of simulating highly complex structures with rigid and deformable components.

Written from the perspective of a numerical analyst, this monograph provides comprehensive information on both the mathematical framework as well as the numerical methods for flexible multibody dynamics. Such a presentation of the subject appears to be needed and should benefit not only graduate students and scientists working in the field but also those interested in time-dependent partial differential equations and heterogeneous problems with multiple time scales. At the same time, a number of open issues at the frontiers of research are addressed by taking a differential-algebraic approach and extending it to the notion of transient saddle point problems.

The results in this book are the product of research work undertaken during the past fifteen years. Parts of the material were also covered in graduate courses offered at the Universität Karlsruhe, the Technische Universität München, and the Technische Universität Kaiserslautern. Though it is self-contained in many respects, the text nonetheless calls for some basic knowledge of discretization schemes for ordinary differential equations and of the finite element method. Simulation examples illustrate the mathematical models and the computational techniques.

The interplay of modeling and numerics is a key feature in flexible multibody dynamics and mainly determines the methodology chosen in this monograph. This is reflected by the organization into two main parts with four chapters each. While the first part develops a detailed mathematical framework for systems of rigid and deformable bodies, the second introduces discretization methods in space and time and concentrates in particular on the problem of different time scales.

Though this outline seems to indicate a purely mathematical treatment of the subject, this book is definitely not intended for the exclusive use of specialists in numerical analysis. On the contrary, it should promote the communication between the engineer and the mathematician, thus leading to a truly interdisciplinary cooperation with mutual benefits.

The origin of this book goes back a long way. After I had completed the Ph.D., it was my advisor Peter Rentrop who first drew my attention to the field of flexible multibody dynamics. It is a pleasure to thank him for his continuing interest in the subject and for his support over the years. The next colleague and friend I would like to thank is Claus Führer, with whom I had various inspiring discussions and who carefully read an earlier version of the manuscript. Many other colleagues have also contributed by providing suggestions and valuable remarks on specific issues: Martin Arnold, Peter Betsch, Severiano Gonzalez Pinto, Ernst Hairer, Domingo Hernández Abreu, Laurent Jay, Christian Lubich, Panos Papadopoulos, Linda Petzold, Werner C. Rheinboldt, John Strain, and Barbara Wohlmuth. Moreover, I wish to sincerely thank all my master and Ph.D. students who worked in this or related fields for their collaboration, their effort, and their patience. Special thanks go to Klaus Dressler and his coworkers at the Fraunhofer ITWM in Kaiserslautern for their hints on practical aspects and for supplying the realistic example in Sect. 8.5. Last but not least, I express my gratitude to Doris Hemmer-Kolb and Kirsten Höffler for the final proofreading and to Mario Aigner from Springer Verlag for the efficient handling of the manuscript.

Kaiserslautern   Bernd Simeon
April 2013

Contents

Part I   Mathematical Models

1   A Point of Departure 3
   1.1   Elastic Pendulum 4
       1.1.1   Small Elastic Displacement 5
       1.1.2   Floating Reference Frame 6
       1.1.3   Galerkin Projection 7
       1.1.4   Time Integration 8
   1.2   Objectives 8
       1.2.1   Treatment of Constraints 9
       1.2.2   Discretization in Space and Time 10
   1.3   Overview 10

2   Rigid Multibody Dynamics 13
   2.1   Equations of Motion 13
       2.1.1   Kinematics 14
       2.1.2   Dynamics 16
       2.1.3   Relation to Hamiltonian Systems 20
       2.1.4   Remarks 20
   2.2   Examples 21
       2.2.1   Slider Crank 21
       2.2.2   Newton–Euler Equations 23
       2.2.3   Wheel Suspension 26
   2.3   Differential-Algebraic Equations 28
       2.3.1   Basic Types of DAEs 28
       2.3.2   The Index 30
       2.3.3   Constraint Manifold and Local State Space Form 35
       2.3.4   Solution Invariants versus Constraints 36
       2.3.5   Differential-Algebraic Equations in Saddle Point Form . . . 38
       2.3.6   Remarks and Further References 39

2.4		Analysis of the Equations of Constrained Mechanical Motion	39
	2.4.1	Index and Existence of Solutions	40
	2.4.2	Minimax Characterization of Constraints	42
	2.4.3	Influence of Perturbations	43
	2.4.4	Alternative Formulations	44
	2.4.5	Remarks	51

3 Elastic Motion 53

3.1	Basic Equations of an Elastic Body	53
	3.1.1 Variational Formulation	53
	3.1.2 Equations of Motion	58
	3.1.3 Smoothness and Appropriate Function Spaces	59
	3.1.4 The Trace Space	64
3.2	Constraints in Linear Elasticity	67
	3.2.1 Variational Formulation	68
	3.2.2 Appropriate Function Spaces	69
3.3	Mathematical Analysis	70
	3.3.1 Existence of Solutions	71
	3.3.2 Influence of Perturbations	76
	3.3.3 Remarks	79
3.4	Related Mathematical Models	80
	3.4.1 Domain Decomposition	80
	3.4.2 Dynamic Contact	82
	3.4.3 Incompressible Elastic Body	84

4 Flexible Multibody Dynamics 87

4.1	Floating Reference Frame	87
	4.1.1 Variational Formulation	88
	4.1.2 Equations of Unconstrained Motion	90
	4.1.3 Extensions and Special Cases	93
4.2	Constraints	97
	4.2.1 Weak Form of Equations of Constrained Motion	97
	4.2.2 Modeling Joints	99
	4.2.3 Remarks	103
4.3	Flexible Multibody System	104
	4.3.1 Variational Formulation	104
	4.3.2 Equations of Motion	106
4.4	Special Bodies	109
	4.4.1 Plane Strain and Plane Stress	109
	4.4.2 Beam Model	110
4.5	Examples	112
	4.5.1 Slider Crank	112
	4.5.2 Truck Model	115
	4.5.3 Pantograph and Catenary	117
4.6	Summary of Key Formulas in Part I	120

Part II Numerical Methods

5 Spatial Discretization 125
  5.1 Finite Element Approximation of Elastic Body 125
    5.1.1 Unconstrained Equations 126
    5.1.2 Constrained Equations of Motion 129
    5.1.3 Remarks 138
  5.2 Spatial Discretization of Flexible Multibody System 139
    5.2.1 Galerkin Projection for Floating Reference Frame 139
    5.2.2 Flexible Multibody System 143
    5.2.3 Model Reduction 146
  5.3 Examples 150
    5.3.1 Slider Crank with Elastic Connecting Rod 150
    5.3.2 Loading Area of Planar Truck Model 153
    5.3.3 Pantograph and Catenary 156

6 Stiff Mechanical System 159
  6.1 Elastic Pendulum Revisited 160
  6.2 General Framework 161
  6.3 State Space Form 163
    6.3.1 Asymptotic Expansion 163
    6.3.2 Elastic Pendulum 167
    6.3.3 Remarks 168
  6.4 The Differential-Algebraic Case 169
    6.4.1 Structure-Preserving Local Parametrization ... 170
    6.4.2 Computational Method 172

7 Time Integration Methods 173
  7.1 Overview on Time Integration Methods 173
    7.1.1 Alternative Formulations of the Equations of Motion 174
    7.1.2 Basic Discretization Schemes 175
  7.2 BDF and Implicit Runge–Kutta Methods 180
    7.2.1 Backward Differentiation Formulas 180
    7.2.2 Error Analysis 182
    7.2.3 Implicit Runge–Kutta Methods of Collocation Type 186
    7.2.4 Application to Differential-Algebraic Equations 188
    7.2.5 Solving Constrained Mechanical Systems in Practice 190
  7.3 Order Reduction for Stiff Mechanical Systems 193
    7.3.1 Beam under Point Load 193
    7.3.2 A Model Problem 195
    7.3.3 Runge–Kutta Methods 196
    7.3.4 Rosenbrock–Wanner Methods 202
    7.3.5 BDF Methods 206
    7.3.6 A Nonlinear Test Equation 207
    7.3.7 Remarks 208

		7.4	Special Integration Methods 208

    7.4 Special Integration Methods 208
        7.4.1 Generalized-α Method 209
        7.4.2 A Variant of the Implicit Midpoint Rule 212

8 Numerical Case Studies 217
    8.1 Slider Crank I 217
    8.2 Slider Crank II 222
    8.3 Pantograph and Catenary 228
    8.4 Planar Truck Model 230
    8.5 3D Trailer Frame 233

References 239

Index 247

Part I
Mathematical Models

Chapter 1
A Point of Departure

What is a flexible multibody system? How can we derive an adequate mathematical model? And what are the major computational challenges that we are facing here? This introductory chapter gives some preliminary answers and, at the same time, illustrates the objectives pursued by this monograph.

A multibody system is defined to be a collection of bodies and interconnection elements. Joints constrain the relative motion of pairs of bodies while springs and dampers act as compliant elements. Furthermore, the bodies possess a certain mass and geometry whereas the interconnections are treated as massless. These basic modeling assumptions apply to a large class of mechanical and structural systems such as vehicles, robots, mechanisms, and air- and spacecrafts. Even the field of biomechanics makes extensive use of multibody models.

Variational principles dating back to Euler and Lagrange characterize the motion of a multibody system. More precisely, these principles provide a methodology, also called *formalism*, to generate equations of motion for arbitrary systems that comprise rigid bodies and interconnection elements. With these formalisms at the core of today's sophisticated simulation software, the automatic generation of models for highly complex structures has become a straightforward task. Depending on the choice of coordinates for the position and orientation of each individual body, such rigid multibody models form a system of ordinary differential equations (ODEs) or, if constraints are present, a system of differential-algebraic equations (DAEs).

In addition to rigid bodies, flexible multibody systems contain elastic components. These systems are used in connection with applications in which the rigid body assumption is no longer valid, a field that has increasingly been in demand in recent years due to a strong trend towards lightweight and high-precision mechanical systems. Accordingly, flexible multibody systems combine models and simulation methods both from rigid body mechanics and from structural analysis. Since elastic bodies are governed by partial differential equations (PDEs) and rigid bodies by ordinary differential or differential-algebraic equations, the mathematical model of a flexible multibody system is naturally heterogeneous. One could also speak of a coupled system of discrete (rigid bodies) and continuous (elastic bodies) components where the interface or coupling conditions deserve particular attention. An-

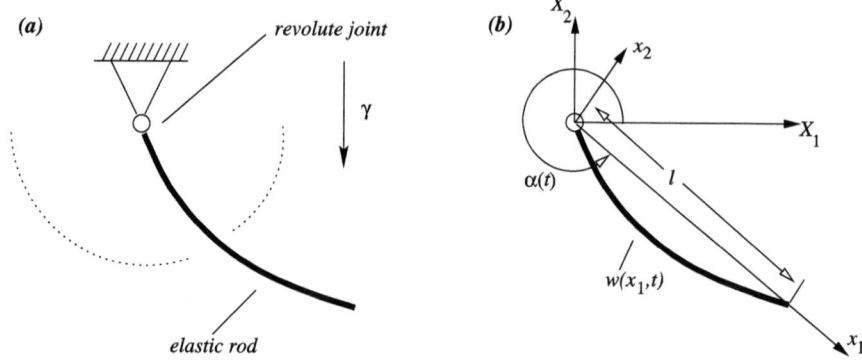

Fig. 1.1 Elastic pendulum with variables α: angle between inertial and body-fixed reference frame, w: displacement of neutral fiber, and constants γ: gravity, l: length

other crucial point are the multiple time scales that typically arise when applying space discretization methods like the finite element method (FEM) to elastic components.

The literature on flexible multibody systems is rich, and various simulation codes now offer corresponding features. To give some basic references, we mention the monographs of Bauchau [Bau10], Bremer [Bre08], Géradin and Cardona [GC00], Schwertassek and Wallrapp [SW99], and Shabana [Sha98]. An overview on simulation codes can be found in Pereia and Ambrosio [PA94] and Schiehlen [Sch90]. We omit an extensive discussion of the existing literature at this point and refer the reader to the following chapters where more references will be given throughout the text. Instead, it is time to get started with an example that illustrates this class of problems.

1.1 Elastic Pendulum

Consider the elastic pendulum displayed in Fig. 1.1(a). It consists of a rod that moves in the plane and a revolute joint at the rod's upper end. Though being a single body system, this example demonstrates the typical modeling steps and, after discretization in space, features two different time scales.

We assume that the rod may undergo a large rotational motion but a small deformation only, which is a typical requirement in flexible multibody dynamics. Figure 1.1(b) shows two reference frames or coordinate systems that we use to express the *kinematics* or geometry of the motion. Coordinates with respect to the inertial reference frame are denoted by (X_1, X_2) whereas the notation (x_1, x_2) stands for points with respect to the *floating* or *body-fixed reference frame* that moves with the rod. A rotation by the scalar angle $\alpha(t)$ transforms quantities from the inertial to the floating frame and vice versa. With respect to the floating frame, the rod's deformation is measured in terms of the displacement of its neutral fiber $w(x_1, t)$.

1.1 Elastic Pendulum

We analyze the elastic pendulum in four steps. To keep the presentation concise, we deliberately leave some loose ends and state the key formulas only. If not indicated otherwise, all variables and parameters are real scalars in this introductory example and written as lightface italic characters.

1.1.1 Small Elastic Displacement

If both reference frames coincide, i.e., $\alpha(t) \equiv 0$, standard beam theory [SW99] describes the motion of material points (x_1, x_2) by the *deformation*

$$\begin{aligned} X_1 &= \varphi_1(x_1, x_2, t) = x_1 - x_2 w'(x_1, t), \\ X_2 &= \varphi_2(x_1, x_2, t) = x_2 + w(x_1, t). \end{aligned} \quad (1.1)$$

Here and in the following, a prime denotes the partial derivative $w' = \partial w / \partial x_1$ with respect to the spatial variable while a dot stands for the derivative $\dot{w} = \partial w / \partial t$ with respect to time. The kinematics of the rod still lacks appropriate boundary conditions. We postulate fixed ends

$$w(0, t) = w(l, t) = 0 \quad (1.2)$$

where l is the pendulum's length.

Hamilton's principle of least action characterizes the motion by

$$\int_{t_0}^{t_1} (T - U) \, dt \to \text{stationary!} \quad (1.3)$$

with kinetic and potential energies T and U, respectively, and arbitrary points of time t_0 and t_1. The kinetic energy can be calculated from the general expression

$$T = \frac{1}{2} \rho \int_\Omega (\dot{\varphi}_1^2 + \dot{\varphi}_2^2) \, dx \quad (1.4)$$

where ρ is the constant mass density and Ω stands for the domain that is occupied by the rod in its undeformed state. In case of a constant rectangular cross section, we have

$$\Omega = (0, l) \times (-\tfrac{h}{2}, \tfrac{h}{2}) \times (-\tfrac{b}{2}, \tfrac{b}{2})$$

with height h and width b. Moreover, integration is to be understood in three dimensions as $dx = d(x_1, x_2, x_3)$, but due to the planar motion assumption and the beam model, we can carry out the integration with respect to x_2 and x_3 directly. The same holds for the derivation of the potential energy U.

Omitting these straightforward calculations, we obtain the energy expressions

$$T = \frac{1}{2} \rho A \int_0^l \dot{w}^2 \, dx_1$$

and
$$U = \frac{1}{2}\text{EI}\int_0^l (w'')^2\, dx_1 + \rho A \gamma \int_0^l w\, dx_1.$$

Here, $A = hb$ is the cross section area, E the modulus of elasticity, I the axial moment of inertia, and γ the gravity constant.

The principle of least action (1.3) combined with the calculus of variations provides a general methodology to derive the partial differential equation that governs the dynamic behavior of an elastic body. In this case, the stationarity condition (1.3) for the action integral leads to the *dispersive wave equation*

$$\rho A \ddot{w} + \text{EI} w'''' = -\rho A \gamma \tag{1.5}$$

with Dirichlet or geometric boundary conditions $w(0,t) = w(l,t) = 0$ and Neumann or natural boundary conditions $w''(0,t) = w''(l,t) = 0$.

We point out that the *strong form* (1.5) is not an appropriate basis for the following steps. Instead, the *weak form* (1.3) or equivalent variational principles are more advantageous, as will become clear in the subsequent chapters.

1.1.2 Floating Reference Frame

The next step combines small displacement with respect to the body-fixed frame and large rotation. We add the angle α as additional variable and describe the total motion of material points (x_1, x_2) in terms of the deformation

$$\begin{pmatrix} \varphi_1(x_1, x_2, t) \\ \varphi_2(x_1, x_2, t) \end{pmatrix} = A(\alpha) \begin{pmatrix} x_1 - x_2 w'(x_1, t) \\ x_2 + w(x_1, t) \end{pmatrix} \tag{1.6}$$

with rotation matrix

$$A(\alpha) = \begin{pmatrix} \cos\alpha & -\sin\alpha \\ \sin\alpha & \cos\alpha \end{pmatrix}.$$

In addition, the fixed end boundary conditions (1.2) still hold and eliminate rigid body motion with respect to the body-fixed frame.

Again, the least action principle (1.3) characterizes the motion of the pendulum. Obviously, the expression for the kinetic energy becomes more complicated as we have to insert the deformation (1.6) in the integral (1.4) and thus get both α and w as variables in our model. The potential energy changes to

$$U = \frac{1}{2}\text{EI}\int_0^l (w'')^2\, dx_1 + \rho A \gamma \int_0^l (x_1 \sin\alpha + w \cos\alpha)\, dx_1.$$

Here, we observe that the first term, the strain energy, is not affected by the rotation.

Based on the expressions for T and U, the principle of least action generates a coupled system that consists of an ordinary differential equation with respect to the

1.1 Elastic Pendulum

angle α and a partial differential equation with respect to the axis displacement w. We skip this system for the moment as its rather complex structure would distract our attention from the main points of interest. In Chap. 4, however, we come back to this loose end.

1.1.3 Galerkin Projection

The displacement of the neutral fiber is now projected to a finite dimensional subspace. To keep the model simple, we employ the rod's first eigenfunction for this purpose and approximate w by

$$w(x_1, t) \doteq \sin\left(\frac{\pi x_1}{l}\right) d(t) \qquad (1.7)$$

with an unknown, time-dependent coefficient d. This approximation of the displacement leads in turn to modified energy expressions where all integrations with respect to the spatial variables can be carried out directly.

Applying one last time the least action principle, this time to the modified energy expressions from the approximation (1.7), we derive the *semi-discretized equations of motion*

$$\left(\frac{J}{\rho Al} + \frac{d^2}{2}\right)\ddot{\alpha} + \frac{l}{\pi}\ddot{d} = -\dot{\alpha} d\dot{d} - \frac{l}{2}\gamma\cos\alpha + \frac{2}{\pi}\gamma d\sin\alpha, \qquad (1.8a)$$

$$\frac{l}{\pi}\ddot{\alpha} + \frac{1}{2}\ddot{d} = \frac{1}{2}\dot{\alpha}^2 d - \frac{2}{\pi}\gamma\cos\alpha - \frac{\omega^2}{2}d. \qquad (1.8b)$$

The additional constants appearing here are $\omega = \pi^2/l^2 \sqrt{EI/(\rho A)}$, the rod's first eigenfrequency, and $J = \rho Al^3/3$, the moment of inertia for the rigid rod.

The structure of the coupled system (1.8a), (1.8b) becomes clearer if we consider two special cases. For vanishing elastic displacement, i.e., $d(t) \equiv 0$, (1.8a) simplifies to the pendulum equation

$$\ddot{\alpha} = -\frac{3\gamma}{2l}\cos\alpha. \qquad (1.9)$$

On the other hand, for vanishing rotation $\alpha(t) \equiv 0$, we obtain from (1.8b) the linear oscillator equation

$$\ddot{d} + \omega^2 d = -\frac{4\gamma}{\pi}. \qquad (1.10)$$

If the frequency ω is large, there is certainly a slow time scale in (1.8a), (1.8b) for the rotational motion and a fast time scale for the elastic vibrations. Nevertheless, in a typical simulation code both equations in (1.8a), (1.8b) are treated as general second order system

$$M(p)\ddot{p} = f(p, \dot{p}) \qquad (1.11)$$

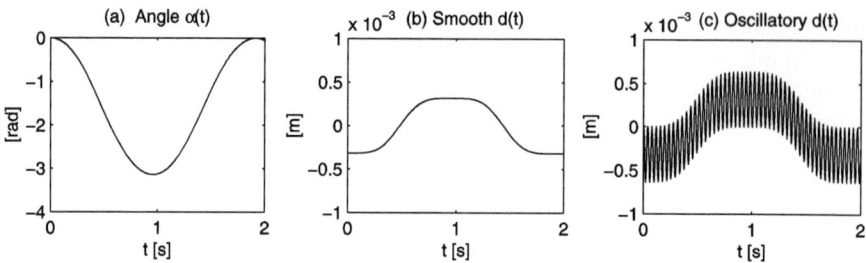

Fig. 1.2 Time integration of (1.8a), (1.8b). In (**b**), the initial value $d(t_0)$ takes the static deformation into account whereas in (**c**) $d(t_0) = 0$

in variables $p(t) := (\alpha(t), d(t))^T \in \mathbb{R}^2$ where the mass matrix $M(p) \in \mathbb{R}^{2 \times 2}$ stands for the coefficients on the left of (1.8a), (1.8b) and the force vector $f(p, \dot{p}) \in \mathbb{R}^2$ for the expressions on the right.

1.1.4 Time Integration

We complete the discussion of the elastic pendulum by a time integration of the equations of motion (1.8a), (1.8b). Figure 1.2 displays the results for the angle α and the displacement d. In a first run, the initial value is chosen as $\alpha(t_0) = 0$ and $d(t_0) = -4\gamma/(\pi\omega^2)$ with zero velocities. The initial displacement with respect to d takes the static deformation due to the gravity into account, and in turn we obtain a *smooth solution* with respect to both variables. On the other hand, if we set $d(t_0) = 0$, the elastic displacement shows high frequency oscillations.

The choice of an appropriate time integration method depends strongly on the solution behavior. If we are solely interested in smooth solutions, the high frequency part can be viewed as perturbation, and methods with *numerical dissipation* will damp out spurious oscillations. One could also say that we are dealing with a *stiff mechanical system*, which is a special kind of a *singularly perturbed system*. However, if the high frequencies need to be resolved, numerical dissipation may lead to erroneous results.

1.2 Objectives

The pendulum example illustrates some of the main features of flexible multibody systems. In fact, most of today's simulation software in the field makes use of the steps outlined above and applies them to far more complex situations. Overall, the models are reasonably settled and computationally, great strides have been made in the last two decades. Yet from a mathematical point of view, there is a number of open issues. This monograph addresses them and, at the same time, tries to provide a unified exposition of models and numerical methods. Two key problems will be covered:

1.2 Objectives

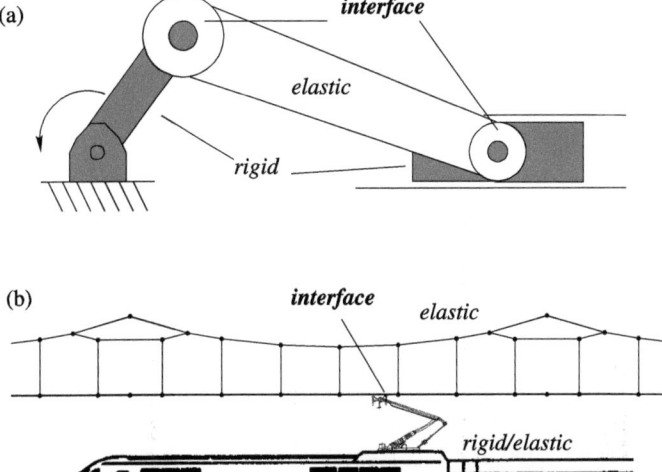

Fig. 1.3 Coupling rigid and elastic bodies at the examples of (**a**) slider crank, (**b**) pantograph and catenary

1.2.1 Treatment of Constraints

The revolute joint in the pendulum example constrains the motion of the rod. However, owing to the adept choice of the angle α as rigid motion coordinate, the governing equations are unconstrained. From rigid body dynamics it is well-known that unconstrained equations of motion are easy to establish for certain standard problems but hard or even impossible to derive in more general situations. For this reason, the Lagrange multiplier technique plays a crucial role in modeling mechanical systems with constraints.

In case of rigid bodies, Lagrange multipliers lead to differential-algebraic equations (DAEs), and due to extensive research efforts, a firm mathematical theory as well as excellent numerical methods have been developed for this problem class. But what happens if we consider both rigid and elastic bodies in combination with joints that constrain their motion? Figure 1.3 shows two examples, a slider crank mechanism and the system of pantograph and catenary. In both examples, rigid and elastic bodies interact via joints and contact conditions. We are thus concerned with coupled systems of discrete (rigid bodies) and continuum (elastic bodies) components where the interface or coupling conditions require particular attention

Since elastic bodies are governed by partial differential equations, we can expect an infinite-dimensional system with some kind of constraints as basic model. What is the relation to standard DAE theory, and where are the differences to systems of rigid bodies? Moreover, what are the implications for numerical analysis? Based on the notion of a transient saddle point problem, the answers given below show which models are well-defined and which are questionable. Additionally, estimates on the influence of perturbations are given. The slider crank and other examples accompany the discussion and illustrate some of the key points. Last but not least, it should be pointed out that the mathematical framework developed here is related to and sheds new light on the theory of dynamic contact and domain decomposition problems.

1.2.2 Discretization in Space and Time

As explained at the pendulum example, the numerical simulation of a flexible multibody system consists of two steps. First, the elastic components are discretized in space, which reduces the governing equations to an extended system of ordinary or differential-algebraic equations. Thereafter, time integration methods solve an initial value problem for the resulting semi-discretized system that involves two types of state variables, namely, those for the rigid body motion, on the one hand, and those for the small elastic displacement, on the other.

Clearly, convergence in space and time is an important issue here, in particular if we use a transient saddle point formulation as model. What are the requirements for the finite element discretization, and what is the structure of the resulting semi-discretized equations? Furthermore, though standard discretization techniques used for contact problems such as node-to-surface algorithms can be carried over to the treatment of joints, it is not clear whether such an effort pays off. As an alternative, rigid body constraints expressed in local joint coordinate systems are widespread but seem to contradict assumptions on the elastic body model.

Last but not least, rigid body motion and elastic deformation may have widely different time scales, and this turns the time integration into a challenging problem. A singularly perturbed system must be solved where the perturbation consists of latent high frequency oscillations. As an asymptotic expansion shows, the perturbed system tends in the limit to a system with additional constraint equations. This in turn leads to order reduction and further numerical difficulties when applying standard DAE solvers to flexible multibody systems. What kind of remedies can we apply, and are there alternative approaches to the time integration problem?

Since the kind of space discretization mainly determines whether a singular perturbation occurs or not, both of the above steps are intertwined. In certain cases, asymptotic techniques can eliminate the stiff modes, which may lead to a reduced model with better properties. In this context, approaches typical in commercial software packages need to be scrutinized since mostly no error estimation procedure is applied and the decision about the quality of the reduced model is left to engineering judgment.

1.3 Overview

After the objectives of this monograph have been stated, we give next a brief overview of the following chapters.

1.3 Overview

Systems of rigid bodies are discussed in Chap. 2. Particular emphasis is placed on the differential-algebraic system of constrained mechanical motion that results from the Lagrange equations of the first kind since this model represents the most general approach to handle complex technical applications and is also widespread in commercial software tools. For those readers that are not familiar with the theory of differential-algebraic equations, the chapter includes furthermore a short introduction to this topic.

Chapter 3 is taken up with the development of a mathematical framework for the constrained motion of a single elastic body. The resulting transient saddle point problem can be viewed as infinite-dimensional differential-algebraic equation and reflects the requirements in flexible multibody dynamics where the elastic body is, like the rigid one, subject to constraints since joints restrict its motion. Moreover, we analyze in this chapter the saddle point model using the inf-sup condition, derive an estimate on the influence of perturbations, and briefly touch upon related mathematical models in computational mechanics.

We introduce next in Chap. 4 the method of floating reference frames in order to express large rotations and translations in combination with small deformation theory. Thus, a set of rigid motion variables plus the displacement field describe the overall motion of the body. The corresponding dynamic equations are transformed into weak form and a suitable formal setting is provided. Several examples illustrate the modeling approach. Part I closes with a summary of key formulas.

Part II covers the numerical methods in the field and begins with Chap. 5 on discretization methods in space. In case of unconstrained motion, the usual projection of the displacement field onto a finite dimensional subspace leads to a system of ordinary differential equations for the elastic motion. If constraints are present, however, the discretization of the saddle point problem results in a differential-algebraic system with certain properties. Convergence results are established in both cases, and the methodology is extended to floating reference frames. Additionally, the treatment of joints is addressed and an overview on model reduction techniques is given.

After discretization in space, the governing equations take the form of a partitioned differential-algebraic system that features two types of state variables, namely, those for the rigid body motion and those for the elastic deformation. As mentioned above, this heterogeneous structure quite often involves widely differing time scales. Chapter 6 is devoted to this important and challenging aspect of flexible multibody dynamics. Employing asymptotic expansions, we analyze standard engineering approaches like the linear theory of elastodynamics and extend the results to the differential-algebraic case by means of suitable local parametrizations.

Chapter 7 deals with time integration methods. Owing to the singular perturbation caused by latent high frequency modes, even robust codes may suffer from effects like order reduction and excessive failures of Newton's method. A thorough investigation for the class of implicit Runge–Kutta methods reveals which variables are most severely affected and leads to practical remedies. Even more, alternative methods that are common in structural dynamics applications are generalized to flexible multibody systems and compared with established differential-algebraic solvers.

Finally, the last Chap. 8 presents several numerical case studies where different models as well as different discretization methods are evaluated. As is the case in many computational methods, theory and numerical practice are closely related here and benefit from each other. Simulations of a slider crank mechanism, a model of a truck, and of the system of pantograph and catenary demonstrate the power of the methods. But deliberately, we also point out possible pitfalls and open problems.

Notational Conventions As far as possible, fairly standard notation has been adopted throughout the text. Vectors $\boldsymbol{v} \in \mathbb{R}^n$ and matrices $\boldsymbol{M} \in \mathbb{R}^{n \times n}$ are denoted by boldface italic characters while scalars $s \in \mathbb{R}$ and components of vectors $v_i \in \mathbb{R}$ or matrices are written as lightface italic characters. An exception to that rule is the notation $x \in \mathbb{R}^3$ for the independent spatial variable of an elastic body, which stresses the analogy to $t \in \mathbb{R}$ as independent variable for time. Finally, second order tensors are represented as symmetric matrices. Unavoidably, there are some minor conflicts of notation, but surrounding explanations and context should make the intended meaning clear.

Chapter 2
Rigid Multibody Dynamics

In this chapter, we give an overview on the mathematical models for the dynamics of systems of rigid bodies. Depending on the choice of coordinates for the position and orientation of each body, the governing equations form either a system of ordinary differential equations or, if constraints are present, a system of differential-algebraic equations. We analyze the structure of these equations, discuss practical aspects, and present several examples. Since differential-algebraic equations are the recurrent theme in this book, we furthermore summarize their most important theoretical properties, which includes the index concept and alternative formulations for the equations of constrained mechanical motion.

Rigid multibody dynamics has experienced a significant development over the last decades. In recent years, however, the main focus of research has shifted towards more refined models that include elastic bodies or mechatronic components. General references are, among others, the monographs of Garcia de Jalón and Bayo [GB94], Magnus [Mag78], Haug [Hau89], Roberson and Schwertassek [RS88], Shabana [Sha98], and Wittenburg [Wit77]. An extensive exposition of both mathematical models and numerical methods is found in Eich-Soellner and Führer [ESF98].

2.1 Equations of Motion

We start with an informal definition.

A multibody system is defined to be a set of rigid bodies and interconnection elements. The latter are joints that constrain the relative motion of pairs of bodies and springs and dampers that act as compliant elements. Moreover, the bodies possess a specific mass and geometry whereas the interconnections are treated as massless, Fig. 2.1.

Variational principles dating back to Euler and Lagrange [Lag88] characterize the motion of a multibody system. These principles constitute a methodology to generate the governing equations and are the basis of today's simulation software. In this context, one speaks also of a *multibody formalism*.

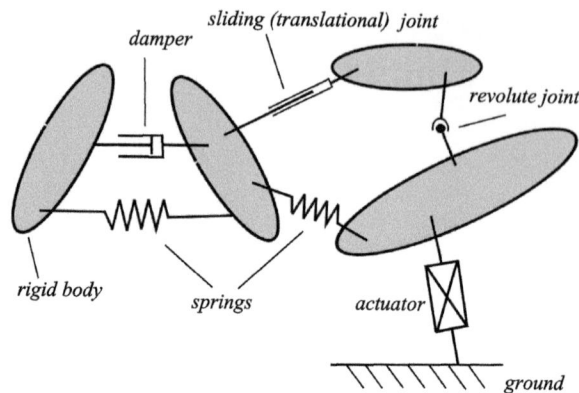

Fig. 2.1 Sketch of a multibody system with rigid bodies and typical interconnections

2.1.1 Kinematics

Let $q(t) \in \mathbb{R}^{n_q}$ denote a vector that comprises the coordinates for position and orientation of all bodies in the system depending on time t. We leave the specifics of the chosen coordinates open at this point but will come back to this issue when discussing standard approaches such as absolute and relative coordinates below. As in the preceding chapter, differentiation with respect to time is expressed by a dot and thus we write $\dot{q}(t)$ and $\ddot{q}(t)$ for the corresponding velocity and acceleration vectors.

The term *kinematics* describes the geometry of the motion q without taking the forces into account that produce it. Revolute, translational, universal, and spherical joints are examples of bondings in a multibody system. They may constrain the motion q and hence determine its kinematics.

If constraints are present, we express the resulting conditions on q in terms of n_λ constraint equations

$$\mathbf{0} = g(q). \tag{2.1}$$

Obviously, a meaningful model requires $n_\lambda < n_q$. Equations (2.1) that restrict the motion q are called *holonomic constraints*, and the rectangular matrix

$$G(q) := \frac{\partial g(q)}{\partial q} \in \mathbb{R}^{n_\lambda \times n_q}$$

is called the *constraint Jacobian*.

We remark that there exist constraints, e.g., driving constraints, that may explicitly depend on time t and that are written as $\mathbf{0} = g(q, t)$. For notational simplicity, however, we omit this dependence in (2.1).

A standard assumption on the constraint Jacobian is the full rank condition

$$\operatorname{rank} G(q) = n_\lambda, \tag{2.2}$$

which means that the constraint equations are linearly independent. In this case, the difference $n_s := n_q - n_\lambda$ is the number of *degrees of freedom* (*DOF*) in the system.

2.1 Equations of Motion

Fig. 2.2 Simple pendulum in the plane

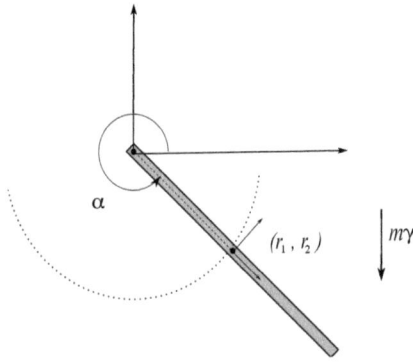

One can also say that the degrees of freedom represent the independent kinematical possibilities of the motion.

Two routes branch off at this point. The first modeling approach expresses the governing equations in terms of the *redundant variables* q and uses additional *Lagrange multipliers* $\lambda(t) \in \mathbb{R}^{n_\lambda}$ to take the constraint equations into account. Alternatively, the second approach introduces *minimal coordinates* $s(t) \in \mathbb{R}^{n_s}$ such that the redundant variables q can be written as a function $q(s)$ and that the constraints are satisfied for all choices of s,

$$g(q(s)) \equiv 0. \tag{2.3}$$

As a consequence, by differentiation of the identity (2.3) with respect to s we get the orthogonality relation

$$G(q(s))\, N(s) = 0 \tag{2.4}$$

with the *null space matrix* $N(s) := \partial q(s)/\partial s \in \mathbb{R}^{n_q \times n_s}$.

Simple Planar Pendulum

An instructive example for the two modeling approaches is given by the simple pendulum of Fig. 2.2. The pendulum is pivoted at the origin of the inertial reference frame and performs a circular motion. One choice for q are absolute or Cartesian coordinates

$$q(t) = \begin{pmatrix} r_1(t) \\ r_2(t) \\ \alpha(t) \end{pmatrix} \tag{2.5}$$

where r_1, r_2 are the coordinates of the centroid and α is the angle between the inertial reference frame and a body-fixed frame placed in the centroid. The $n_q = 3$ coordinates completely specify the position and orientation of the pendulum in the plane. Due to the revolute joint at the tip, the kinematic constraint is prescribed by

$n_\lambda = 2$ equations

$$0 = \begin{pmatrix} r_1 - \frac{l}{2}\cos\alpha \\ r_2 - \frac{l}{2}\sin\alpha \end{pmatrix} =: g(q) \quad (2.6)$$

where l denotes the length of the pendulum. The constraint Jacobian reads

$$G(q) = \begin{pmatrix} 1 & 0 & \frac{l}{2}\sin\alpha \\ 0 & 1 & -\frac{l}{2}\cos\alpha \end{pmatrix}$$

and is obviously of full rank 2.

On the other hand, we can directly eliminate $r_1 = \frac{l}{2}\cos\alpha$ and $r_2 = \frac{l}{2}\sin\alpha$ from the constraints and select the angle α as minimal coordinate for the constrained motion. The number of DOF is hence $n_s = 1$, and the redundant coordinates can be written as

$$q(\alpha) = \left(\frac{l}{2}\cos\alpha, \frac{l}{2}\sin\alpha, \alpha\right)^T \quad (2.7)$$

while

$$N(\alpha) = \left(-\frac{l}{2}\sin\alpha, \frac{l}{2}\cos\alpha, 1\right)^T \quad (2.8)$$

is the corresponding null space matrix.

The pendulum example might suggest that an adept choice of variables in terms of minimal coordinates is clearly preferable. We postpone the discussion of this issue to the following section and study next the dynamics of a multibody system.

2.1.2 Dynamics

Above we have sketched two approaches to model the kinematics of a multibody system. These two approaches are also fundamental for the *dynamics* of the system, which is the study of forces and how the motion changes under their influence.

Lagrange Equations of the First Kind

Using both the redundant position variables q and additional Lagrange multipliers λ to describe the dynamics leads to the *equations of constrained mechanical motion*, also called the *Lagrange equations of the first kind*,

$$M(q)\ddot{q} = f(q, \dot{q}, t) - G(q)^T \lambda, \quad (2.9a)$$
$$0 = g(q), \quad (2.9b)$$

where $M(q) \in \mathbb{R}^{n_q \times n_q}$ stands for the *mass matrix* and $f(q, \dot{q}, t) \in \mathbb{R}^{n_q}$ for the vector of *applied and internal forces*. In Chap. 1, we employed Hamilton's principle

2.1 Equations of Motion

of least action (1.3) to derive the equations for the pendulum example. The same principle holds for *conservative multibody systems*, i.e., systems where the applied forces can be written as the gradient of a potential U. The equations of motion (2.9a), (2.9b) then follow from

$$\int_{t_0}^{t_1} \left(T - U - g(q)^T \lambda \right) dt \to \text{stationary !} \qquad (2.10)$$

where the kinetic energy possesses a representation as quadratic form

$$T(q, \dot{q}) = \frac{1}{2} \dot{q}^T M(q) \dot{q}.$$

In the least action principle (2.10), we observe the fundamental Lagrange multiplier technique for coupling constraints and dynamics [Bri08].

Extensions of the multiplier technique exist in various more general settings such as dissipative systems or even inequality constraints. The underlying idea is always the same, namely to add an expression $g(q)^T \lambda$ or $G(q)^T \lambda$ to the variational principle or the resulting system of dynamic equations, respectively. In the non-conservative case, the Lagrange equations of the first kind read [RS88, Sha98]

$$\frac{d}{dt} \left(\frac{\partial}{\partial \dot{q}} T(q, \dot{q}) \right) - \frac{\partial}{\partial q} T(q, \dot{q}) = f_a(q, \dot{q}, t) - G(q)^T \lambda, \qquad (2.11)$$
$$0 = g(q)$$

with applied forces f_a. Carrying out the differentiations and defining the force vector f as

$$f(q, \dot{q}, t) := f_a(q, \dot{q}, t) + \frac{\partial}{\partial q} \left(\frac{1}{2} \dot{q}^T M(q) \dot{q} \right) - \left(\frac{d}{dt} M(q) \right) \dot{q},$$

which includes Coriolis and centrifugal terms, results in the equations of motion (2.9a), (2.9b). In the conservative case, $f_a = -\partial U / \partial q$ holds, and the principle (2.10) is equivalent to (2.11).

It should be remarked that for ease of presentation, we omit momentarily the treatment of generalized velocities resulting from three-dimensional rotation matrices. For that case, an additional kinematic equation $\dot{q} = S(q) v$ with transformation matrix S and velocity vector v needs to be taken into account [ESF98, RS88]. Section 2.2 will cover that point in more detail.

Lagrange Equations of the Second Kind

The Lagrange equations of the first kind lead to a system of second order differential equations with additional constraints (2.9a), (2.9b), which is a special form of a differential-algebraic equation. Applying minimal coordinates s, on the other hand, eliminates the constraints and allows generating a system of ordinary differential equations.

If we insert the coordinate transformation $q = q(s(t))$ into the principle (2.10) or apply it directly to Eqs. (2.9a), (2.9b) of the first kind, the constraints and Lagrange multipliers cancel due to the property (2.3). The equations of motion, the *Lagrange equations of the second kind*, then take the form

$$C(s)\ddot{s} = h(s, \dot{s}, t). \tag{2.12}$$

This system of second order ordinary differential equations bears also the name *state space form*.

For a closer look at Eqs. (2.12) of the second kind, we recall the null space matrix N from (2.4) and derive the relations

$$\frac{d}{dt}q(s) = N(s)\dot{s}, \qquad \frac{d^2}{dt^2}q(s) = N(s)\ddot{s} + \frac{\partial N(s)}{\partial s}(\dot{s}, \dot{s})$$

for the velocity and acceleration vectors. Inserting these relations into the dynamic equations (2.9a), we conclude that the mass matrix C in (2.12) has the form

$$C(s) = N(s)^T M(q(s)) N(s) \in \mathbb{R}^{n_s \times n_s},$$

which means that M is mapped onto the space of minimal coordinates. Furthermore, the force vector $h(s, \dot{s}, t) \in \mathbb{R}^{n_s}$ is given by the expression

$$h(s, \dot{s}, t) = N(s)^T f(q(s), N(s)\dot{s}, t) - N(s)^T M(q(s)) \frac{\partial N(s)}{\partial s}(\dot{s}, \dot{s}).$$

If the mapping from minimal coordinates s to redundant coordinates q is globally defined, then it is obviously possible to transform the equations of the first kind (2.12) back to the differential-algebraic equation (2.9a), (2.9b), and the two modeling approaches turn out to be equivalent. However, the existence of a globally valid set of minimal coordinates is a crucial point that cannot be guaranteed in general.

Dynamics of Simple Pendulum

To illustrate the two models for the dynamics, we take up the simple pendulum of Fig. 2.2 and state the resulting equations of motion. The Cartesian coordinates (2.5) lead to the differential-algebraic equation

$$\begin{pmatrix} m & 0 & 0 \\ 0 & m & 0 \\ 0 & 0 & m\frac{l^2}{12} \end{pmatrix} \begin{pmatrix} \ddot{r}_1 \\ \ddot{r}_2 \\ \ddot{\alpha} \end{pmatrix} = \begin{pmatrix} 0 \\ -m\gamma \\ 0 \end{pmatrix} - \begin{pmatrix} 1 & 0 \\ 0 & 1 \\ \frac{l}{2}\sin\alpha & -\frac{l}{2}\cos\alpha \end{pmatrix} \begin{pmatrix} \lambda_1 \\ \lambda_2 \end{pmatrix}, \tag{2.13a}$$

$$0 = \begin{pmatrix} r_1 - \frac{l}{2}\cos\alpha \\ r_2 - \frac{l}{2}\sin\alpha \end{pmatrix}, \tag{2.13b}$$

where m stands for the pendulum's mass and γ for the gravitation constant.

2.1 Equations of Motion

Expressing the motion in the minimal coordinate α, however, results in the state space form

$$J\ddot{\alpha} = -m\gamma \frac{l}{2} \cos\alpha \qquad (2.14)$$

with moment of inertia $J = l^2 m/3$.

The pendulum equation (2.14) should look familiar to the reader. In fact, the special case (1.9) of rigid motion as part of the equations of the elastic pendulum (1.8a), (1.8b) follows from (2.14) by simply canceling the parameters m and l on both sides.

Comparing the Approaches

So far, our discussion of the two modeling approaches has excluded an important issue: the question which of the two approaches should be preferred.

At first sight, and in particular when looking at simple models such as the planar pendulum (2.14), the state space form (2.12) seems a more appropriate basis for the numerical treatment than the differential-algebraic system (2.9a), (2.9b) in redundant coordinates. In practice, however, the state space form suffers from serious drawbacks.

The analytical complexity of the constraint equations (2.1) makes it in various applications impossible to obtain a set of minimal coordinates that is valid for all configurations of the multibody system. Moreover, although we know from the theorem on implicit functions that such a set exists in a neighborhood of the current configuration, it might lose its validity when the configuration changes. This holds in particular for multibody systems with so-called *closed kinematic loops*.

Even more, the modeling of subsystems like electrical and hydraulic feedback controls, which are essential for the performance of modern mechanical systems, is limited. The differential-algebraic model, on the other hand, bypasses topological analysis and offers the choice of using a set of coordinates q that possess physical significance.

A look at the leading software tools in the field underlines this reasoning. Some of the codes generate a differential-algebraic model whenever a constraint is present, while others try to generate a state space form as long as it is convenient. But the majority of the commercial products relies on the differential-algebraic approach as the most general way to handle complex technical applications [GB94, Sch90].

In Sect. 2.4 below, we will formulate our assessment of the differences between the two modeling approaches more rigorously by resorting to a differential-geometric point of view where the constraints (2.1) define a *submanifold* of $\mathbb{R}^{n_q}$, the so-called *constraint manifold*. For the moment, however, we continue with additional aspects of rigid multibody dynamics.

2.1.3 Relation to Hamiltonian Systems

In the conservative case, the Lagrange equations of the first and second kind can be reformulated by the transformation to Hamilton's canonical equations. Though this relation is of minor relevance in engineering applications, we mention it here for completeness.

We define the Lagrange function

$$L(q, \dot{q}) := T(q, \dot{q}) - U(q) \qquad (2.15)$$

as the difference of kinetic and potential energy. The conjugate momenta $p(t) \in \mathbb{R}^{n_q}$ are then given by

$$p := \frac{\partial}{\partial \dot{q}} L(q, \dot{q}) = M(q)\dot{q}, \qquad (2.16)$$

and for the Hamiltonian we set

$$H := p^T \dot{q} - L(q, \dot{q}). \qquad (2.17)$$

Since the velocity $\dot{q}$ can be expressed as $\dot{q}(p, q)$ due to (2.16) if the mass matrix is invertible, we view the Hamiltonian as a function $H = H(p, q)$. Moreover, we observe that H is the total energy of the system because

$$H = p^T M(q)^{-1} p - \frac{1}{2} \dot{q}^T M(q) \dot{q} + U(q) = T + U.$$

Using the least action principle (2.10) in the new coordinates p and q and applying the Lagrange multiplier technique as above in the presence of constraints, we can express the equations of motion as

$$\begin{aligned} \dot{q} &= \frac{\partial}{\partial p} H(p, q), \\ \dot{p} &= -\frac{\partial}{\partial q} H(p, q) - G(q)^T \lambda, \\ 0 &= g(q). \end{aligned} \qquad (2.18)$$

The *Hamiltonian equations* (2.18) possess a rich mathematical structure. Most technical multibody applications, however, are not conservative as they contain dampers and actuators, and for this reason we do not pursue this model any further. An extensive treatment of the corresponding subject of geometric numerical integration is given in the monograph by Hairer, Lubich and Wanner [HLW02].

2.1.4 Remarks

The differential-algebraic model (2.9a), (2.9b) does not cover all aspects of rigid multibody dynamics. In particular, features such as control laws, non-holonomic constraints, and substructures require a more general formulation. Corresponding extensions are discussed in [ESF98, SFR91, Sim95], while systems with nonholonomic constraints are treated in depth by Rabier and Rheinboldt [RR00].

Several different methodologies for the derivation of the governing equations are successfully applied in multibody dynamics. Examples are the principle of virtual work, the principle of Jourdain, and the Newton–Euler equations in combination with the principle of D'Alembert. As discussed in [BP92], these approaches are basically equivalent and lead to the same mathematical model. The crucial point is not the selection of the variational principle but the choice of coordinates and how the equations are implemented in a computer program.

With respect to the choice of coordinates, one primarily distinguishes between absolute and relative coordinates. Absolute or Cartesian coordinates describe the motion of each body with respect to an inertial reference frame while relative or joint coordinates are based on relative motions between interacting bodies. Using absolute coordinates results in a large number of equations which have a clear and sparse structure and are inexpensive to compute. Furthermore, constraints always imply a differential-algebraic model [Hau89].

Relative coordinates, on the other hand, lead to a reduced number of equations and are the basis of the recursive $O(n_b)$ formalisms that evaluate the equations of motion with a complexity that is linear in the number n_b of bodies [AC03, BJO86, EFS94]. Moreover, in case of systems with a tree structure, all kinematic bondings due to joints can be eliminated and a state space form is obtained. In general, the system matrices are full and require more complicated computations than for absolute coordinates.

A further approach, the natural coordinates as described in [GB94], employs orientation vectors for expressing the rotation of individual bodies. Overall, this leads again to a differential-algebraic model in redundant coordinates where the constraint equations, at least for typical joints, are quadratic functions and thus particularly easy to evaluate.

2.2 Examples

In this section, we look at three examples of multibody systems and discuss several aspects, in particular the choice of coordinates and the structure of the resulting equations of motion.

2.2.1 Slider Crank

The planar slider crank shown in Fig. 2.3 consists of three rigid bodies: the crank, the connecting rod, and the sliding block. Three revolute joints and one translational joint connect the bodies and constrain their motion. We use this mechanism

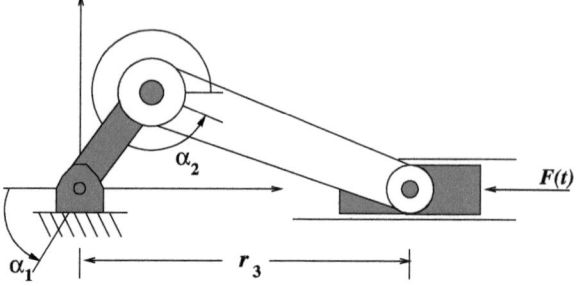

Fig. 2.3 Slider crank with crank, connecting rod, and sliding block

as example for the structure of the constrained equations of motion (2.9a), (2.9b). In Chap. 4, the rigid connecting rod will be replaced by an elastic beam.

For this particular model, $n_q = 3$ coordinates $q = (\alpha_1, \alpha_2, r_3)^T$ describe the motion of the bodies. Note that in case of unconstrained motion, each body has three degrees of freedom in the plane, which would yield a total of nine absolute coordinates. Owing to the joints, however, this system has solely $n_s = 1$ degree of freedom.

More specifically, by choosing angles α_1 and α_2 as variables, the first two revolute joints between ground and crank and between crank and connecting rod are directly taken into account. The corresponding constraints are thus automatically satisfied, and the resulting kinematics of crank and connecting rod corresponds to the motion of a double pendulum. The translational joint prevents rotation and vertical translation of the sliding block, which means that the horizontal displacement r_3 suffices as variable for the motion of this body.

Although this adept choice of coordinates eliminates the constraints that stem from two revolute and one translational joint, the third revolute joint between sliding block and connecting rod still leads to $n_\lambda = 2$ constraint equations.

Summarizing, the equations of motion (2.9a), (2.9b) read in this case

$$M \begin{pmatrix} \ddot{\alpha}_1 \\ \ddot{\alpha}_2 \\ \ddot{r}_3 \end{pmatrix} = \begin{pmatrix} -1/2 l_1 l_2 m_2 \dot{\alpha}_2^2 \sin(\alpha_1 - \alpha_2) \\ 1/2 l_1 l_2 m_2 \dot{\alpha}_1^2 \sin(\alpha_1 - \alpha_2) \\ -F(t) \end{pmatrix} - \begin{pmatrix} l_1 \cos \alpha_1 & l_1 \sin \alpha_1 \\ l_2 \cos \alpha_2 & l_2 \sin \alpha_2 \\ 0 & 1 \end{pmatrix} \begin{pmatrix} \lambda_1 \\ \lambda_2 \end{pmatrix},$$

$$\mathbf{0} = \begin{pmatrix} l_1 \sin \alpha_1 + l_2 \sin \alpha_2 \\ r_3 - l_1 \cos \alpha_1 - l_2 \cos \alpha_2 \end{pmatrix},$$

with mass matrix

$$M(\alpha_1, \alpha_2) = \begin{pmatrix} J_1 + m_2 l_1^2 & 1/2 l_1 l_2 m_2 \cos(\alpha_1 - \alpha_2) & 0 \\ 1/2 l_1 l_2 m_2 \cos(\alpha_1 - \alpha_2) & J_2 & 0 \\ 0 & 0 & m_3 \end{pmatrix}.$$

The constants m_1, m_2, m_3 are the masses of the bodies, J_1, J_2 are the moments of inertia and l_1, l_2 the lengths of the rods. More details on this and related slider crank models can be found in [Hau89, Sim96].

2.2 Examples

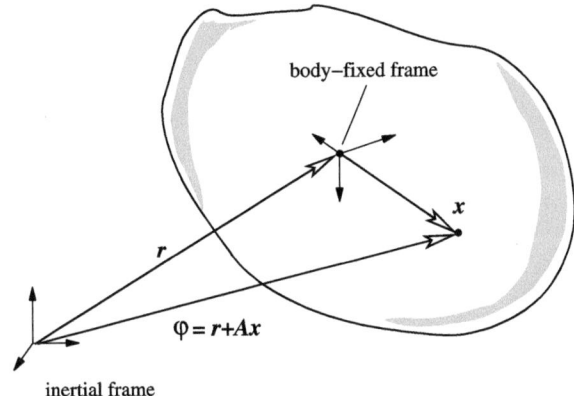

Fig. 2.4 Kinematics of a rigid body in three-dimensional space

A closer look at the constraint equations shows that we can easily eliminate the variable r_3 and write it as $r_3 = l_1 \cos\alpha_1 + l_2 \cos\alpha_2$. In contrast, the other constraint equation is a nonlinear relation

$$l_1 \sin\alpha_1 = -l_2 \sin\alpha_2$$

between the angles α_1 and α_2. One of the angles could be chosen as minimal coordinate of the system. E.g., if $l_1 < l_2$, we could express the first angle by $\alpha_1 = -\arcsin(l_1/l_2 \sin\alpha_2)$, but obviously the complexity of the resulting ordinary differential equation in the minimum coordinate α_2 would grow strongly.

2.2.2 Newton–Euler Equations

As second example, we consider a single rigid body in three-dimensional space. The well-known Newton–Euler equations describe the body's motion, and the question is how they are related to the framework of the Lagrange equations above.

Figure 2.4 depicts the body and two reference frames, a time-invariant orthonormal (inertial) frame and a body-fixed frame, whose origin is placed in the body's centroid. Let $x \in \mathbb{R}^3$ denote the coordinates of a point with respect to the body-fixed frame and $\boldsymbol{r}(t) \in \mathbb{R}^3$ the position of the centroid with respect to the inertial frame. The kinematics of the point is then given by

$$\boldsymbol{\varphi}(x,t) = \boldsymbol{r}(t) + A(t)x \qquad (2.19)$$

where $A(t) \in \mathbb{R}^{3\times 3}$ is the matrix that describes the rotation of the body-fixed frame. More precisely, we have $A(t) \in SO(3)$, the special orthogonal group that consists of all orthonormal and orientation-preserving matrices,

$$SO(3) = \{A \in \mathbb{R}^{3\times 3} : A^T A = AA^T = I \text{ and } \det A = 1\},$$

where I stands for the 3×3 identity matrix.

The body's kinetic energy reads

$$T = \frac{1}{2}\int_\Omega \rho \dot{\boldsymbol{\varphi}}^T \dot{\boldsymbol{\varphi}}\,\mathrm{d}x \qquad (2.20)$$

with ρ standing for the mass density and Ω for the domain that is occupied by the body. By differentiation of (2.19) we obtain

$$\frac{\mathrm{d}}{\mathrm{d}t}\boldsymbol{\varphi}(x,t) = \frac{\mathrm{d}}{\mathrm{d}t}\boldsymbol{r}(t) + \frac{\mathrm{d}}{\mathrm{d}t}\boldsymbol{A}(t)x = \dot{\boldsymbol{r}}(t) + \boldsymbol{A}(t)\tilde{\boldsymbol{\omega}}(t)x. \qquad (2.21)$$

Here, the angular velocity $\boldsymbol{\omega}(t) \in \mathbb{R}^3$ with respect to the body-fixed frame has been introduced from Poisson's equation

$$\frac{\mathrm{d}}{\mathrm{d}t}\boldsymbol{A}(t)^T \boldsymbol{A}(t) + \boldsymbol{A}(t)^T \frac{\mathrm{d}}{\mathrm{d}t}\boldsymbol{A}(t) = \boldsymbol{0}, \qquad (2.22)$$

which is a consequence of the orthogonality $\boldsymbol{A}^T \boldsymbol{A} = \boldsymbol{I}$. Poisson's equation (2.22) implies the skew-symmetry of the product

$$\boldsymbol{A}(t)^T \frac{\mathrm{d}}{\mathrm{d}t}\boldsymbol{A}(t) = \begin{pmatrix} 0 & -\omega_3(t) & \omega_2(t) \\ \omega_3(t) & 0 & -\omega_1(t) \\ -\omega_2(t) & \omega_1(t) & 0 \end{pmatrix} =: \tilde{\boldsymbol{\omega}}(t) \qquad (2.23)$$

and leads to the formal definition of the angular velocity $\boldsymbol{\omega}$. Notice that we employed here the compact notation $\tilde{\boldsymbol{a}}$ for a skew symmetric matrix based on the vector $\boldsymbol{a} \in \mathbb{R}^3$. This "tilde" notation is also a convenient means to express the vector product $\boldsymbol{a} \times \boldsymbol{b}$ as $\tilde{\boldsymbol{a}}\boldsymbol{b}$. Among others, the rules

$$\tilde{\boldsymbol{a}}\boldsymbol{b} = -\tilde{\boldsymbol{b}}\boldsymbol{a}, \qquad \tilde{\boldsymbol{a}}\boldsymbol{b} + \boldsymbol{a}\boldsymbol{b}^T = \tilde{\boldsymbol{b}}\boldsymbol{a} + \boldsymbol{b}\boldsymbol{a}^T$$

are frequently used. Premultiplying (2.23) by $\boldsymbol{A}$ and exploiting one more time the orthogonality, we deduce the relation

$$\frac{\mathrm{d}}{\mathrm{d}t}\boldsymbol{A}(t) = \boldsymbol{A}(t)\tilde{\boldsymbol{\omega}}(t)$$

between the rotation matrix $\boldsymbol{A}$ and the angular velocity $\boldsymbol{\omega}$, which leads to the representation (2.21).

The matrix $\boldsymbol{A}$ depends on the angles that are actually chosen to express the rotation. In case of Euler angles $\boldsymbol{\alpha}(t) = (\alpha_1(t), \alpha_2(t), \alpha_3(t))^T$ we have, omitting the time-dependence,

$$\boldsymbol{A}(\boldsymbol{\alpha}) = \begin{pmatrix} \cos\alpha_1 & -\sin\alpha_1 & 0 \\ \sin\alpha_1 & \cos\alpha_1 & 0 \\ 0 & 0 & 1 \end{pmatrix} \begin{pmatrix} 1 & 0 & 0 \\ 0 & \cos\alpha_2 & -\sin\alpha_2 \\ 0 & \sin\alpha_2 & \cos\alpha_2 \end{pmatrix} \begin{pmatrix} \cos\alpha_3 & -\sin\alpha_3 & 0 \\ \sin\alpha_3 & \cos\alpha_3 & 0 \\ 0 & 0 & 1 \end{pmatrix}.$$

2.2 Examples

Inserting this expression into the kinematic equation (2.21) and integrating, we arrive at the kinetic energy

$$T = \frac{1}{2} m \dot{r}(t)^T \dot{r}(t) + \frac{1}{2} \omega(t)^T \mathbf{J} \omega(t) \qquad (2.24)$$

where $m = \int \rho \, dx$ is the body's mass and $\mathbf{J} = \int \rho \tilde{x}^T \tilde{x} \, dx \in \mathbb{R}^{3 \times 3}$ its inertia tensor. The dynamics of a single rigid body in three-dimensional space is then described by the Newton–Euler equations [RS88]

$$\begin{aligned} m \ddot{r} &= f_f(t), \\ \mathbf{J} \dot{\omega} + \tilde{\omega} \mathbf{J} \omega &= f_m(t) \end{aligned} \qquad (2.25)$$

with applied forces $f_f(t) \in \mathbb{R}^3$ and applied moments $f_m(t) \in \mathbb{R}^3$.

Obviously, the Newton–Euler equations (2.25) do not fully comply with the second order structure presented in the Lagrange equations (2.9a), (2.9b) and (2.12). One way out is the introduction of a kinematic equation $\dot{\alpha} = S(\alpha) \omega$ where the transformation matrix $S(\alpha) \in \mathbb{R}^{3 \times 3}$ can be determined from the relation (2.23). Formally, we thus obtain for the law of angular momentum

$$\dot{\alpha} = S(\alpha) \omega, \qquad \mathbf{J} \dot{\omega} = f_m(t) - \tilde{\omega} \mathbf{J} \omega. \qquad (2.26)$$

This natural extension of the second order formulations, however, suffers from the drawback that the matrix $S(\alpha)$ might become singular due to the use of the Euler angles. The same drawback arises when using Rodrigues parameters or Cardan angles instead [RS88].

The other way out, which also circumvents the singularity problem, is the usage of four Euler parameters $\theta(t) = (\theta_0(t), \theta_1(t), \theta_2(t), \theta_3(t))^T$ for the description of spatial rotation. Based on this particular application of the quaternion algebra, it is possible to express the direction cosine matrix as

$$A(\theta) = \begin{pmatrix} \theta_0^2 + \theta_1^2 - \theta_2^2 - \theta_3^2 & 2(\theta_1 \theta_2 - \theta_0 \theta_3) & 2(\theta_1 \theta_3 + \theta_0 \theta_2) \\ 2(\theta_1 \theta_2 + \theta_0 \theta_3) & \theta_0^2 - \theta_1^2 + \theta_2^2 - \theta_3^2 & 2(\theta_2 \theta_3 - \theta_0 \theta_1) \\ 2(\theta_1 \theta_3 - \theta_0 \theta_2) & 2(\theta_2 \theta_3 + \theta_0 \theta_1) & \theta_0^2 - \theta_1^2 - \theta_2^2 + \theta_3^2 \end{pmatrix}.$$

The four Euler parameters are required to satisfy the normalizing condition

$$0 = \theta^T \theta - 1. \qquad (2.27)$$

Now the relation between angular velocity and Euler parameters reads

$$\omega = Q(\theta) \dot{\theta}, \qquad Q(\theta) := 2 \begin{pmatrix} -\theta_1 & \theta_0 & \theta_3 & -\theta_2 \\ -\theta_2 & -\theta_3 & \theta_0 & \theta_1 \\ -\theta_3 & \theta_2 & -\theta_1 & \theta_0 \end{pmatrix}, \qquad (2.28)$$

and the equations of motion preserve the second order structure. However, the normalizing condition (2.27) leads to a differential-algebraic formulation

$$m\ddot{r} = f_f(t),$$
$$Q(\theta)^T \mathbf{J} Q(\theta) \ddot{\theta} = Q(\theta)^T f_m(t) - Q(\theta)^T \widetilde{Q(\theta)\dot{\theta}} \mathbf{J} Q(\theta) \dot{\theta} - \theta \lambda, \quad (2.29)$$
$$0 = \theta^T \theta - 1,$$

which can be subsumed under the Lagrange equations of the first kind (2.9a), (2.9b).

Observe that the 4×4 inertia matrix $Q(\theta)^T \mathbf{J} Q(\theta)$ in (2.29) is singular while the augmented 5×5 matrix

$$\begin{pmatrix} Q(\theta)^T \mathbf{J} Q(\theta) & \theta \\ \theta^T & 0 \end{pmatrix} \quad (2.30)$$

is invertible for all θ that satisfy (2.27).

Besides the second order system (2.29), an unconstrained first order formulation for the equations of motion based on the Euler parameters is also common [RS10]. To this end, we premultiply the relation (2.28) between angular velocity ω and velocity $\dot{\theta}$ by $Q(\theta)^T$, yielding

$$Q(\theta)^T \omega = Q(\theta)^T Q(\theta) \dot{\theta} = 4(I - \theta \theta^T) \dot{\theta} = 4 \dot{\theta}.$$

Here, the normalization condition (2.27) has been differentiated to obtain the orthogonality property $\theta^T \dot{\theta} = 0$. The law of angular momentum is then expressed by

$$\dot{\theta} = \frac{1}{4} Q(\theta)^T \omega, \qquad \mathbf{J}\dot{\omega} = f_m(t) - \tilde{\omega} \mathbf{J} \omega. \quad (2.31)$$

In contrast to (2.26), the seven equations (2.31) are free of singularities. An interesting side aspect of (2.31) is that the properties $0 = \theta^T \theta - 1$ and $0 = \theta^T \dot{\theta}$ of the Euler parameters have now turned into *invariants*, and preserving them under discretization requires extra care. This issue will be taken up again in Sect. 2.3.4.

For more details on the use of Euler parameters and the underlying theory, we refer to Haug [Hau89] and Rabier and Rheinboldt [RR00].

2.2.3 Wheel Suspension

The above discussion of the Newton–Euler equations might suggest that in each spatial multibody system the description of rotations leads either to singularities or requires the use of Euler parameters. But this is not the case, since in practice joints constrain the motion, which can be exploited by reducing the rotational degrees of freedom a priori. In the following example, this approach is realized by means of relative coordinates.

2.2 Examples

Fig. 2.5 Five-link wheel suspension

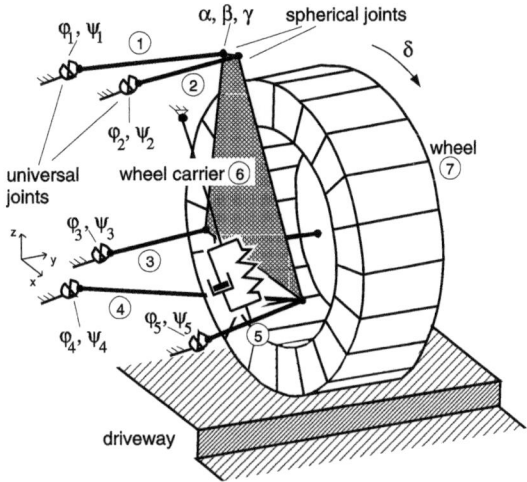

The spatial multibody system in Fig. 2.5 is a so-called *five-link wheel suspension*. It serves as a high comfort rear suspension system in upper class automobiles. The model consists of seven bodies, i.e., wheel, wheel carrier, and five rods connecting wheel carrier and car body. Besides several universal and spherical joints, it additionally features a spring-damper element, which represents a coil spring with shock absorber between rod number 5 and car body.

Using absolute coordinates, i.e., three rotations and three translations for each body, to describe the motion results in $7 \cdot 6 = 42$ position variables for this multibody system. A universal joint implies four constraint equations while a spherical joint is expressed in terms of three constraints. Overall, this leads to 40 additional constraints that determine the kinematics.

Alternatively, it is possible to partially eliminate the constraint equations by using the $n_q = 14$ relative coordinates

$$\boldsymbol{q} = (\varphi_1, \psi_1, \alpha, \beta, \gamma, \varphi_2, \psi_2, \varphi_3, \psi_3, \varphi_4, \psi_4, \varphi_5, \psi_5, \delta)^T,$$

which describe the relative angles in the joints and the rotation of the wheel. Yet, since the model has four closed kinematic loops, there exists no global state space form (2.12) in minimal coordinates, and the loop closing conditions still imply $n_\lambda = 12$ constraints. In total, despite taking advantage of relative coordinates, the equations of motion have the second order differential-algebraic form (2.9a), (2.9b) with solely $n_q - n_\lambda = 2$ degrees of freedom.

Hiller and Frik [HF92] proposed this wheel suspension as a benchmark problem in vehicle system dynamics. The generation of the governing equations requires the aid of a multibody formalism as the suspension model spreads over several thousand lines of code when written in symbolic form. For further details see [Sim95].

2.3 Differential-Algebraic Equations

The equations of constrained mechanical motion (2.9a), (2.9b) represent a special type of a differential-algebraic equation (DAE). As the analysis of (2.9a), (2.9b) and the design of appropriate numerical methods require a deeper understanding of the corresponding DAE theory, we include at this point a brief summary of the most important facts. For an in-depth exposition on differential-algebraic equations, the reader is referred to the monographs of Brenan, Campbell and Petzold [BCP96], Hairer and Wanner [HW96], Kunkel and Mehrmann [KM06], Lamour, März and Tischendorf [LMT13], and to the survey of Rabier and Rheinboldt [RR02].

2.3.1 Basic Types of DAEs

The most general type of a time-dependent differential equation is the fully implicit system

$$F(\dot{x}, x, t) = 0 \qquad (2.32)$$

with state variables $x(t) \in \mathbb{R}^{n_x}$ and a nonlinear, vector-valued function F of corresponding dimension. If the $n_x \times n_x$ Jacobian $\partial F / \partial \dot{x}$ is invertible, then by the implicit function theorem, it is, at least formally, possible to transform (2.32) to an explicit system of ordinary differential equations. If $\partial F / \partial \dot{x}$ is singular, however, (2.32) constitutes a *system of differential-algebraic equations*.

In most applications one actually has substantially more structural information available than in (2.32). An important class are *linear-implicit systems* of the form

$$E \dot{x} = \phi(x, t) \qquad (2.33)$$

with singular matrix $E \in \mathbb{R}^{n_x \times n_x}$ and right hand side function ϕ.

In some cases, e.g., in electric circuit simulation, the matrix E depends on the unknown states x, but here we will assume that E is constant. By applying Gaussian elimination with total pivoting or the singular value decomposition, it is then possible to transform E into

$$UEV = \begin{pmatrix} I & 0 \\ 0 & 0 \end{pmatrix} \qquad (2.34)$$

with invertible matrices U, V. The block matrix on the right has the same rank as E and features an identity matrix I and a zero block on the diagonal, which can be exploited to introduce new variables

$$V^{-1} x =: \begin{pmatrix} y \\ z \end{pmatrix} \qquad (2.35)$$

of appropriate dimensions. Multiplying (2.33) from the left by U, we obtain

$$\begin{pmatrix} \dot{y} \\ 0 \end{pmatrix} = U\phi \left(V \begin{pmatrix} y \\ z \end{pmatrix}, t \right) =: \begin{pmatrix} a(y, z, t) \\ b(y, z, t) \end{pmatrix}.$$

2.3 Differential-Algebraic Equations

It is convenient to further convert this system to autonomous form by adding the equation $\dot{t} = 1$ and appending time t as variable to the vector y. Keeping the notation unchanged, this yields finally the *semi-explicit system*

$$\dot{y} = a(y, z), \tag{2.36a}$$

$$0 = b(y, z). \tag{2.36b}$$

The differential-algebraic system (2.36a), (2.36b) shows a clear separation into n_y differential equations (2.36a) for the *differential variable* $y(t) \in \mathbb{R}^{n_y}$ and n_z constraints (2.36b), which define the *algebraic variables* $z \in \mathbb{R}^{n_z}$.

For the convergence analysis of numerical time integration methods, the system (2.36a), (2.36b) is usually the easiest starting point. If the method is invariant under a transformation from the linear-implicit system (2.33) to (2.36a), (2.36b), the convergence results then also hold for (2.33).

Van der Pol's Equation

To study an example for a semi-explicit system, we consider Van der Pol's equation

$$\epsilon \ddot{q} + (q^2 - 1)\dot{q} + q = 0 \tag{2.37}$$

with parameter $\epsilon > 0$. This is an oscillator equation with a nonlinear damping term that acts as a controller. For large amplitudes $q^2 > 1$, the damping term introduces dissipation into the system while for small values $q^2 < 1$, the sign changes and the damping term is replaced by an excitation, leading thus to a self-exciting oscillator. Introducing Liénhard's coordinates [HNW93]

$$z := q, \quad y := \epsilon \dot{z} + (z^3/3 - z),$$

we transform (2.37) into the first order system

$$\dot{y} = -z, \tag{2.38a}$$

$$\epsilon \dot{z} = y - \frac{z^3}{3} + z. \tag{2.38b}$$

The case $\epsilon \ll 1$ is of special interest. In the limit $\epsilon = 0$, Eq. (2.38b) turns into a constraint and we arrive at the semi-explicit system

$$\dot{y} = -z, \tag{2.39a}$$

$$0 = y - \frac{z^3}{3} + z. \tag{2.39b}$$

In other words, Van der Pol's equations (2.38a), (2.38b) in Liénhard's coordinates is an example of a *singularly perturbed system* which tends to the semi-explicit DAE (2.39a), (2.39b) when $\epsilon \to 0$.

Such a close relation between a singularly perturbed system and a differential-algebraic equation is quite common and can be found in various application fields. Often, the parameter ϵ stands for an almost negligible physical quantity or the presence of strongly different time scales. Analyzing the *reduced system*, in this case (2.39a), (2.39b), usually proves successful to gain a better understanding of the original perturbed equation [O'M74]. In the context of *regularization methods*, this relation is also exploited, but in reverse direction [Han90]. One starts with a DAE such as (2.39a), (2.39b) and replaces it by a singularly perturbed ODE, in this case (2.38a), (2.38b). We will take up this connection in Chap. 6 when discussing stiff mechanical systems.

2.3.2 The Index

The index of a differential-algebraic equation measures its well-posedness when compared to an ordinary differential equation. This key concept has evolved over several decades, and today a number of definitions with different emphasis exist. In this short survey, we focus first on linear constant coefficient systems and the nilpotency index, continue with the differentiation index and finally include also the perturbation index.

Linear Constant Coefficient DAEs

The analysis of a linear constant coefficient DAE reveals already a number of fundamental properties. Consider the initial value problem

$$E\dot{x} + Hx = c, \qquad x(t_0) = x_0, \qquad (2.40)$$

with constant matrices $E, H \in \mathbb{R}^{n_x \times n_x}$ and some time-dependent excitation $c(t) \in \mathbb{R}^{n_x}$. While the matrix E is regular in the ODE case and can be brought to the right hand side by formal inversion, it is singular in the DAE case. The singularity, however, may not be arbitrary. We assume that the *matrix pencil* (E, H) is regular, i.e., that there exists $\mu \in \mathbb{C}$ such that the matrix $\mu E + H$ is regular. Otherwise, the pencil is singular, and (2.40) has either no or infinitely many solutions [Cam82].

If (E, H) is a regular pencil, there exist non-singular matrices U and V such that

$$UEV = \begin{pmatrix} I & 0 \\ 0 & N \end{pmatrix}, \quad UHV = \begin{pmatrix} C & 0 \\ 0 & I \end{pmatrix} \qquad (2.41)$$

where N is a nilpotent matrix, I an identity matrix, and C a matrix that can be assumed to be in Jordan canonical form. The transformation (2.41) is called the *Kronecker canonical form* [Kro90]. It is a generalization of the Jordan canonical form and contains the essential structure of the linear system (2.40).

2.3 Differential-Algebraic Equations

The proof of (2.41) is particularly easy for $\mu = 0$, i.e., when the matrix H is regular. Then, the Jordan canonical form of $H^{-1}E$ yields

$$H^{-1}E = TJT^{-1}, \quad J = \begin{pmatrix} R & 0 \\ 0 & N \end{pmatrix},$$

where R corresponds to the Jordan blocks with nonzero eigenvalues and N to the zero eigenvalues. Thus, setting $C := R^{-1}$ and

$$U := T^{-1}H^{-1}, \quad V := T\begin{pmatrix} R^{-1} & 0 \\ 0 & I \end{pmatrix}$$

results in the decomposition (2.41). For the derivation in the general case see Gantmacher [Gan59].

In the Kronecker canonical form (2.41), the singularity of the DAE is represented by the nilpotent matrix N. Its degree of nilpotency, i.e., the smallest positive integer k such that $N^k = 0$, plays a key role when studying closed-form solutions of the linear system (2.40).

In passing we note that the transformation matrices U and V in the Kronecker form (2.41) are, in general, not equivalent to the matrices in the decomposition (2.34). An exception is the case $k = 1$ where N^1 is the zero matrix. Moreover, while standard numerical algorithms compute the decomposition (2.34) in a stable way, the computation of the Jordan canonical form and consequently also of the Kronecker form are notoriously ill-conditioned problems.

To construct a solution of (2.40), we introduce new variables and right hand side vectors

$$V^{-1}x =: \begin{pmatrix} y \\ z \end{pmatrix}, \quad Uc =: \begin{pmatrix} \delta \\ \theta \end{pmatrix}. \tag{2.42}$$

Premultiplying (2.40) by U then leads to the *decoupled system*

$$\dot{y} + Cy = \delta, \tag{2.43a}$$

$$N\dot{z} + z = \theta. \tag{2.43b}$$

While the solution of the ODE (2.43a) follows by integrating and results in an expression based on the matrix exponential $\exp(-C(t - t_0))$, Eq. (2.43b) for z can be solved recursively by differentiating. More precisely, we have

$$N\ddot{z} + \dot{z} = \dot{\theta} \quad \Rightarrow \quad N^2\ddot{z} = -N\dot{z} + N\dot{\theta} = z - \theta + N\dot{\theta}.$$

Repeating the differentiation and multiplication by N, we can eventually exploit the nilpotency and get

$$0 = N^k z^{(k)} = (-1)^k z + \sum_{\ell=0}^{k-1} (-1)^{k-1-\ell} N^\ell \theta^{(\ell)}.$$

This implies the explicit representation

$$z = \sum_{\ell=0}^{k-1}(-1)^\ell N^\ell \theta^{(\ell)}. \tag{2.44}$$

The above solution procedure illustrates several crucial points about differential-algebraic equations and how they differ from ordinary differential equations. We highlight two of these:

(i) The solution of (2.40) rests on $k-1$ differentiation steps. This requires that the derivatives of certain components of θ exist up to $\ell = k-1$. Furthermore, some components of z may only be continuous but not differentiable depending on the smoothness of θ.
(ii) The components of z are directly given in terms of the right hand side data θ and its derivatives. Accordingly, the initial value $z(t_0) = z_0$ is fully determined by (2.44) and, in contrast to y_0, cannot be chosen arbitrarily. Initial values (y_0, z_0) where z_0 satisfies (2.44) are called *consistent*. The same terminology applies to the initial value x_0, which is consistent if, after the transformation (2.42), z_0 satisfies (2.44).

Nilpotency Index

As seen above, constructing a closed-form solution of a linear DAE is based on repeated differentiation until the nilpotency of the singular part comes into play. Since differentiation is an unstable numerical process, the degree of nilpotency k measures the numerical difficulty associated with the system (2.40). The integer k is therefore called the *(nilpotency) index of Eq.* (2.40).

Obviously, the nilpotency index is restricted to linear systems with a regular matrix pencil. Its generalization to the nonlinear case will be discussed next.

Differentiation Index

Both the index concept and the definition of consistent initial values can be generalized for the fully implicit system (2.32). Following Gear [Gea88, Gea90], we define the index k by

$k = 0$: If $\partial F/\partial \dot{x}$ is non-singular, the index is 0.
$k > 0$: Otherwise, consider the system of equations

$$F(\dot{x}, x, t) = 0,$$

$$\frac{d}{dt}F(\dot{x}, x, t) = \frac{\partial}{\partial \dot{x}}F(\dot{x}, x, t)x^{(2)} + \ldots = 0, \tag{2.45}$$

$$\vdots$$

$$\frac{d^s}{dt^s}F(\dot{x}, x, t) = \frac{\partial}{\partial \dot{x}}F(\dot{x}, x, t)x^{(s+1)} + \ldots = 0$$

2.3 Differential-Algebraic Equations

as a system in the separate dependent variables $\dot{x}, x^{(2)}, \ldots, x^{(s+1)}$, with x and t as independent variables. Then the index k is the smallest s for which it is possible, using algebraic manipulations only, to extract an ordinary differential equation $\dot{x} = \psi(x, t)$ from (2.45).

In the case of linear constant coefficient systems (2.40), the differential index is equivalent to the nilpotency index.

As example, we consider the semi-explicit system (2.36a), (2.36b) where the unknown variables are a priori partitioned into differential variables y and algebraic variables z. This allows one to determine the index by simply differentiating the constraint equation $0 = b(y, z)$. We assume that the Jacobian matrix

$$\frac{\partial b}{\partial z}(y, z) \in \mathbb{R}^{n_z \times n_z} \quad \text{is invertible} \tag{2.46}$$

in a neighborhood of the solution. Differentiating (2.36b) then leads to

$$0 = \frac{d}{dt}b(y, z) = \frac{\partial b}{\partial y}(y, z)\dot{y} + \frac{\partial b}{\partial z}(y, z)\dot{z}.$$

This implies

$$\dot{z} = -\left(\frac{\partial b}{\partial z}(y, z)\right)^{-1} \frac{\partial b}{\partial y}(y, z) \cdot a(y, z), \tag{2.47}$$

which is the desired ordinary differential equation for the variable z. In other words, (2.36a), (2.36b) is of index $k = 1$ if the assumption (2.46) holds. The combination of (2.47) with (2.36a) is called the *underlying ODE* of the differential-algebraic system. An initial value (y_0, z_0) of (2.36a), (2.36b) is consistent if the constraint is satisfied, $0 = b(y_0, z_0)$.

As second example, we analyze the semi-explicit equation

$$\dot{y} = a(y, z), \tag{2.48a}$$

$$0 = b(y) \tag{2.48b}$$

under the assumption

$$\frac{\partial b}{\partial y}(y) \cdot \frac{\partial a}{\partial z}(y, z) \in \mathbb{R}^{n_z \times n_z} \quad \text{is invertible} \tag{2.49}$$

in a neighborhood of the solution. Now, differentiating twice and setting $B(y) := \partial b(y)/\partial y$, we get

$$0 = B(y)\dot{y} = B(y)a(y, z) \tag{2.50}$$

and

$$0 = B(y)\frac{\partial a}{\partial z}(y, z)\dot{z} + B(y)\frac{\partial a}{\partial y}(y, z)a(y, z) + \frac{d}{dt}B(y) \cdot a(y, z). \tag{2.51}$$

Due to assumption (2.49), the index of (2.48a), (2.48b) is thus $k = 2$.

As a rule of thumb, differential-algebraic equations of index 2 or higher are generally more difficult to analyze and to solve numerically than ordinary differential equations or DAEs of index 1. One reason for this is the presence of *hidden constraints* such as (2.50). Note that a consistent initial value for the system (2.48a), (2.48b) must satisfy both the original and the hidden constraint, i.e.,

$$0 = b(y_0), \quad 0 = B(y_0) a(y_0, z_0). \tag{2.52}$$

In practice, finding such consistent initial values may constitute a challenging problem of its own [BCP96, ST00].

Perturbation Index

The index is the standard approach to classify differential-algebraic equations. While the differential index is based on successively differentiating the original DAE until the obtained system can be solved for $\dot{x}$, the *perturbation index* introduced by Hairer, Lubich and Roche [HLR89] measures the sensitivity of the solutions to perturbations in the equation:

The system $F(\dot{x}, x, t) = 0$ has perturbation index $k \geq 1$ along a solution $x(t)$ on $[t_0, t_1]$ if k is the smallest integer such that, for all functions $\hat{x}$ having a defect

$$F(\dot{\hat{x}}, \hat{x}, t) = \delta(t),$$

there exists on $[t_0, t_1]$ an estimate

$$\|\hat{x}(t) - x(t)\| \leq c \left(\|\hat{x}(t_0) - x(t_0)\| + \max_{t_0 \leq \xi \leq t} \|\delta(\xi)\| + \ldots + \max_{t_0 \leq \xi \leq t} \|\delta^{(k-1)}(\xi)\| \right)$$

whenever the expression on the right hand side is sufficiently small. Note that the constant c depends only on F and on the length of the interval, but not on the perturbation δ. The perturbation index is $k = 0$ if

$$\|\hat{x}(t) - x(t)\| \leq c \left(\|\hat{x}(t_0) - x(t_0)\| + \max_{t_0 \leq \xi \leq t} \left\| \int_{t_0}^{\xi} \delta(\tau) \, d\tau \right\| \right),$$

which is satisfied for ordinary differential equations.

If the perturbation index exceeds $k = 1$, derivatives of the perturbation show up in the estimate and indicate a certain degree of ill-posedness. E.g., if δ contains a small high frequency term $\epsilon \sin \omega t$ with $\epsilon \ll 1$ and $\omega \gg 1$, the resulting derivatives will induce a severe amplification in the bound for $\hat{x}(t) - x(t)$.

Unfortunately, the differential and the perturbation index are not equivalent in general and may even differ substantially [CG95]. As an example, consider the linear-implicit system

$$\begin{pmatrix} 0 & y_3 & 0 \\ 0 & 0 & y_3 \\ 0 & 0 & 0 \end{pmatrix} \begin{pmatrix} \dot{y}_1 \\ \dot{y}_2 \\ \dot{y}_3 \end{pmatrix} + \begin{pmatrix} y_1 \\ y_2 \\ y_3 \end{pmatrix} = \mathbf{0}. \tag{2.53}$$

2.3 Differential-Algebraic Equations

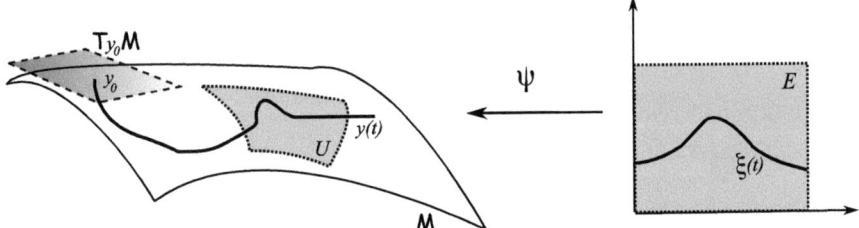

Fig. 2.6 Manifold $\mathcal{M}$, tangent space $\mathcal{T}_y\mathcal{M}$, and local parametrization

The last equation is $y_3 = 0$, which immediately implies $y_1 = 0$ and $y_2 = 0$. Differentiating these equations once yields the underlying ordinary differential equation, and accordingly the differential index equals 1. If the right hand side, on the other hand, is perturbed by $\boldsymbol{\delta} = (\delta_1, \delta_2, \delta_3)^T$, we can compute the perturbed solution in a way similar to the derivation of (2.44), obtaining eventually an expression for $\hat{y}_1$ that involves the second derivative $\delta_3^{(2)}$. The perturbation index is hence 3.

The example (2.53) extends easily to arbitrary dimension n_y. While the perturbation index equals n_y and grows with the dimension, the differential index stays at 1. In case of semi-explicit systems (2.36a), (2.36b), however, such an inconsistence does not arise, and both indices can be shown to be equivalent.

2.3.3 Constraint Manifold and Local State Space Form

So far, our discussion of differential-algebraic equations has been mainly inspired by differential calculus and algebraic considerations. A fundamentally different aspect comes into play by adopting the viewpoint of *differential equations on manifolds*, as introduced by Rheinboldt [Rhe84].

To illustrate this approach, we consider again the semi-explicit system (2.48a), (2.48b) of index 2 where the constraint $\mathbf{0} = \boldsymbol{b}(\boldsymbol{y})$, assuming sufficient differentiability, defines the manifold

$$\mathcal{M} := \left\{ \boldsymbol{y} \in \mathbb{R}^{n_y} : \boldsymbol{b}(\boldsymbol{y}) = \mathbf{0} \right\}. \tag{2.54}$$

The full rank condition (2.49) for the matrix product $\partial \boldsymbol{b}/\partial \boldsymbol{y} \cdot \partial \boldsymbol{a}/\partial \boldsymbol{z}$ implies that the Jacobian $\boldsymbol{B}(\boldsymbol{y}) = \partial \boldsymbol{b}(\boldsymbol{y})/\partial \boldsymbol{y} \in \mathbb{R}^{n_z \times n_y}$ possesses also full rank n_z. Hence, for fixed $\boldsymbol{y} \in \mathcal{M}$, the tangent space

$$\mathcal{T}_y\mathcal{M} := \left\{ \boldsymbol{v} \in \mathbb{R}^{n_y} : \boldsymbol{B}(\boldsymbol{y})\boldsymbol{v} = \mathbf{0} \right\} \tag{2.55}$$

spans the kernel of $\boldsymbol{B}$ and has the same dimension $n_y - n_z$ as the manifold $\mathcal{M}$. Figure 2.6 depicts $\mathcal{M}$, $\mathcal{T}_y\mathcal{M}$, and a solution of the DAE (2.48a), (2.48b), which, starting from a consistent initial value, is required to proceed on the manifold.

The differential equation on the manifold $\mathcal{M}$ that is equivalent to the DAE (2.48a), (2.48b) is obtained as follows: The hidden constraint

$$0 = B(y)a(y,z)$$

can be solved for $z(y)$ according to the rank condition (2.49) and the implicit function theorem. Moreover, for $y \in \mathcal{M}$ we have $a(y,z(y)) \in T_y\mathcal{M}$, which defines a vector field on the manifold $\mathcal{M}$ [AMR88]. Overall,

$$\dot{y} = a(y,z(y)) \quad \text{for } y \in \mathcal{M} \tag{2.56}$$

represents then a differential equation on the manifold [Arn81, Rhe84].

In theory, and also computationally [Rhe96], it is possible to transform the differential equation (2.56) from the manifold to an ordinary differential equation in a linear space of dimension $n_y - n_z$. For this purpose, one introduces a *local parametrization*

$$\psi : E \to \mathcal{U} \tag{2.57}$$

where E is an open subset of $\mathbb{R}^{n_y - n_z}$ and $\mathcal{U} \subset \mathcal{M}$, see Fig. 2.6. Such a parametrization is not unique and holds only locally in general. It is, however, possible to extend it to a family of parametrizations such that the whole manifold is covered. For $y \in \mathcal{U}$ and local coordinates $\xi \in E$ we thus get the relations

$$y = \psi(\xi), \quad \dot{y} = \Psi(\xi)\dot{\xi}, \quad \Psi(\xi) := \frac{\partial \psi}{\partial \xi}(\xi) \in \mathbb{R}^{n_y \times (n_y - n_z)}.$$

Premultiplying (2.56) by the transpose of the Jacobian $\Psi(\xi)$ of the parametrization and substituting y by $\psi(\xi)$, we arrive at

$$\Psi(\xi)^T \Psi(\xi) \dot{\xi} = \Psi(\xi)^T a(\psi(\xi), z(\psi(\xi))). \tag{2.58}$$

Since the Jacobian Ψ has full rank for a valid parametrization, the matrix $\Psi^T \Psi$ is invertible, and (2.58) constitutes the desired ordinary differential equation in the local coordinates ξ. In analogy to the mechanical system (2.12) in minimal coordinates, we call (2.58) a *local state space form*.

The process of transforming a differential equation on a manifold to a local state space form constitutes a *push forward* operator, while the reverse mapping is called a *pull back* operator [AMR88]. It is important to realize that the previously defined concept of an index does not appear in the theory of differential equations on manifolds. Finding hidden constraints by differentiation, however, is also crucial for the classification of DAEs from a geometric point of view.

2.3.4 Solution Invariants versus Constraints

Differential equations with invariants are sometimes confused with differential-algebraic equations. In fact, there exists a close relation between both concepts, and this connection plays a major role for *index reduction* and *stabilization techniques*.

2.3 Differential-Algebraic Equations

Assume the solution of the ordinary differential equation

$$\dot{y} = \phi(y), \tag{2.59}$$

satisfies the invariant

$$0 = c(y). \tag{2.60}$$

An example are the Newton–Euler equations with quaternions in the formulation (2.31).

If (2.60) is even valid in a neighborhood of the solution, the invariant is called a *first integral*. Since (2.60) holds for all t, we may differentiate the invariant and obtain the relation

$$0 = C(y)\dot{y} = C(y)\phi(y) \tag{2.61}$$

with $C(y) := \partial c(y)/\partial y$.

Linear invariants are preserved by standard time integration methods, but other types of invariants require either specific integrators, e.g., symplectic methods in case of Hamiltonian systems, or additional measures to enforce (2.60). In this context, reformulating the equations as a differential-algebraic system

$$\dot{y} = \phi(y) - C(y)^T z, \tag{2.62a}$$
$$0 = c(y) \tag{2.62b}$$

is a tempting option. The DAE (2.62a), (2.62b) is semi-explicit, with the algebraic variables z playing the role of Lagrange multipliers. Moreover, the invariant (2.60) has turned into the constraint (2.62b).

If the Jacobian $C(y)$ has full rank, the condition (2.49) is satisfied, and the DAE (2.62a), (2.62b) is thus of index 2. Even more, assume y solves (2.59) and satisfies (2.60) as well as (2.61). Inserting y into the DAE (2.62a), (2.62b), we have by differentiating the constraint

$$0 = C(y)\phi(y) - C(y)C(y)^T z = -C(y)C(y)^T z.$$

Since the square matrix $C(y)C(y)^T$ is invertible if $C(y)$ has full rank, we conclude that $z = 0$. In other words, the extra multipliers vanish, and the pair (y, z) where $z \equiv 0$ solves the DAE (2.62a), (2.62b).

Appending an invariant as a constraint by means of suitable Lagrange multipliers is not always a good remedy. Properties such as symplecticity require a specific discretization and cannot be simply enforced in this way [HW96, §VII.2]. For higher index problems, however, this approach can be used to lower the index without losing the information of the original constraint. In Sect. 2.4.4 below, we will explain this in more detail when studying alternative formulations of the equations of constrained mechanical motion.

2.3.5 Differential-Algebraic Equations in Saddle Point Form

At the end of this brief exposition on differential-algebraic equations, we discuss a special structure that is widespread in applications and that plays a key role throughout this book. Consider the linear semi-explicit system

$$\dot{y} + Ay + Bz = \ell(t), \qquad (2.63a)$$

$$By = m(t), \qquad (2.63b)$$

with a positive semi-definite matrix $A \in \mathbb{R}^{n_y \times n_y}$ and a rectangular matrix $B \in \mathbb{R}^{n_z \times n_y}$, $n_z < n_y$, of full rank. As the DAE (2.63a), (2.63b) can be subsumed under the class of semi-explicit systems (2.48a), (2.48b), its index is obviously 2. Written as general linear system $E\dot{x} + Hx = c$ as in (2.40) in the unknowns $x = (y, z)$, we have

$$E = \begin{pmatrix} I & 0 \\ 0 & 0 \end{pmatrix}, \qquad H = \begin{pmatrix} A & B^T \\ B & 0 \end{pmatrix}. \qquad (2.64)$$

While the matrix E reflects the singularity of the system, the matrix H comprises the constraints and the coupling of differential and algebraic variables.

We call the symmetric matrix H a *saddle point matrix*. The name is inspired by optimization theory, as we explain in the following.

A function $h : \mathbb{R}^2 \to \mathbb{R}$ has a saddle point in $x^* \in \mathbb{R}^2$ if the Hessian matrix $\nabla^2 h(x^*)$ possesses one positive and one negative eigenvalue. A particular example is the case where

$$\nabla^2 h(x^*) = \begin{pmatrix} a & b \\ b & 0 \end{pmatrix} \quad \text{with } a \geq 0, b \neq 0. \qquad (2.65)$$

A straightforward calculation shows that the two eigenvalues are here given by

$$\mu_1 = \frac{a}{2} + \sqrt{a^2/4 + b^2} > a \geq 0, \quad \mu_2 = \frac{a}{2} - \sqrt{a^2/4 + b^2} < 0.$$

Thus, for a positive and b different from zero, (2.65) satisfies the saddle point criterion.

The generalization of this little example replaces a by the positive semi-definite matrix A and b by the constraint matrix B, which is required to be of full rank. It can then be shown that the matrix H from (2.64) is indefinite as well, which gives rise to the term saddle point matrix. Moreover, if we study the optimization problem

$$y^T A y - \ell \to \text{min!} \quad \text{subject to } By = m$$

and apply the Lagrange multiplier theorem, we obtain the necessary condition that

$$\begin{pmatrix} A & B^T \\ B & 0 \end{pmatrix} \begin{pmatrix} y \\ z \end{pmatrix} = \begin{pmatrix} \ell \\ m \end{pmatrix}.$$

Once again, the fundamental saddle point structure becomes evident here.

In the transient case, appending constraints by Lagrange multipliers also implants a natural saddle point structure into the differential-algebraic system. The treatment of invariants as analyzed in (2.62a), (2.62b) is another, in this case nonlinear example where we encounter the same structural feature.

2.3.6 Remarks and Further References

Historically, the subject of differential-algebraic equations has emerged from two application fields, namely multibody dynamics and electric circuit analysis. Though both fields seem completely different at first sight, there are some analogies in the mathematical models. In particular, the idea of viewing the technical system as a *network* of different components, whose interaction is described by the laws of physics, opened the door for a generalized mathematical framework. References for differential-algebraic equations in circuit simulation are, e.g., Günther and Feldmann [GF99] and März and Tischendorf [MT97].

Unstructured problems of higher index and their numerical solution based on the *derivative array* (2.45) are discussed, among others, in Campbell [Cam93] and Kunkel and Mehrmann [KM98], while a general existence theory of DAEs is presented in Rabier and Rheinboldt [RR02]. As references for linear differential-algebraic equations with variable coefficients we mention finally Kunkel and Mehrmann [KM96] and März [Mär96].

2.4 Analysis of the Equations of Constrained Mechanical Motion

The state space form (2.12) is a system of second order ordinary differential equations. Consequently, all results from standard theory on the existence and uniqueness of solutions carry over. In contrast, the equations of constrained mechanical motion (2.9a), (2.9b) constitute a differential-algebraic system of index 3, as we will see in the following. For this purpose, it is convenient to rewrite the equations as a system of first order

$$\dot{q} = v, \tag{2.66a}$$

$$M(q)\dot{v} = f(q, v, t) - G(q)^T \lambda, \tag{2.66b}$$

$$0 = g(q) \tag{2.66c}$$

with additional velocity variables $v(t) \in \mathbb{R}^{n_q}$.

2.4.1 Index and Existence of Solutions

In order to analyze the differential-algebraic equation (2.66a)–(2.66c), we apply a step-by-step procedure that identifies the hidden constraints and allows to solve for the Lagrange multiplier λ, which is the algebraic variable.

By differentiating the constraints (2.66c) with respect to time, we obtain the *constraints at velocity level*

$$0 = \frac{\mathrm{d}}{\mathrm{d}t} g(q) = G(q)\dot{q} = G(q)v. \tag{2.67}$$

A second differentiation step yields the *constraints at acceleration level*

$$0 = \frac{\mathrm{d}^2}{\mathrm{d}t^2} g(q) = G(q)\dot{v} + \kappa(q,v), \quad \kappa(q,v) := \frac{\partial G(q)}{\partial q}(v,v), \tag{2.68}$$

where the two-form κ comprises additional derivative terms.

Combining the dynamic equation (2.66b) and the acceleration constraints (2.68), we finally arrive at the linear system

$$\begin{pmatrix} M(q) & G(q)^T \\ G(q) & 0 \end{pmatrix} \begin{pmatrix} \dot{v} \\ \lambda \end{pmatrix} = \begin{pmatrix} f(q,v,t) \\ -\kappa(q,v) \end{pmatrix}. \tag{2.69}$$

The matrix on the left hand side has a saddle point structure. We presuppose that

$$\begin{pmatrix} M(q) & G(q)^T \\ G(q) & 0 \end{pmatrix} \text{ is invertible} \tag{2.70}$$

in a neighborhood of the solution. A necessary but not sufficient condition for (2.70) is the full rank of the constraint Jacobian G as stated in (2.2). If in addition the mass matrix M is symmetric positive definite, (2.70) obviously holds. We remark, however, that there are applications where the mass matrix is singular but the prerequisite (2.70) nevertheless is satisfied. An example is given by the formulation (2.29) for the Newton–Euler equations with quaternions.

Assuming (2.70) and a symmetric positive definite mass matrix, we can solve the linear system (2.69) for the acceleration $\dot{v}$ and the Lagrange multiplier λ by block Gaussian elimination. This leads to the explicit expressions

$$\dot{v} = M(q)^{-1}\left(f(q,v,t) - G^T(q)\lambda\right), \tag{2.71}$$

$$\lambda = \left(G(q)M(q)^{-1}G^T(q)\right)^{-1}\left(G(q)M(q)^{-1}f(q,v,t) + \kappa(q,v)\right). \tag{2.72}$$

The representation $\lambda = \lambda(q,v,t)$ from (2.72) is now inserted into (2.71), which leads to an ordinary differential equation for the velocity variables v. Under the usual assumption of Lipschitz continuity of the corresponding right hand side, the

2.4 Analysis of the Equations of Constrained Mechanical Motion

unique solution (q, v) of (2.66a) and (2.71) is guaranteed and in turn, the multiplier λ is uniquely determined as well.

At the same time, the above differentiation steps also define the index of the DAE (2.66a)–(2.66c). Two differentiation steps result in the linear system (2.69) that allows to solve for the Lagrange multiplier as a function $\lambda(q, v, t)$. A final third differentiation step yields an ordinary differential equation for this algebraic variable, which implies that the differentiation index is 3.

Note that the solution of the ODE (2.66a) and (2.71) does not necessarily fulfill the differential-algebraic equation (2.66a)–(2.66c) since the differentiation steps involve a loss of integration constants. However, if the initial values (q_0, v_0) are consistent, i.e., if they satisfy the original constraints and the velocity constraints,

$$0 = g(q_0), \qquad 0 = G(q_0) v_0, \tag{2.73}$$

the solution of (2.66a) and (2.71) also fulfills the original system (2.66a)–(2.66c). Above all, depending on q_0 and v_0, the initial value λ_0 for the Lagrange multiplier is completely determined by (2.72).

Hidden Constraints of Planar Pendulum

As example for the hidden constraints of the equations of motion, we consider the planar pendulum (2.13a), (2.13b) with velocity variables

$$v(t) = \begin{pmatrix} v_1(t) \\ v_2(t) \\ v_3(t) \end{pmatrix} := \begin{pmatrix} \dot{r}_1(t) \\ \dot{r}_2(t) \\ \dot{\alpha}(t) \end{pmatrix}.$$

Differentiating the constraints (2.13b) at position level, we obtain the constraints at velocity level

$$0 = \frac{d}{dt} \begin{pmatrix} r_1 - \frac{l}{2} \cos\alpha \\ r_2 - \frac{l}{2} \sin\alpha \end{pmatrix} = \begin{pmatrix} 1 & 0 & \frac{l}{2}\sin\alpha \\ 0 & 1 & -\frac{l}{2}\cos\alpha \end{pmatrix} \begin{pmatrix} v_1 \\ v_2 \\ v_3 \end{pmatrix}.$$

The constraints at acceleration level, finally, are given by

$$0 = \frac{d^2}{dt^2} \begin{pmatrix} r_1 - \frac{l}{2}\cos\alpha \\ r_2 - \frac{l}{2}\sin\alpha \end{pmatrix} = \begin{pmatrix} 1 & 0 & \frac{l}{2}\sin\alpha \\ 0 & 1 & -\frac{l}{2}\cos\alpha \end{pmatrix} \begin{pmatrix} \dot{v}_1 \\ \dot{v}_2 \\ \dot{v}_3 \end{pmatrix} + v_3^2 \begin{pmatrix} \frac{l}{2}\cos\alpha \\ \frac{l}{2}\sin\alpha \end{pmatrix}.$$

For completeness, we furthermore state Eq. (2.72) for the two arising Lagrange multipliers. It reads

$$\begin{pmatrix} \lambda_1 \\ \lambda_2 \end{pmatrix} = -\frac{m\gamma}{4} \begin{pmatrix} 3\sin\alpha\cos\alpha \\ 1 + 3\sin^2\alpha \end{pmatrix} + m v_3^2 \begin{pmatrix} \frac{l}{2}\cos\alpha \\ \frac{l}{2}\sin\alpha \end{pmatrix}.$$

We point out that, in practice, the complexity of the equations of motion prohibits such explicit expressions as given here for the planar pendulum example. Instead, the multibody formalisms provide automatic procedures for numerically evaluating constraints and differentiated constraints.

2.4.2 Minimax Characterization of Constraints

As a prelude to the topic of transient saddle point problems in Chap. 3, we reconsider next the full rank criterion (2.2). The above analysis steps demonstrate that the full rank of the constraint Jacobian G is a fundamental property. We can reformulate this criterion in terms of the singular values of G, which are given by the factorization [GL96, §2.5]

$$U^T G(q) V = \text{diag}(\sigma_1, \ldots, \sigma_{n_\lambda}) \in \mathbb{R}^{n_\lambda \times n_q}$$

with orthogonal matrices $U \in \mathbb{R}^{n_\lambda \times n_\lambda}$ and $V \in \mathbb{R}^{n_q \times n_q}$. The singular values are ordered as

$$\sigma_1 \geq \sigma_2 \geq \ldots \geq \sigma_{n_\lambda} \geq 0,$$

and for the full rank of G we require $\sigma_{\min} := \sigma_{n_\lambda} > 0$.

Omitting the argument q of G for brevity, we can reformulate this criterion by observing that, for any vector $0 \neq \lambda \in \mathbb{R}^{n_\lambda}$ and $\mu := U^T \lambda$,

$$\frac{\lambda^T G G^T \lambda}{\lambda^T \lambda} = \frac{\mu^T \text{diag}(\sigma_1^2, \ldots, \sigma_{\min}^2) \mu}{\mu^T \mu} \geq \sigma_{\min}^2.$$

In case of $\mu = (0, \ldots, 0, 1)^T \in \mathbb{R}^{n_\lambda}$, this inequality is sharp and we conclude

$$\sigma_{\min}^2 = \min_\lambda \frac{\lambda^T G G^T \lambda}{\lambda^T \lambda} \iff \sigma_{\min} = \min_\lambda \frac{\|G^T \lambda\|_2}{\|\lambda\|_2}.$$

Moreover, using the definition of the operator norm we get the identity

$$\|G^T \lambda\|_2 = \max_v \frac{\|v^T G^T \lambda\|_2}{\|v\|_2} = \max_v \frac{\lambda^T G v}{\|v\|_2}$$

since $\|v^T G^T \lambda\|_2 = |\lambda^T G v|$.

Overall, we have thus derived the *minimax characterization*

$$\sigma_{\min}(G) = \min_\lambda \max_v \frac{\lambda^T G v}{\|v\|_2 \|\lambda\|_2} > 0. \tag{2.74}$$

In Chap. 3, the essential minimax condition (2.74) for the regularity of rigid body constraints will be generalized to an inf-sup condition for constraints on elastic bodies.

2.4 Analysis of the Equations of Constrained Mechanical Motion

Constraint Jacobian of Planar Pendulum

To illustrate the minimax criterion (2.74), we analyze the constraint Jacobian

$$G(q) = \begin{pmatrix} 1 & 0 & \frac{l}{2}\sin\alpha \\ 0 & 1 & -\frac{l}{2}\cos\alpha \end{pmatrix}$$

of the planar pendulum example (2.13a), (2.13b). Computing the eigenvalues of the symmetric matrix

$$G(q)G(q)^T = \begin{pmatrix} 1 + \frac{l^2}{4}\sin^2\alpha & -\frac{l^2}{4}\sin\alpha\cos\alpha \\ -\frac{l^2}{4}\sin\alpha\cos\alpha & 1 + \frac{l^2}{4}\cos^2\alpha \end{pmatrix}$$

yields the squared singular values of $G(q)$, and in this way we find that

$$\sigma_1 = \sqrt{2}, \quad \sigma_2 = 1$$

for all configurations q of the system. The minimax condition (2.74) is hence guaranteed.

2.4.3 Influence of Perturbations

The previous analysis has shown that the equations of motion (2.66a)–(2.66c) possess a unique solution provided that the full rank or minimax criterion (2.74) holds and the right hand side of (2.71) is Lipschitz continuous. For well-posedness of the equations, however, we need to look at small perturbations. As discussed above, the perturbation index measures the sensitivity of the solution of a DAE with respect to perturbations in the equations. The perturbed version of the equations of constrained mechanical motion (2.66a)–(2.66c) takes the form

$$\dot{\hat{q}} = \hat{v} + \gamma(t), \qquad (2.75a)$$

$$M(\hat{q})\dot{\hat{v}} = f(\hat{q}, \hat{v}, t) - G(\hat{q})^T\hat{\lambda} + \delta(t), \qquad (2.75b)$$

$$0 = g(\hat{q}) + \theta(t). \qquad (2.75c)$$

Arnold [Arn95] proved that the position variables q and the velocity variables v are not sensitive to such perturbations but the Lagrange multipliers λ indeed are. We have the sharp estimate

$$\|\hat{\lambda}(t) - \lambda(t)\| \leq c \Big(\Delta_0 + \max_{\tau \in [t_0,t]} \|\delta(\tau)\| + \max_{\tau \in [t_0,t]} \|\theta(\tau)\| \qquad (2.76)$$

$$+ \max_{\tau \in [t_0,t]} \|\dot{\theta}(\tau)\| + \max_{\tau \in [t_0,t]} \|\ddot{\theta}(\tau)\| \Big).$$

In this estimate,

$$\Delta_0 := \|\hat{q}(t_0) - q(t_0)\| + \|S(\hat{q}(t_0))\hat{v}(t_0) - S(q(t_0))v(t_0)\|$$

denotes the difference in the initial values. Moreover,

$$S(q) := I - \left(M^{-1}G^T\left(GM^{-1}G^T\right)^{-1}G\right)(q)$$

is a projector onto the kernel of the constraint Jacobian G, and c is a constant that may depend on the size of $\max_\tau \|\delta(\tau)\|$.

What are the consequences of this result? Perturbations in the constraints strongly influence the Lagrange multipliers. In particular, the second time derivative $\ddot{\theta}$ is typical for DAEs of index 3. If θ is a small but high frequency signal, we get an amplification of the perturbation and the problem is not well-posed in this formulation.

The amplification of perturbations is not the only obstacle to the numerical treatment of the equations of motion (2.66a)–(2.66c). In a numerical time integration scheme, the differentiation step is replaced by a difference quotient, which implies a division by the stepsize. Therefore, the approximation properties of the numerical scheme deteriorate and we observe phenomena like order reduction, ill-conditioning, or even loss of convergence. Most severely affected are typically the Lagrange multipliers. In Chap. 7, this issue will be studied in more detail.

2.4.4 Alternative Formulations

In view of the above discussion, it is mostly not advisable to tackle DAEs of index 3 directly. Instead, it has become standard in multibody dynamics to lower the index first by introducing alternative formulations.

Formulation of Index 1

The differentiation process for determining the index revealed the hidden constraints at velocity and at acceleration level. It is a straightforward idea to replace now the original position constraint (2.66c) by one of the hidden constraints. Selecting the acceleration equation (2.68) for this purpose, one obtains the *formulation of index 1*

$$\begin{aligned}\dot{q} &= v,\\ M(q)\dot{v} &= f(q,v,t) - G(q)^T\lambda,\\ 0 &= G(q)\dot{v} + \kappa(q,v).\end{aligned} \quad (2.77)$$

This system is of index 1, and at first sight one could expect much less difficulties here. But a closer view shows that (2.77) lacks the information of the original position and velocity constraints, which have become *invariants of the system*. In

2.4 Analysis of the Equations of Constrained Mechanical Motion

general, these invariants are not preserved under discretization, and the numerical solution may thus turn unstable, which is called the *drift off phenomenon*.

To give a first idea of the drift off, we write the position constraint as

$$\boldsymbol{w}(t) := \boldsymbol{g}(\boldsymbol{q}(t))$$

and differentiate twice. The differential equation $\ddot{\boldsymbol{w}} = \boldsymbol{0}$ corresponds then to the acceleration constraint (2.68), but we assume now small errors in the right hand side and the initial values and consider instead the initial value problem

$$\ddot{\boldsymbol{w}} = \boldsymbol{\zeta}_a, \quad \dot{\boldsymbol{w}}(t_0) = \boldsymbol{\zeta}_v, \quad \boldsymbol{w}(t_0) = \boldsymbol{\zeta}_p \qquad (2.78)$$

with constants $\boldsymbol{\zeta}_a, \boldsymbol{\zeta}_v, \boldsymbol{\zeta}_p \in \mathbb{R}^{n_\lambda}$. By integrating twice, we obtain

$$\boldsymbol{w}(t) = \frac{1}{2}(t - t_0)^2 \boldsymbol{\zeta}_a + (t - t_0)\boldsymbol{\zeta}_v + \boldsymbol{\zeta}_p. \qquad (2.79)$$

We thus observe a violation of the position constraint that grows quadratically with time. In Chap. 7, this intuitive reasoning will be made more precise by studying the discrete error propagation in the constraints when numerically solving the formulation of index 1. For the moment, we conclude that there is an inherent quadratic instability in this formulation.

Baumgarte Stabilization

A very early cure for the drift off goes back to Baumgarte [Bau72]. The idea is to combine original and differentiated constraints and form the new constraint

$$\boldsymbol{0} = \boldsymbol{G}(\boldsymbol{q})\dot{\boldsymbol{v}} + \kappa(\boldsymbol{q}, \boldsymbol{v}) + 2\alpha \boldsymbol{G}(\boldsymbol{q})\boldsymbol{v} + \beta^2 \boldsymbol{g}(\boldsymbol{q}) \qquad (2.80)$$

with scalar parameters α and β. In (2.77), the acceleration constraint is then replaced by (2.80), which leaves the index unchanged. The free parameters α and β should be chosen in such a way that

$$\boldsymbol{0} = \ddot{\boldsymbol{w}} + 2\alpha \dot{\boldsymbol{w}} + \beta^2 \boldsymbol{w} \qquad (2.81)$$

becomes an asymptotically stable equation.

E.g., if $\alpha = \beta$, the solution of the perturbed system

$$\ddot{\boldsymbol{w}} + 2\alpha \dot{\boldsymbol{w}} + \alpha^2 \boldsymbol{w} = \boldsymbol{\zeta}_a, \quad \dot{\boldsymbol{w}}(t_0) = \boldsymbol{\zeta}_v, \quad \boldsymbol{w}(t_0) = \boldsymbol{\zeta}_p,$$

is given by

$$\boldsymbol{w}(t) = \left(\boldsymbol{\zeta}_p + (t - t_0)(\boldsymbol{\zeta}_p + \alpha \boldsymbol{\zeta}_v)\right) \exp(-\alpha(t - t_0)) + \frac{\boldsymbol{\zeta}_a}{\alpha^2}.$$

For $\alpha > 0$, the exponential function decays and damps out the initial deviations $\boldsymbol{\zeta}_p$ and $\boldsymbol{\zeta}_v$ in the position and velocity constraints, respectively.

The crucial point in Baumgarte's approach is the choice of the parameters. Also, extraneous eigenvalues are introduced in this way. For a detailed analysis of this stabilization and related techniques we refer to Ascher et al. [ACPR95, AL97].

Formulation of Index 2

Instead of the acceleration constraints, one can also use the velocity constraints (2.67) to replace (2.66c). This leads to the formulation

$$\dot{q} = v,$$
$$M(q)\dot{v} = f(q, v, t) - G(q)^T \lambda, \qquad (2.82)$$
$$0 = G(q)v.$$

Now the index is 2, but similar to the index 1 case, the information of the position constraint is lost. The resulting drift off is noticeable but stays linear, which means a significant improvement compared to (2.77). Nevertheless, additional measures such as stabilization by projection are often applied when discretizing (2.82).

The evaluation of the acceleration constraint (2.68) requires expressions with second derivatives of the constraints and is thus computationally rather expensive. For the formulation (2.82), however, there is almost no extra effort when evaluating the velocity constraints since the constraint Jacobian $G(q)$ needs to be evaluated anyway to form the product $G(q)^T \lambda$.

GGL Formulation

On the one hand, we have seen that it is desirable for the governing equations to have an index as small as possible. On the other hand, though simple differentiation lowers the index, it may lead to drift off. An elegant way out of this dilemma is due to Gear, Gupta and Leimkuhler [GGL85]. This formulation starts with the kinematic and dynamic equations (2.66a), (2.66b) combined with the constraints at velocity level (2.67). The position constraints (2.9b) are interpreted as invariants and appended by means of extra Lagrange multipliers, cf. Sect. 2.3.4.

In this way, one obtains an enlarged system

$$\dot{q} = v - G(q)^T \mu,$$
$$M(q)\dot{v} = f(q, v, t) - G(q)^T \lambda, \qquad (2.83)$$
$$0 = G(q)v,$$
$$0 = g(q)$$

with additional multipliers $\mu(t) \in \mathbb{R}^{n_\lambda}$. A straightforward calculation shows

$$0 = \frac{d}{dt} g(q) = G(q)\dot{q} = G(q)v - G(q)G^T(q)\mu = -G(q)G^T(q)\mu$$

2.4 Analysis of the Equations of Constrained Mechanical Motion

and one concludes $\mu = 0$ since $G(q)$ is of full rank and hence $G(q)G^T(q)$ invertible. With the additional multipliers μ vanishing, (2.83) and the original equations of motion (2.9a), (2.9b) coincide along any solution. Yet, the index of the *GGL formulation* (2.83) is 2 instead of 3. Some authors refer to (2.83) also as *stabilized index-2 system*.

A scaled variant of the GGL formulation (2.83) is also widespread where the kinematic equation is replaced by

$$S\dot{q} = Sv - G(q)^T \mu. \tag{2.84}$$

The scaling matrix $S \in \mathbb{R}^{n_q \times n_q}$ should be symmetric positive definite so that $G(q)S^{-1}G^T(q)$ is invertible and the above conclusion $\mu = 0$ remains valid. Specific choices are $S = M(q)$ or, better with respect to efficiency, $S = M_d(q)$ where M_d is a diagonal mass matrix obtained from mass lumping.

Overdetermined Formulation

From an analytical point of view, one could drop the extra multiplier μ in (2.83) and consider instead the *overdetermined system*

$$\begin{aligned} \dot{q} &= v, \\ M(q)\dot{v} &= f(q,v,t) - G(q)^T \lambda, \\ 0 &= G(q)v, \\ 0 &= g(q). \end{aligned} \tag{2.85}$$

Though there are more equations than unknowns in (2.85), the solution is unique and, given consistent initial values, coincides with the solution of the original system (2.66a)–(2.66c). Even more, one could add the acceleration constraint (2.68) to (2.85) so that all hidden constraints are explicitly stated.

Under discretization, however, an overdetermined system such as (2.85) can only be solved in a least squares sense. As investigated in Führer [Füh88] and Führer and Leimkuhler [FL91] for the BDF methods, it is possible to construct a least squares objective function that inherits certain properties of the state space form (2.12) and that defines an integration scheme equivalent to the discretization of (2.83).

Local State Space Form

In Sect. 2.1.2, we have contrasted the constrained equations of motion (2.9a), (2.9b) with the state space form (2.12), which is a system of second order ordinary differential equations. The reasoning about the restrictions of the state space form contained a loose end that we now take up. It is very helpful in this context to view the

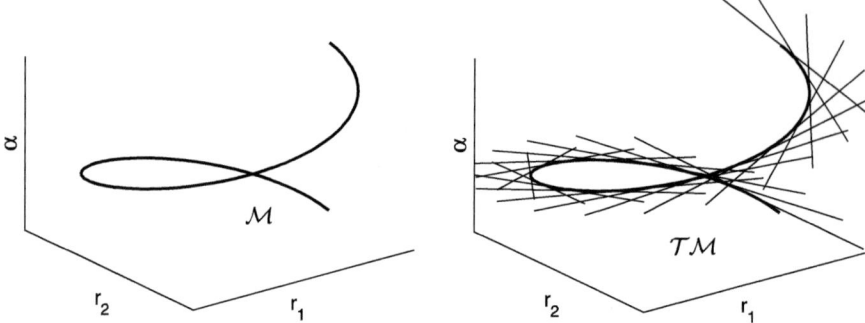

Fig. 2.7 Constraint manifold (*left*) and tangent bundle (*right*) for the planar pendulum (2.13a), (2.13b). The constraint manifold is the helix defined by (2.7), and the disjoint union of all tangents to the helix forms the tangent bundle

equations of constrained mechanical motion as differential equations on a manifold, compare Sect. 2.3.3.

The position constraint (2.66c) defines the *constraint manifold*

$$\mathcal{M} := \{q \in \mathbb{R}^{n_q} : \mathbf{0} = g(q)\}. \tag{2.86}$$

Furthermore, combining (2.66c) and the velocity constraint (2.67) leads to the *tangent bundle*

$$\mathcal{TM} := \{(q, v) \in \mathbb{R}^{n_q} \times \mathbb{R}^{n_q} : \mathbf{0} = g(q), \mathbf{0} = G(q)v\}. \tag{2.87}$$

Figure 2.7 illustrates the constraint manifold and the tangent bundle for the planar pendulum example (2.13a), (2.13b).

Given $(q_0, v_0) \in \mathcal{TM}$, our task is now to construct a parametrization such that the resulting analogue of the state space form (2.12) can be evaluated. As discussed in Sect. 2.3.3, such a parametrization in general holds only locally, and this restriction will become clearer when we discuss two established approaches in the following.

We start with the method of *coordinate partitioning* by Wehage and Haug [WH82]. The full rank of the constraint Jacobian $G(q_0)$ implies that, after an appropriate permutation of the columns of $G(q_0)$ which we omit here for simplicity, the coordinate vector q can be partitioned into *independent coordinates* $q_I(t) \in \mathbb{R}^{n_q - n_\lambda}$ and *dependent coordinates* $q_D(t) \in \mathbb{R}^{n_\lambda}$ such that

$$q = \begin{pmatrix} q_I \\ q_D \end{pmatrix} \quad \text{and} \quad G_D(q_0) := \frac{\partial g(q_0)}{\partial q_D} \in \mathbb{R}^{n_\lambda \times n_\lambda} \text{ is invertible.} \tag{2.88}$$

Recall that $n_s = n_q - n_\lambda$ is the number of DOF, and these degrees of freedom are momentarily identified with q_I.

2.4 Analysis of the Equations of Constrained Mechanical Motion

By the implicit function theorem, the constraint $0 = g(q_I, q_D)$ defines formally a function $q_D = \eta(q_I)$ in a neighborhood of q_0, and by computing the derivative and observing the chain rule, we obtain

$$0 = \frac{\partial}{\partial q_I} g(q_I, q_D) = G_I(q_I, q_D) + G_D(q_I, q_D) \frac{\partial \eta(q_I)}{\partial q_I}$$

where $G_I(q) := \partial g(q)/\partial q_I \in \mathbb{R}^{n_\lambda \times n_s}$. Overall, we have

$$\frac{\partial \eta(q_I)}{\partial q_I} = -(G_D^{-1} G_I)(q_I, \eta(q_I)) \tag{2.89}$$

for the relation between dependent and independent coordinates.

In the language of differential geometry, the mapping

$$\psi : E \to \mathcal{U}, \quad q_I \mapsto \begin{pmatrix} q_I \\ \eta(q_I) \end{pmatrix} \tag{2.90}$$

constitutes a local parametrization of the manifold $\mathcal{M}$, with E being an open subset of $\mathbb{R}^{n_s}$ and $\mathcal{U} \subset \mathcal{M}$. The same partitioning into dependent and independent coordinates can be applied to the velocity vector $v = (v_I, v_D)$, and the pair (ψ, Ψ) with

$$\Psi(q_I) = \frac{\partial \psi(q_I)}{\partial q_I}$$

parametrizes the tangent bundle (2.87).

Looking back to the state space form (2.12), it is now evident that the independent coordinates q_I are a special choice of minimal coordinates, and furthermore, the null space matrix N is given by

$$N(q_I) = \frac{\partial \psi(q_I)}{\partial q_I} = \Psi(q_I) = \begin{pmatrix} I \\ -(G_D^{-1} G_I)(q_I, \eta(q_I)) \end{pmatrix}. \tag{2.91}$$

Thus, all ingredients for evaluating the equations of motion (2.12) are defined. One should be aware, however, that the construction of the local parametrization outlined so far is not suitable for a computer implementation. One reason lies in the second fundamental form $\partial N(q_I)/\partial q_I (v_I, v_I)$, which is required in (2.12) and which is not given explicitly.

To this end, we use the coordinate partitioning to express the acceleration constraint (2.68) as

$$0 = G_I(q)\dot{v}_I + G_D(q)\dot{v}_D + \kappa(q, v),$$

which implies that

$$\dot{v}_D = -(G_D^{-1} G_I)(q)\dot{v}_I - G_D^{-1}(q)\kappa(q, v). \tag{2.92}$$

Due to the parametrization, we have on the other hand

$$\dot{v} = \frac{d}{dt}(N(q_I)v_I) = N(q_I)\dot{v}_I + \frac{\partial N(q_I)}{\partial q_I}(v_I, v_I)$$

$$= \begin{pmatrix} I \\ -G_D^{-1}G_I \end{pmatrix} \dot{v}_I + \begin{pmatrix} 0 \\ -\frac{\partial(G_D^{-1}G_I)}{\partial q_I}(v_I, v_I) \end{pmatrix}. \quad (2.93)$$

Comparing the second row of (2.93) with (2.92), we conclude that the evaluation of the second fundamental form $\partial N(q_I)/\partial q_I(v_I, v_I)$ can be accomplished by forming the acceleration term $G_D^{-1}(q)\kappa(q,v)$ where $(q,v) \in \mathcal{T}\mathcal{M}$.

In practice, this procedure is implicitly performed. Given (q_I, v_I), the evaluation of the local state space form starts by computing

$$q = \psi(q_I), \quad v = N(q_I)v_I,$$

which involves the solution of the nonlinear system $0 = g(q_I, q_D)$ for q_D. Next, one solves the linear system

$$\begin{pmatrix} M(q) & G(q)^T \\ G(q) & 0 \end{pmatrix} \begin{pmatrix} w \\ \lambda \end{pmatrix} = \begin{pmatrix} f(q,v,t) \\ -\kappa(q,v) \end{pmatrix} \quad (2.94)$$

for the acceleration w and the Lagrange multiplier λ. Finally, the local state space form, written as a system of first order, is given by

$$\dot{q}_I = v_I, \quad \dot{v}_I = w_I \quad (2.95)$$

where w_I denotes the entries of the acceleration vector that correspond to the independent coordinates.

An alternative approach for constructing a local state space form, introduced by Potra and Rheinboldt [PR91], is based on a *tangent space parametrization*. The relation between redundant coordinates and local minimal coordinates $s(t) \in E \subset \mathbb{R}^{n_s}$ is written as

$$q - q_b = Q_1 u + Q_2 s \quad (2.96)$$

where q_b stands for the base point of the parametrization. Moreover, the columns of the matrix $Q_2 \in \mathbb{R}^{n_q \times n_s}$ form a basis of the tangent space

$$\mathcal{T}_{q_b}\mathcal{M} = \{v \in \mathbb{R}^{n_q} : G(q_b)v = 0\},$$

and the columns of $Q_1 \in \mathbb{R}^{n_q \times n_\lambda}$ complement Q_2 such that the square matrix (Q_1, Q_2) is invertible. The local parametrization reads now

$$\psi : E \to \mathcal{U} \subset \mathcal{M}, \quad s \mapsto q = Q_1 u + Q_2 s + q_b, \quad (2.97)$$

and is again implicitly defined by solving the nonlinear system $g(q) = 0$ and (2.96) for q and u. Moreover, the evaluation of the state space form (2.12) can then be implemented in a fashion similar to the coordinate partitioning method.

2.4 Analysis of the Equations of Constrained Mechanical Motion

While the coordinate partitioning method uses local state variables with physical significance, the tangent space parametrization can be computed in such a way that the corresponding transformations are particularly stable from a numerical point of view. More precisely, the QR factorization with column pivoting applied to G^T gives [GL96]

$$G(q_b)^T = Q \begin{pmatrix} R \\ 0 \end{pmatrix} P$$

with orthogonal matrix $Q \in \mathbb{R}^{n_q \times n_q}$, upper triangular matrix $R \in \mathbb{R}^{n_\lambda \times n_\lambda}$, and permutation matrix $P \in \mathbb{R}^{n_\lambda \times n_\lambda}$. By partitioning Q into n_λ and n_s columns,

$$Q = (Q_1, Q_2),$$

one obtains the matrices Q_1 and Q_2 of the parametrization (2.96). The property $G(q_b) Q_2 = 0$ is immediately verified, and the basis of the tangent space and its complement are orthonormal by construction. For more details see [Rhe96].

Both the coordinate partitioning and the tangent space parametrization are local in nature, and the reason for this lies in the implicit function theorem, which is directly linked to topological properties of the multibody system. During a time integration process, the validity of a given set of minimal coordinates needs to be checked, and frequent changes of the parametrization might be necessary if the system contains closed kinematic loops.

2.4.5 Remarks

We close this chapter with a few further remarks on the equations of constrained mechanical motion.

Some of the aspects discussed above will play an important role when dealing with time integration methods for differential-algebraic systems in Chap. 7. There we will mostly concentrate on the formulations (2.82) and (2.83) that are of index 2 and avoid the evaluation of the acceleration constraints. The local state space form, on the other hand, will be the main tool to derive an asymptotic expansion and a corresponding computational method for stiff mechanical systems with constraints in Chap. 6.

Recently, physical invariants of mechanical systems, e.g., energy and momentum, and their preservation under numerical discretization have found much interest. We have already touched this topic in Sect. 2.1.3 in the context of Hamiltonian systems and mention here additionally the discrete null space method of Betsch [Bet05] and Betsch and Leyendecker [BL06], which is closely related to the local state space form approaches.

Chapter 3
Elastic Motion

In the following chapters, we turn our attention to the more general class of flexible multibody systems that include rigid as well as elastic bodies. While an elastic body is described by a partial differential equation, the rigid body motion, on the other hand, satisfies an ordinary differential equation or, in the presence of Euler parameters or joints, a differential-algebraic equation. Thus, the overall mathematical model consists of a coupled system with a subtle structure.

We apply a step-by-step procedure in order to derive the underlying models. This chapter is mainly devoted to the motion of a single elastic body under the assumption of linear elasticity. Large rotations and translations will be treated by the method of floating reference frames in the subsequent Chap. 4. As in Chap. 2, Hamilton's principle of stationary action is the starting point to generate the equations of motion. Constraints are appended by means of Lagrange multipliers and lead to the notion of a time-dependent saddle point formulation that can be viewed as the continuous analogue of the constrained system for rigid bodies.

3.1 Basic Equations of an Elastic Body

This section gives a brief outline of elasticity theory. For more details and the mathematical background, the reader is referred to Braess [Bra01], Ciarlet [Cia88], Duvaut and Lions [DL75], and Marsden and Hughes [MH83].

Consider the elastic body in Fig. 3.1. It is subjected to volume forces like the gravity and to surface forces such as spring forces. Additionally, some part of the surface is linked to a rigid body and performs a prescribed motion.

3.1.1 Variational Formulation

For the mathematical model, we identify each of the body's material points by its position $x \in \overline{\Omega}$ where the domain $\Omega \subset \mathbb{R}^3$ is the open interior of a *reference configuration* that stands for the undeformed state. The surface or boundary $\partial \Omega$ of Ω

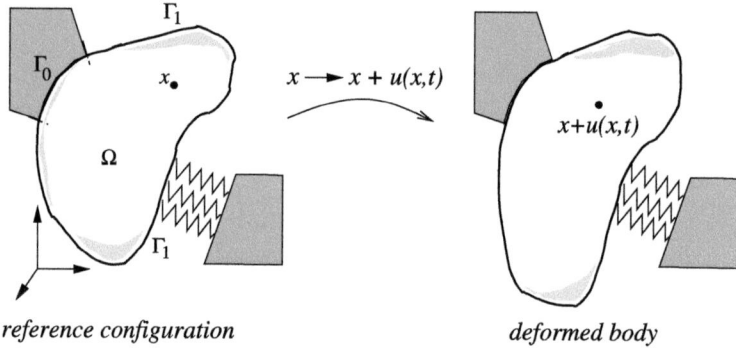

Fig. 3.1 Sketch of an elastic body with reference frame and displacement field u

consists of two parts. On the segment Γ_0 the motion is prescribed, and on Γ_1 a surface traction is given.

Kinematics

We introduce next an orthonormal reference frame that does not change in time. With respect to this inertial reference frame, the *deformation* φ maps each material point $x \in \Omega$ onto its deformed state $\varphi(x,t)$ at time t. Commonly, the deformation is expressed in terms of the *displacement field*

$$\varphi(x,t) = x + u(x,t), \quad u(x,t) = \begin{pmatrix} u_1(x_1,x_2,x_3,t) \\ u_2(x_1,x_2,x_3,t) \\ u_3(x_1,x_2,x_3,t) \end{pmatrix} \in \mathbb{R}^3. \tag{3.1}$$

With respect to the displacement field u, the prescribed motion on the boundary Γ_0 represents a *geometric* or *Dirichlet boundary condition*

$$u(x,t) = u_0(x,t) \quad \text{on } \Gamma_0 \tag{3.2}$$

with given function u_0.

Strain measures changes of local length elements in the deformed body. One choice is the 3×3 Jacobian $\nabla \varphi = \partial \varphi / \partial x$, which is called the deformation gradient (observe the potentially misleading terminology). A standard requirement is that

$$\det \nabla \varphi > 0, \tag{3.3}$$

which means that the deformation is orientation-preserving. The deformation gradient as strain measure suffers from the drawback that it is not invariant under rigid motion.

A better choice is the *Green–Lagrangian strain tensor*

$$E = \frac{1}{2}(\nabla \varphi^T \nabla \varphi - I) = \frac{1}{2}(\nabla u + \nabla u^T + \nabla u^T \nabla u) \tag{3.4}$$

3.1 Basic Equations of an Elastic Body

since it is invariant under translation and rotation. By a tensor E we mean here and in the following a second order tensor in matrix representation:

$$E(x,t) = (E_{ij}(x,t))_{i,j=1,2,3} \in \mathbb{R}^{3\times 3}.$$

In *linear elasticity theory* only small displacements are considered, and instead of the strain tensor E one uses

$$\varepsilon = \varepsilon(u) = \frac{1}{2}(\nabla u + \nabla u^T). \tag{3.5}$$

The strain tensor ε consists thus of the symmetric part of the displacement gradient ∇u. Notice that the arguments x and t have been suppressed here and in (3.4) for better readability. In the following, we will proceed with this notational convention and solely point out special dependencies if required.

Material Law

The two most important quantities in elasticity theory are strain and *stress*. While strain is a kinematic quantity, its counterpart stress is defined as the density of surface force per area element. In other words, stress describes the pressure caused by inner forces acting on the surface of infinitely small volume elements in the deformed body. In a famous theorem, Cauchy proved existence, uniqueness and symmetry of the stress tensor in the deformed configuration, see, e.g., [Cia88]. Expressed in the coordinates of the undeformed body, the so-called *Lagrange representation*, we denote the symmetric stress tensor by $\Sigma(x,t) \in \mathbb{R}^{3\times 3}$.

Material laws relate strain E and stress Σ. We assume here a homogeneous and isotropic material such that the linear law of St. Venant–Kirchhoff

$$\Sigma = \Lambda_1(\text{trace } E)I + 2\Lambda_2 E, \tag{3.6}$$

holds, with 3×3 identity matrix I and real Lamé constants Λ_1 and Λ_2 as material parameters. Relation (3.6) constitutes a special case of a hyperelastic material [Cia88, §4.4]. Such materials possess the *strain energy*

$$W_{nl} = \frac{1}{2}\int_\Omega \Sigma : E \, dx \tag{3.7}$$

where the expression $\Sigma : E := \text{trace}(\Sigma E)$ stands for the inner product of symmetric tensors and the subscript of W_{nl} indicates that (3.7) is valid in the geometrically nonlinear case.

In linear elasticity, we write σ instead of Σ for the stress tensor. The material law (3.6) reduces here to *Hooke's law*

$$\sigma = \sigma(u) = \Lambda_1(\text{trace } \varepsilon(u))I + 2\Lambda_2 \varepsilon(u), \tag{3.8}$$

and the strain energy is given by

$$W = \frac{1}{2} \int_\Omega \sigma : \varepsilon \, dx. \tag{3.9}$$

Frequently, one makes use of an alternative notation for Hooke's law that is based on a vector representation of the tensors σ and ε. Set

$$\underline{\varepsilon} := (\varepsilon_{11}, \varepsilon_{22}, \varepsilon_{33}, 2\varepsilon_{12}, 2\varepsilon_{23}, 2\varepsilon_{13})^T, \quad \underline{\sigma} := (\sigma_{11}, \sigma_{22}, \sigma_{33}, \sigma_{12}, \sigma_{23}, \sigma_{13})^T$$

and replace the Lamé constants by *Poisson's number* $\nu = \Lambda_1/(2(\Lambda_1 + \Lambda_2))$ and *modulus of elasticity* $E = \Lambda_2(3\Lambda_1 + 2\Lambda_2)/(\Lambda_1 + \Lambda_2)$. Hooke's law (3.8) then reads $\underline{\sigma} = C \underline{\varepsilon}$ with constant 6×6 matrix

$$C = \frac{E}{(1+\nu)(1-2\nu)} \begin{pmatrix} 1-\nu & \nu & \nu & & & \\ \nu & 1-\nu & \nu & & & \\ \nu & \nu & 1-\nu & & & \\ & & & (1-2\nu)/2 & & \\ & & & & (1-2\nu)/2 & \\ & & & & & (1-2\nu)/2 \end{pmatrix}. \tag{3.10}$$

Principle of Least Action

We concentrate now on linear elasticity and provide the energy expressions required for Hamilton's principle. The kinetic energy is given by

$$T = \frac{1}{2} \int_\Omega \rho \dot{u}^T \dot{u} \, dx \tag{3.11}$$

with constant mass density ρ. The potential energy consists of strain energy (3.9) minus volume and surface force terms and reads

$$U = \frac{1}{2} \int_\Omega \sigma(u) : \varepsilon(u) \, dx - \int_\Omega u^T \beta \, dx - \int_{\Gamma_1} u^T \tau \, ds. \tag{3.12}$$

Here, $\beta(x, t) \in \mathbb{R}^3$ denotes the density of volume forces such as the gravity and $\tau(x, t) \in \mathbb{R}^3$ the given surface traction on the boundary Γ_1.

Like in the rigid body case, the principle of least action characterizes the motion u by the requirement

$$\int_{t_0}^{t_1} (T - U) \, dt \to \text{stationary!} \tag{3.13}$$

We use (3.13) in order to derive both the weak and the strong form of the equations of motion. For brevity, we assume sufficient smoothness and postpone the introduction of appropriate function spaces to the next subsection, which will deal with the weak form in more detail.

3.1 Basic Equations of an Elastic Body

Suppose there exists a displacement field u that satisfies (3.13). We perturb u by the *variation* θv and analyse $u + \theta v$ where θ is a real parameter and $v = v(x,t)$. The perturbed displacement $u + \theta v$ must also satisfy the boundary condition (3.2) on Γ_0, and hence we require $v(\cdot, t) = \mathbf{0}$ on Γ_0 at any time $t_0 \leq t \leq t_1$. Additionally, the two trajectories are assumed to coincide at arbitrary but fixed points in time t_0 and t_1, i.e., $v(\cdot, t_0) = v(\cdot, t_1) = \mathbf{0}$ in Ω. This means that the variation does neither affect the starting point nor the end point of the deformation path. We call v then *admissible* or *consistent*.

In the first step, we define the function

$$J(\theta) := \int_{t_0}^{t_1} (T(u + \theta v) - U(u + \theta v))\,dt \tag{3.14}$$

and postulate due to the stationarity condition (3.13)

$$0 = \frac{d}{d\theta} J(\theta) \bigg|_{\theta=0}.$$

This leads to

$$0 = \int_{t_0}^{t_1} \left(\int_\Omega \rho \dot{v}^T \dot{u}\,dx - \int_\Omega \sigma(u) : \varepsilon(v)\,dx + \int_\Omega v^T \beta\,dx + \int_{\Gamma_1} v^T \tau\,ds \right) dt.$$

Secondly, integration by parts with respect to time t results in

$$\int_{t_0}^{t_1} \int_\Omega \rho \dot{v}^T \dot{u}\,dx\,dt = \int_\Omega \rho v^T \dot{u}\,dx \bigg|_{t_0}^{t_1} - \int_{t_0}^{t_1} \int_\Omega \rho v^T \ddot{u}\,dx\,dt.$$

The first term on the right hand side vanishes due to $v(\cdot,t_0) = v(\cdot,t_1) = \mathbf{0}$, and it follows

$$0 = \int_{t_0}^{t_1} \left(\int_\Omega \left(\rho v^T \ddot{u} + \sigma(u) : \varepsilon(v) - v^T \beta \right) dx - \int_{\Gamma_1} v^T \tau\,ds \right) dt, \tag{3.15}$$

which must hold for all admissible functions v. As third step, we apply the Gauss theorem

$$\int_\Omega \operatorname{div} F(x)\,dx = \int_{\partial\Omega} F(s)^T n(s)\,ds$$

to the vector field $F(x) := \sigma(u(x,\cdot))v(x,\cdot)$, which yields the relation, also known as Green's formula,

$$\int_\Omega \sigma(u) : \varepsilon(v)\,dx = \int_\Omega \sigma : \nabla v\,dx = \int_{\partial\Omega} v^T \sigma n\,ds - \int_\Omega v^T \operatorname{div} \sigma\,dx \tag{3.16}$$

for the variation of the strain energy. Note that the **div** operator for the tensor $\boldsymbol{\sigma}$ is defined as the vector whose components are the divergences of the rows of $\boldsymbol{\sigma}$,

$$\boldsymbol{\sigma} = \begin{pmatrix} \sigma_{11} & \sigma_{12} & \sigma_{13} \\ \sigma_{21} & \sigma_{22} & \sigma_{23} \\ \sigma_{31} & \sigma_{32} & \sigma_{33} \end{pmatrix} \Rightarrow \mathbf{div}\,\boldsymbol{\sigma} = \begin{pmatrix} \partial\sigma_{11}/\partial x_1 + \partial\sigma_{12}/\partial x_2 + \partial\sigma_{13}/\partial x_3 \\ \partial\sigma_{21}/\partial x_1 + \partial\sigma_{22}/\partial x_2 + \partial\sigma_{23}/\partial x_3 \\ \partial\sigma_{31}/\partial x_1 + \partial\sigma_{32}/\partial x_2 + \partial\sigma_{33}/\partial x_3 \end{pmatrix}.$$

Furthermore, $\boldsymbol{n}(x)$ denotes the outward normal vector on the boundary $\partial\Omega$. We make finally use of the relation (3.16) and obtain from (3.15)

$$0 = \int_{t_0}^{t_1} \int_{\Omega} \boldsymbol{v}^T \left(\rho \ddot{\boldsymbol{u}} - \mathbf{div}\,\boldsymbol{\sigma} - \boldsymbol{\beta}\right) \mathrm{d}x\,\mathrm{d}t + \int_{t_0}^{t_1} \int_{\Gamma_1} \boldsymbol{v}^T (\boldsymbol{\sigma}\boldsymbol{n} - \boldsymbol{\tau})\,\mathrm{d}s\,\mathrm{d}t, \qquad (3.17)$$

which must again hold for all admissible $\boldsymbol{v}$.

3.1.2 Equations of Motion

The variational formulation (3.17) holds in particular if $\boldsymbol{v}$ vanishes not only on Γ_0 but also on Γ_1. Then, however, the second integral vanishes, and the fundamental lemma of the calculus of variations [CH68, §4.3] yields immediately that the integrand of the first integral must vanish, too. In turn, if $\boldsymbol{v}$ is not equal to zero on Γ_1, it follows that the integrand of the surface integral must be equal to zero, which results in a second boundary condition, the so-called *natural* or *Neumann boundary condition*.

Summing up, we have derived the *strong form of linear elasticity*

$$\rho \ddot{\boldsymbol{u}}(x, t) = \mathbf{div}\,\boldsymbol{\sigma}(\boldsymbol{u}(x, t)) + \boldsymbol{\beta}(x, t) \qquad \text{in } \Omega, \qquad (3.18\mathrm{a})$$

$$\boldsymbol{u}(x, t) = \boldsymbol{u}_0(x, t) \qquad \text{on } \Gamma_0, \qquad (3.18\mathrm{b})$$

$$\boldsymbol{\sigma}(\boldsymbol{u}(x, t))\boldsymbol{n}(x) = \boldsymbol{\tau}(x, t) \qquad \text{on } \Gamma_1. \qquad (3.18\mathrm{c})$$

The strong form consists of the *balance of momentum* (3.18a), a hyperbolic system of PDEs which can be understood as generalized wave equation, and corresponding boundary conditions (3.18b) and (3.18c). Hooke's law (3.8), which defines the stress as function $\boldsymbol{\sigma}(\boldsymbol{u})$, completes the strong form. We remark that (3.18a)–(3.18c) can be extended to more general cases like large deformation, nonlinear material laws or dissipation where the above derivation does not apply. Since we focus on the weak form in the sequel, however, we do not pursue the strong form any further.

Weak Form of the Equations of Motion

In order to pass to a weak formulation or work principle, we have to go back half the way in the above derivation. We consider now *test functions* that are time-independent, $\boldsymbol{v} = \boldsymbol{v}(x)$, and that satisfy the homogeneous boundary condition $\boldsymbol{v} = \boldsymbol{0}$

3.1 Basic Equations of an Elastic Body

on Γ_0. Instead of v, the *virtual displacement* $\delta u = \theta v$ is frequently used in the engineering literature.

If we multiply the strong form (3.18a) from the left by v and integrate over the whole body, we can apply Green's formula (3.16) in reverse. This leads directly to the *weak form* of the balance equations or the so-called *principle of virtual work*,

$$\int_\Omega \rho v^T \ddot{u}\, dx + \int_\Omega \sigma(u) : \varepsilon(v)\, dx = \int_\Omega v^T \beta\, dx + \int_{\Gamma_1} v^T \tau\, ds, \qquad (3.19)$$

which must hold for all test functions v with $v = \mathbf{0}$ on Γ_0, compare (3.15). The next subsection will concentrate on the weak form, in particular on appropriate function spaces and smoothness requirements. Before, we present an important extension of (3.19) for which Hamilton's principle does not apply.

Dissipation is not covered by the above derivation since Hamilton's principle is restricted to conservative systems. For the weak as well as the strong form, extensions exist, though. We restrict the discussion to *Rayleigh damping*, which is a very common model in structural dynamics. Instead of Hooke's law (3.8), one assumes here the stress-strain relation [Hug87]

$$\sigma = \sigma(u, \dot{u}) = \Lambda_1 (\text{trace } \varepsilon(u + \zeta_1 \dot{u}))I + 2\Lambda_2 \varepsilon(u + \zeta_1 \dot{u}). \qquad (3.20)$$

The real parameter $\zeta_1 \geq 0$ specifies damping depending on the body's frequencies. With a second parameter $\zeta_2 \geq 0$ determining viscous damping, the weak form of an elastic body with Rayleigh damping reads

$$\int_\Omega \rho v^T (\ddot{u} + \zeta_2 \dot{u})\, dx + \int_\Omega \sigma(u, \dot{u}) : \varepsilon(v)\, dx = \int_\Omega v^T \beta\, dx + \int_{\Gamma_1} v^T \tau\, ds. \qquad (3.21)$$

This dissipative model of the elastic body features a velocity proportional damping: If $\zeta_1 > 0$, higher frequencies are more strongly damped than lower frequencies, which agrees with experimental data.

3.1.3 Smoothness and Appropriate Function Spaces

In the preceding derivation of the equations of motion, the displacement u and the test function v were assumed to be "smooth enough". We will now make this statement more precise for the weak form (3.19). The reader is referred to Brenner/Scott [BS91] for the details from functional analysis.

A closer look at (3.19) shows that the term $\sigma(u) : \varepsilon(v)$ with stress $\sigma(u)$ and strain $\varepsilon(v)$ essentially consists of products of the form

$$\frac{\partial u_i}{\partial x_j} \cdot \frac{\partial v_k}{\partial x_l},$$

which are multiplied by material parameters, summed up and then integrated over the domain Ω. Thus, the obvious question is: Which kind of continuity do we require for u and v to obtain a well-defined model? The class of C^1-functions would be a first option, but in the well-established finite element method, the corresponding piecewise polynomial functions are smooth on element interiors but have only C^0-continuity across element boundaries. In order to deal with such functions, we hence need a different notion of smoothness than the one offered by the class of differentiable functions.

To simplify the discussion, we consider first scalar functions $f, g : \Omega \to \mathbb{R}$ where $\Omega \subset \mathbb{R}^d$ is a domain in d-dimensional space before we come back to the vector fields u and v and the weak form (3.19). The domain Ω is assumed to possess a sufficiently smooth boundary. If the boundary can be split into segments that are parametrized by continuously differentiable functions, this assumption will be satisfied.

Weak Derivatives

The space $L_2(\Omega)$ consists of all functions f whose square integral over Ω exists in the sense of Lebesgue, i.e., for which

$$\int_\Omega f(x)^2 \, dx < \infty.$$

As example, consider in one space dimension the open interval $\Omega = (0, 1)$ and the functions $f(x) = x^{-1/4}$ and $g(x) = x^{-1/2}$. While f is in $L_2(\Omega)$, the function g is not due to the singularity of $\ln x = \int x^{-1} \, dx$ at $x = 0$. With inner product

$$(f, g)_{L_2(\Omega)} := \int_\Omega f(x) g(x) \, dx \tag{3.22}$$

and norm

$$\|f\|_{L_2(\Omega)} = \sqrt{(f, f)_{L_2(\Omega)}},$$

the vector space $L_2(\Omega)$ obtains a special structure with important properties. In particular, each Cauchy sequence in $L_2(\Omega)$ converges to an element of $L_2(\Omega)$, which means that the space is complete. Such a complete space with inner product and corresponding norm is called a Hilbert space.

Two functions f and g are identical in L_2 if $\|f - g\|_{L_2(\Omega)} = 0$. Due to the subtleties of the Lebesgue integral, this allows f and g to differ in sets of measure zero, e.g., in isolated points, and we speak of equivalence classes of functions. For our purposes, however, it is sufficient to keep in mind that if $\|f - g\|_{L_2(\Omega)} = 0$, both f and g represent the same function except for trivial differences. Moreover, the space $L_2(\Omega)$ may contain rather rough functions for which the evaluation in a single point makes no sense.

3.1 Basic Equations of an Elastic Body

For $f \in L_2(\Omega)$ and $\Omega \subset \mathbb{R}$, the classical finite difference definition of the first derivative

$$f'(x) = \lim_{h \to 0} \frac{f(x+h) - f(x)}{h}$$

is not meaningful. As we have seen above for the weak form, the derivatives actually appear always inside an integral, and this inspires a more general concept for a derivative where rather rough functions in L_2 are combined with very smooth functions in C^∞. More precisely, we define

$$C_0^\infty(\Omega) := \{\phi \in C^\infty(\Omega) : \Omega \supset \operatorname{supp}\phi \text{ is compact}\}$$

where $\operatorname{supp}\phi$ stands for the support of ϕ, i.e., for the set where $\phi \neq 0$. Clearly, for $\phi \in C_0^\infty(\Omega)$ the partial derivatives

$$D^\alpha \phi = \frac{\partial^{|\alpha|}\phi}{\partial x_1^{\alpha_1} \partial x_2^{\alpha_2} \ldots \partial x_d^{\alpha_d}}$$

exist for the multi-index $\alpha = (\alpha_1, \ldots, \alpha_d)$ with $|\alpha| = \sum \alpha_i$ and spatial dimension d. We say that a function $f \in L_2(\Omega)$ possesses a *weak derivative* $D_w^\alpha f$ if a function $g \in L_2(\Omega)$ exists with the property

$$\int_\Omega g(x)\phi(x)\,dx = (-1)^{|\alpha|} \int_\Omega f(x) D^\alpha \phi(x)\,dx \qquad \forall \phi \in C_0^\infty(\Omega).$$

Then we set $D_w^\alpha f = g$.

In one spatial dimension and for $f \in C^1(\bar{\Omega})$, integration by parts yields directly

$$-\int_\Omega f(x)\phi'(x)\,dx = -f(x)\phi(x)\Big|_0^1 + \int_\Omega f'(x)\phi(x)\,dx = \int_\Omega f'(x)\phi(x)\,dx,$$

which implies $g = f'$ and shows the equivalence of weak and classical derivative in this case.

Note that the concept of weak derivatives can be defined in an analogous way for all differential operators that are in common use in connection with vector fields.

Sobolev Spaces

For $m \geq 0$ we denote by $H^m(\Omega)$ the set of all functions $f \in L_2(\Omega)$ which possess weak derivatives $D_w^\alpha f \in L_2(\Omega)$ for all $|\alpha| \leq m$. In $H^m(\Omega)$ we introduce the inner product

$$(f, g)_{H^m(\Omega)} := \sum_{|\alpha| \leq m} (D_w^\alpha f, D_w^\alpha g)_{L^2(\Omega)}$$

and the corresponding norm

$$\|f\|_{H^m(\Omega)} := \sqrt{(f, f)_{H^m(\Omega)}}.$$

The space $H^m(\Omega)$ is again a Hilbert space, which motivates the naming convention based on the letter "H", and it is called a Sobolev space.

It can be shown that $C^\infty(\Omega) \cap H^m(\Omega)$ is dense in $H^m(\Omega)$, which may be viewed as an alternative definition of the Sobolev spaces. This means that $H^m(\Omega)$ is the completion of $C^\infty(\Omega)$ with respect to the norm $\|\cdot\|_{H^m(\Omega)}$. Besides the norm $\|\cdot\|_{H^m(\Omega)}$, the highest order weak derivative terms define furthermore a semi-norm

$$|f|_{H^m(\Omega)} := \left(\sum_{|\alpha|=m} \|D_w^\alpha f\|_{L_2(\Omega)}^2 \right)^{1/2} \tag{3.23}$$

For constant functions f we have $|f|_{H^m(\Omega)} = 0$ when $m \geq 1$, which immediately implies that a crucial norm property is not satisfied in this case.

The basic setting for the weak formulation of a second order partial differential equation is the Sobolev space $H^1(\Omega)$. In other words, $m = 1$ represents the most important case in practice and provides a powerful framework for theoretical results such as existence and uniqueness theorems and for numerical convergence proofs.

For convenience, we explicitly state the corresponding inner product in the form

$$(f,g)_{H^1(\Omega)} := (f,g)_{L_2(\Omega)} + \sum_{i=1}^d (\partial f/\partial x_i, \partial g/\partial x_i)_{L_2(\Omega)} \tag{3.24}$$

where we, from now on, do not explicitly mark the derivatives as weak derivatives to keep the notation concise. Whenever derivatives appear in a weak formulation in the following, they will be assumed to be weak ones. The norm that is induced by (3.24) is denoted by $\|\cdot\|_{H^1(\Omega)}$.

It should be stressed that the properties of functions in $H^1(\Omega)$ depend on the dimension of Ω. In $d = 1$ space dimension, each function in $H^1(\Omega)$ is bounded and continuous, i.e., $H^1(\Omega) \subset C_b^0(\Omega)$. However, in $d = 2$ or $d = 3$ space dimensions, by the Sobolev embedding theorem we have boundedness and continuity for functions in $H^2(\Omega) \subset H^1(\Omega)$ only. As a consequence, the evaluation of a function $u \in H^1(\Omega)$ in a point x is not well-defined in case of $d = 2$ or $d = 3$ space dimensions.

Regardless of the spatial dimension, however, continuous functions that are piecewise differentiable, such as the functions typically used in a finite element method, always belong to $H^1(\Omega)$.

Extension to Vector Fields

We now come back to the weak form (3.19) and introduce the spaces

$$\mathcal{V} := H^1(\Omega)^3 = \left\{ v = (v_1, v_2, v_3)^T : v_i \in H^1(\Omega), i = 1, 2, 3 \right\}, \tag{3.25}$$

$$\mathcal{V}_0 := \{ v \in \mathcal{V} : v = 0 \text{ on } \Gamma_0 \}. \tag{3.26}$$

3.1 Basic Equations of an Elastic Body

The space $\mathcal{V}$ is just the extension of $H^1(\Omega)$ to three dimensions. Its subspace $\mathcal{V}_0$ contains all test functions which satisfy the homogeneous boundary condition on Γ_0. The Hilbert space $\mathcal{V}$ is equipped with the inner product

$$(\boldsymbol{u}, \boldsymbol{v})_{\mathcal{V}} := \sum_{i=1}^{3} (u_i, v_i)_{H^1(\Omega)}$$

and the corresponding norm $\|\cdot\|_{\mathcal{V}}$.

For $\boldsymbol{u}, \boldsymbol{v} \in \mathcal{V}$, we write the product of stress and strain as

$$a(\boldsymbol{u}, \boldsymbol{v}) := \int_{\Omega} \boldsymbol{\sigma}(\boldsymbol{u}) : \boldsymbol{\varepsilon}(\boldsymbol{v}) \, dx. \tag{3.27}$$

The notation $a(\boldsymbol{u}, \boldsymbol{v})$ on the left stands for a *bilinear form*, which means that it is linear in both arguments and maps $\boldsymbol{u}$ and $\boldsymbol{v}$ to a real number. The strain energy W from (3.9) can therefore be written as $W = \frac{1}{2} a(\boldsymbol{u}, \boldsymbol{u})$. Moreover, due to Korn's inequality [Cia88], the bilinear form a is $\mathcal{V}_0$-*elliptic* or *coercive* on $\mathcal{V}_0$: There is a constant $\alpha > 0$ such that

$$a(\boldsymbol{v}, \boldsymbol{v}) \geq \alpha \|\boldsymbol{v}\|_{\mathcal{V}}^2 \qquad \forall \boldsymbol{v} \in \mathcal{V}_0. \tag{3.28}$$

The right hand side of the weak form (3.19) maps each test function $\boldsymbol{v} \in \mathcal{V}_0$ to a real number. Consequently, it defines a *linear functional* $\boldsymbol{l} \in \mathcal{V}_0'$ where $\mathcal{V}_0'$ is the dual space of $\mathcal{V}_0$. Instead of $l(\boldsymbol{v})$, we use the notation

$$\langle \boldsymbol{l}, \boldsymbol{v} \rangle := \int_{\Omega} \boldsymbol{v}^T \boldsymbol{\beta} \, dx + \int_{\Gamma_1} \boldsymbol{v}^T \boldsymbol{\tau} \, ds \tag{3.29}$$

to indicate the inner product structure. In this context, one calls $\langle \boldsymbol{l}, \boldsymbol{v} \rangle$ the *duality pairing* of $\boldsymbol{l}$ and $\boldsymbol{v}$. The inertia term on the left hand side is treated in a similar fashion by setting

$$\langle \rho \ddot{\boldsymbol{u}}, \boldsymbol{v} \rangle := \int_{\Omega} \boldsymbol{v}^T \rho \ddot{\boldsymbol{u}} \, dx. \tag{3.30}$$

Based on this notation, we can rewrite the weak form (3.19) in the following way, cf. [DL75, §2.2]:

Variational Problem At each instant of time t, find a displacement field $\boldsymbol{u}(\cdot, t) \in \mathcal{V}$ such that $\boldsymbol{u}(\cdot, t) = \boldsymbol{u}_0(\cdot, t)$ on Γ_0 and

$$\langle \rho \ddot{\boldsymbol{u}}, \boldsymbol{v} \rangle + a(\boldsymbol{u}, \boldsymbol{v}) = \langle \boldsymbol{l}, \boldsymbol{v} \rangle \qquad \forall \boldsymbol{v} \in \mathcal{V}_0. \tag{3.31}$$

The initial values are also given in a weak sense. The initial displacement satisfies

$$\langle \boldsymbol{u}(\cdot, t_0), \boldsymbol{v} \rangle = \langle \boldsymbol{u}_{t_0}, \boldsymbol{v} \rangle \qquad \forall \boldsymbol{v} \in \mathcal{V}_0 \tag{3.32}$$

and the initial velocity

$$\langle \dot{\boldsymbol{u}}(\cdot, t_0), \boldsymbol{v} \rangle = \langle \dot{\boldsymbol{u}}_{t_0}, \boldsymbol{v} \rangle \qquad \forall \boldsymbol{v} \in L_2(\Omega)^3, \tag{3.33}$$

Fig. 3.2 Domain $\Omega \subset \mathbb{R}^2$ and parametrization of boundary segment Γ_i

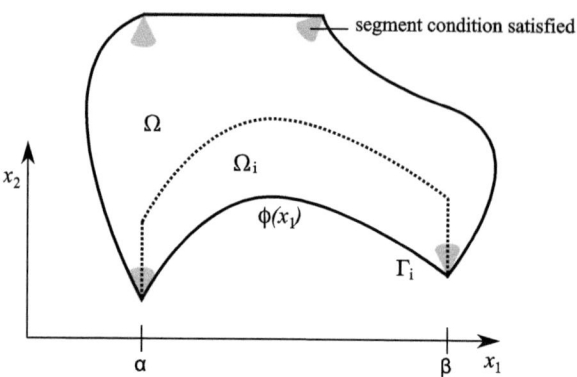

with given initial data $\boldsymbol{u}_{t_0} = \boldsymbol{u}_{t_0}(x)$ and $\dot{\boldsymbol{u}}_{t_0} = \dot{\boldsymbol{u}}_{t_0}(x)$. Note that the velocity $\dot{\boldsymbol{u}}$ is in general an element of $L_2(\Omega)^3$ only and hence less smooth with respect to the spatial variable than the displacement $\boldsymbol{u}$.

3.1.4 The Trace Space

The variational problem (3.31) still contains an inconsistency. While the balance of momentum and the initial values are expressed in weak form, the boundary condition $\boldsymbol{u}(\cdot, t)|_{\Gamma_0} = \boldsymbol{u}_0(\cdot, t)$ is written in strong form. As we have seen above when discussing the Sobolev space $H^1(\Omega)$, the pointwise evaluation of a function in $H^1(\Omega)$ makes only sense for $\Omega \subset \mathbb{R}$. If $\Omega \subset \mathbb{R}^d$ where $d \geq 2$, the space $H^1(\Omega)$ may contain singularities, and the restriction to the boundary is unclear. This issue affects also the surface integral in (3.29) that stems from the surface tractions. To summarize, we cannot interpret boundary conditions in a pointwise sense in case of $d \geq 2$.

The so-called *trace* of a function $f \in H^1(\Omega)$ provides a way out of this dilemma. To explain the idea behind this notion, we consider the planar case $d = 2$ and a smooth function $f \in C^1(\bar{\Omega})$. For such a function, the line integral is defined as

$$\int_\Gamma f \, ds = \int_\alpha^\beta f(\varphi(\tau)) |\varphi'(\tau)| \, d\tau \tag{3.34}$$

where $\varphi : [\alpha, \beta] \to \Gamma$ stands for the parametrization of the boundary segment $\Gamma \subset \partial \Omega \subset \mathbb{R}^2$. Obviously, the evaluation of f is well-defined along the curve that forms Γ. Our goal is now to derive an upper bound for $\|f\|_{L_2(\partial \Omega)}$ in terms of $\|f\|_{H^1(\Omega)}$, which would relate the measuring of f along the boundary to the standard norm in the domain. A density argument will then lead to the formal definition of the trace space [Bra01].

Figure 3.2 illustrates the domain Ω and its boundary $\partial \Omega$, which consists of a finite number of segments Γ_i, $i = 1, \ldots, n_\Gamma$. Each Γ_i is assumed to possess a C^1-parametrization. Picking Γ_i and introducing an appropriate local coordinate system,

3.1 Basic Equations of an Elastic Body

this curve can be written as

$$\Gamma_i = \left\{ x \in \mathbb{R}^2 : x_2 = \phi(x_1), \, \alpha \leq x_1 \leq \beta \right\}.$$

We thus have $\varphi(\tau) = (\tau, \phi(\tau))$, $\alpha \leq \tau \leq \beta$, as the parametrization of Γ_i. Associated with Γ_i, for some $r > 0$ the set

$$\Omega_i = \left\{ x \in \mathbb{R}^2 : \phi(x_1) \leq x_2 \leq \phi(x_1) + r, \, \alpha \leq x_1 \leq \beta \right\}$$

is part of Ω. For this to hold, we require a *segment condition*, which basically states that the angle that can be measured in the transition point between two adjacent boundary segments is greater than zero.

For $f \in C^1(\bar{\Omega})$ and $x \in \Gamma_i$, we have the identity

$$f(x_1, \phi(x_1)) = f(x_1, \phi(x_1) + t) - \int_0^t \frac{\partial f}{\partial x_2}(x_1, \phi(x_1) + s) \, ds, \quad 0 \leq t \leq r.$$

Integrating on both sides from 0 to r, we obtain

$$rf(x_1, \phi(x_1)) = \int_0^r f(x_1, \phi(x_1) + t) \, dt - \int_0^r \frac{\partial f}{\partial x_2}(x_1, \phi(x_1) + t)(r - t) \, dt.$$

Next, we square both sides and apply the inequality $(c + d)^2 \leq 2c^2 + 2d^2$, which yields

$$r^2 f(x_1, \phi(x_1))^2 \leq 2 \left(\int_0^r f(x_1, \phi(x_1) + t) \, dt \right)^2$$

$$+ 2 \left(\int_0^r \frac{\partial f}{\partial x_2}(x_1, \phi(x_1) + t)(r - t) \, dt \right)^2.$$

The Schwarz inequality then gives

$$r^2 f(x_1, \phi(x_1))^2 \leq 2 \int_0^r 1^2 \, dt \int_0^r f(x_1, \phi(x_1) + t)^2 \, dt$$

$$+ 2 \int_0^r (r - t)^2 \, dt \int_0^r \left(\frac{\partial f}{\partial x_2}(x_1, \phi(x_1) + t) \right)^2 dt.$$

Since $\int 1 \, dt = r$ and $\int (r - t)^2 \, dt = r^3/3$, we divide both sides by r^2 and integrate over x_1 to obtain

$$\int_\alpha^\beta f(x_1, \phi(x_1))^2 dx_1 \leq \frac{2}{r} \int_{\Omega_i} f^2 d(x_1, x_2) + \frac{r}{3} \int_{\Omega_i} \left(\frac{\partial f}{\partial x_2} \right)^2 d(x_1, x_2).$$

From the line integral (3.34) we recall that the line element along Γ_i reads $ds = |\varphi'| d\tau = \sqrt{1 + (\phi')^2} d\tau$. Thus, we arrive at the bound

$$\int_{\Gamma_i} f^2 \, ds = \int_\alpha^\beta f(\tau, \phi(\tau))^2 \sqrt{1 + (\phi')^2} d\tau \leq c_i \left(\frac{2}{r} \|f\|_{L_2(\Omega)}^2 + \frac{r}{3} \left| \frac{\partial f}{\partial x_2} \right|_{H^1(\Omega)}^2 \right)$$

where $c_i = \max\{\sqrt{1 + (\phi'(\tau))^2}, \alpha \leq \tau \leq \beta\}$.

By summing up over all boundary segments, we finally get, with a suitable constant c that depends on the domain Ω, the desired estimate

$$\|f\|_{L_2(\partial\Omega)} \leq c \|f\|_{H^1(\Omega)}. \tag{3.35}$$

Though (3.35) holds so far only for $f \in H^1(\Omega) \cap C^1(\bar{\Omega})$, we can extend the reasoning to $f \in H^1(\Omega)$ by considering an appropriate sequence of functions $f_j \in H^1(\Omega) \cap C^1(\bar{\Omega})$ that converge to f in the H^1-norm. The completeness of $L_2(\partial\Omega)$ then guarantees that this limit process is well-defined and results in the same estimate.

The Trace Operator

We are now in a position to introduce the trace as the restriction of Sobolev-class functions to the boundary. The trace is a mapping $\gamma : H^1(\Omega) \to L_2(\partial\Omega)$ such that

$$\begin{cases} \gamma f = f|_{\partial\Omega} & \text{for } f \in H^1(\Omega) \cap C^1(\bar{\Omega}), \\ \|\gamma f\|_{L_2(\partial\Omega)} \leq c \|f\|_{H^1(\Omega)} & \text{for } f \in H^1(\Omega). \end{cases} \tag{3.36}$$

In other words, the trace is a bounded linear operator that coincides with the restriction to $\partial\Omega$ in case of C^1-functions.

Note that this result does not cure the problem with pointwise values of f on $\partial\Omega$. What we have shown is that $f|_{\partial\Omega}$ is well-defined in terms of the trace operator and square integrable on $\partial\Omega$.

The trace γ is continuous but in general not surjective. The image of γ defines the *trace space* $H^{1/2}(\partial\Omega) \subset L_2(\partial\Omega)$. Fractional Sobolev spaces such as $H^{1/2}(\partial\Omega)$ form a continuous scale between integer order Sobolev spaces. Without going into the details, we point out that taking the trace costs $1/2$ of a derivative. Intuitively, this can be explained as follows: For $d = 2$, a function $f \in H^1(\Omega)$ might possess singularities, and if we apply the trace to f, the result is not necessarily a function in $H^1(\partial\Omega)$ because the boundary $\partial\Omega$ is a curve, i.e., one-dimensional. Correspondingly, $H^1(\partial\Omega)$ is equivalent to the class of continuous and bounded functions on $\partial\Omega$ and too small to contain all possible traces. As an example, consider

$$f(x) = \log\log \frac{2}{\|x\|_2} \quad \text{on } \Omega = \{x \in \mathbb{R}^2 : x_1 > 0, x_2 > 0, \|x\|_2 < \tfrac{1}{2}\}.$$

3.2 Constraints in Linear Elasticity

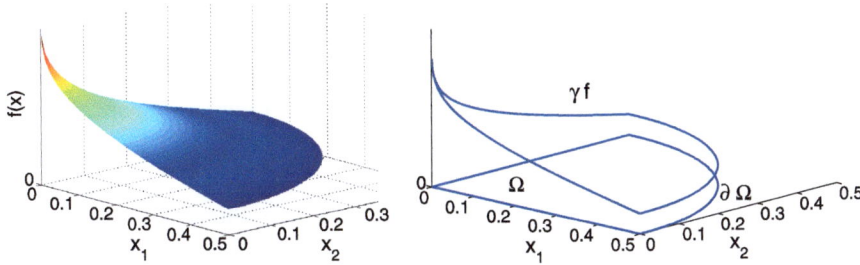

Fig. 3.3 Example of a function $f \in H^1(\Omega)$ whose trace γf is not in $H^1(\partial\Omega)$

The domain Ω and the function f are displayed in Fig. 3.3, and they fit into the framework of Fig. 3.2. Though f possesses a singularity at $x = 0$, it can be shown that $f \in H^1(\Omega)$. However, since $0 \in \partial\Omega$, the singularity precludes that γf is in $H^1(\partial\Omega)$.

Coming back to the weak form (3.31), we state the boundary conditions now more precisely. By simply restricting $\partial\Omega$ to Γ_0, we obtain the Hilbert space $H^{1/2}(\Gamma_0)$. In the sections below, we will also make use of its dual space $H^{1/2}(\Gamma_0)'$, which is the linear space of all bounded linear functionals on $H^{1/2}(\Gamma_0)$. Note that in general, this dual space is not equivalent to $H^{-1/2}(\Gamma_0)$, which is the dual space of $H^{1/2}_{00}(\Gamma_0)$, the space of functions with smooth decay to zero when approaching $\partial\Gamma_0$.

The equality sign in the Dirichlet boundary condition should therefore be understood as equality in $H^{1/2}(\Gamma_0)^3$. In the next section, we will study an alternative formulation of this boundary condition as constraint that is appended by Lagrange multipliers, and in this context, the trace will play an important role.

3.2 Constraints in Linear Elasticity

In the following, we introduce a model with constraints that is based on Lagrange multipliers on the boundary Γ_0. The reason for this new model lies in the structure of flexible multibody systems where the elastic body is, like the rigid one, subject to constraints since joints restrict its motion. The weak form (3.19) or (3.31), respectively, takes such a constraint into account by formulating it as a boundary condition $\boldsymbol{u} = \boldsymbol{u}_0$ on Γ_0. In general, however, the function $\boldsymbol{u}_0$ depends on the motion of neighboring bodies and is hence unknown. Though a certain choice of coordinates may circumvent this difficulty in some cases, there is a need for models with explicitly formulated constraints.

3.2.1 Variational Formulation

The fundamental idea is to append the boundary condition $\boldsymbol{u} = \boldsymbol{u}_0$ as a constraint in Hamilton's principle. With kinetic energy T from (3.11) and potential energy U from (3.12), we require

$$\int_{t_0}^{t_1} \left(T - U - \int_{\Gamma_0} \boldsymbol{\mu}^T (\boldsymbol{u} - \boldsymbol{u}_0) \, \mathrm{d}s \right) \mathrm{d}t \to \text{stationary !} \tag{3.37}$$

Here, $\boldsymbol{\mu} = \boldsymbol{\mu}(x,t)$ is a Lagrange multiplier defined on the boundary Γ_0, and the displacement $\boldsymbol{u}$ is now arbitrary and not a priori equal to $\boldsymbol{u}_0$ on Γ_0.

We postpone the introduction of appropriate function spaces for the moment and proceed in the standard way. Define

$$J(\theta) := \int_{t_0}^{t_1} \left(T(\boldsymbol{u} + \theta \boldsymbol{v}) - U(\boldsymbol{u} + \theta \boldsymbol{v}) - \int_{\Gamma_0} (\boldsymbol{\mu} + \theta \boldsymbol{\vartheta})^T (\boldsymbol{u} + \theta \boldsymbol{v} - \boldsymbol{u}_0) \, \mathrm{d}s \right) \mathrm{d}t$$

with perturbation $\boldsymbol{v} = \boldsymbol{v}(x,t)$ for the displacement and $\boldsymbol{\vartheta} = \boldsymbol{\vartheta}(x,t)$ for the multiplier. As above, we require the properties $\boldsymbol{v}(\cdot, t_0) = \boldsymbol{v}(\cdot, t_1) = \boldsymbol{0}$ in Ω and $\boldsymbol{\vartheta}(\cdot, t_0) = \boldsymbol{\vartheta}(\cdot, t_1) = \boldsymbol{0}$ on Γ_0.

Next, we evaluate the stationarity condition $0 = \mathrm{d}J(\theta)/\mathrm{d}\theta$ at $\theta = 0$. After the same transformations as in Sect. 3.1.1, we get

$$0 = \int_{t_0}^{t_1} \int_{\Omega} \boldsymbol{v}^T (\rho \ddot{\boldsymbol{u}} - \mathbf{div}\,\boldsymbol{\sigma} - \boldsymbol{\beta}) \, \mathrm{d}x \, \mathrm{d}t + \int_{t_0}^{t_1} \int_{\Gamma_1} \boldsymbol{v}^T (\boldsymbol{\sigma} \boldsymbol{n} - \boldsymbol{\tau}) \, \mathrm{d}s \, \mathrm{d}t$$
$$+ \int_{t_0}^{t_1} \int_{\Gamma_0} \boldsymbol{v}^T (\boldsymbol{\sigma} \boldsymbol{n} + \boldsymbol{\mu}) \, \mathrm{d}s \, \mathrm{d}t + \int_{t_0}^{t_1} \int_{\Gamma_0} \boldsymbol{\vartheta}^T (\boldsymbol{u} - \boldsymbol{u}_0) \, \mathrm{d}s \, \mathrm{d}t,$$

which must hold for all $\boldsymbol{v}$ and $\boldsymbol{\vartheta}$. Applying the fundamental lemma of the calculus of variations, we can draw the following conclusions: The first integral leads one more time to the balance equation (3.18a) in strong form, and the second integral to the Neumann boundary condition (3.18c). A new aspect is introduced by the third integral, which did not show up in Sect. 3.1.1 since there we had the property $\boldsymbol{v} = \boldsymbol{0}$ on Γ_0. Now we obtain $\boldsymbol{\mu} = -\boldsymbol{\sigma}\boldsymbol{n}$ on Γ_0, i.e., the Lagrange multiplier can be viewed as negative stress induced by the constraint. The fourth integral yields exactly this constraint, $\boldsymbol{0} = \boldsymbol{u} - \boldsymbol{u}_0$.

Equations of Constrained Linear Elasticity

Summing up, the approach (3.37) leads to the equations

$$\rho \ddot{\boldsymbol{u}}(x,t) = \mathbf{div}\,\boldsymbol{\sigma}(\boldsymbol{u}(x,t)) + \boldsymbol{\beta}(x,t) \quad \text{in } \Omega, \tag{3.38a}$$
$$\boldsymbol{u}(x,t) = \boldsymbol{u}_0(x,t) \quad \text{on } \Gamma_0, \tag{3.38b}$$

3.2 Constraints in Linear Elasticity

$$\sigma(u(x,t))n(x) = -\mu(x,t) \quad \text{on } \Gamma_0, \quad (3.38c)$$

$$\sigma(u(x,t))n(x) = \tau(x,t) \quad \text{on } \Gamma_1, \quad (3.38d)$$

which is equivalent with the strong form (3.18a)–(3.18c). More insight is obtained, though, if one passes to the corresponding weak form.

For this purpose, we premultiply the balance equation (3.38a) by the time-independent but otherwise arbitrary test function v, integrate over Ω, apply Green's formula (3.16) in reverse, and get

$$\int_\Omega \rho v^T \ddot{u}\,dx + \int_\Omega \sigma(u):\varepsilon(v)\,dx - \int_{\Gamma_0} v^T \sigma n\,ds = \int_\Omega v^T \beta\,dx + \int_{\Gamma_1} v^T \sigma n\,ds.$$

In the surface integral on the right, we replace σn by τ due to (3.38d), and in the surface integral on the left, we replace σn by the unknown multiplier $-\mu$ due to (3.38c). Moreover, we multiply the constraint (3.38b) by the time-independent test function ϑ defined on Γ_0 and integrate over the boundary in order to derive a *weak constraint equation*. In total, the weak form of the equations of constrained linear elasticity reads

$$\int_\Omega \rho v^T \ddot{u}\,dx + \int_\Omega \sigma : \varepsilon\,dx + \int_{\Gamma_0} v^T \mu\,ds = \int_\Omega v^T \beta\,dx + \int_{\Gamma_1} v^T \tau\,ds, \quad (3.39a)$$

$$\int_{\Gamma_0} \vartheta^T u\,ds = \int_{\Gamma_0} \vartheta^T u_0\,ds. \quad (3.39b)$$

This formulation is a time-dependent or *transient saddle point problem*. It must hold for all test functions v and ϑ. Note that the equations of motion (3.39a), (3.39b) are not restricted to the conservative case. Extensions like dissipation (3.20) are straightforward to include.

3.2.2 Appropriate Function Spaces

Based on the framework of Sect. 3.1.3, we state now the saddle point problem (3.39a), (3.39b) in a more precise way.

Consider the constraint equation (3.39b) and recall that u_0 is element of the trace space $H^{1/2}(\Gamma_0)^3$. The product $\int_{\Gamma_0} \vartheta^T u_0\,ds$ makes not only sense for ϑ element of $L_2(\Gamma_0)^3$ but moreover for ϑ element of the dual space

$$\mathcal{Q} := (H^{1/2}(\Gamma_0)^3)'. \quad (3.40)$$

The function u_0 is thus an element of the dual dual space $\mathcal{Q}' = H^{1/2}(\Gamma_0)^3$ and can be viewed as linear functional $m \in \mathcal{Q}'$. Instead of $m(\vartheta)$, we make use of the duality pairing $\langle m, \vartheta \rangle$, but this time as pairing on $\mathcal{Q}' \times \mathcal{Q}$. For ϑ element of $L_2(\Gamma_0)^3$ we have in particular

$$\langle m, \vartheta \rangle = \int_{\Gamma_0} \vartheta^T u_0 \, ds.$$

The left hand side of the constraint equation (3.39b) has the same structure since the trace γu of u on Γ_0 is element of the trace space. In this case, however, we view u as element of $\mathcal{V} = H^1(\Omega)^3$ and the integral as *bilinear form b* on the product space $\mathcal{V} \times \mathcal{Q}$. In other words, we set

$$b(u, \vartheta) := \langle \gamma u, \vartheta \rangle \tag{3.41}$$

for all $u \in \mathcal{V}$ and $\vartheta \in \mathcal{Q}$. Note that this bilinear form appears as $b(v, \mu)$ in the balance equation (3.39a), too.

The other integrals in the balance equation can be treated in the same way as in Sect. 3.1.3, i.e., the bilinear form a denotes the variation of the strain energy and the linear functional l the right hand side, with duality pairing $\langle , \rangle$ on $\mathcal{V}' \times \mathcal{V}$ (we omit extra subscripts to distinguish between both pairings). In this way, we finally reformulate the equations of motion and obtain the following abstract setting:

Variational Problem At each instant of time t, find a displacement field $u(\cdot, t) \in \mathcal{V}$ and a multiplier $\mu(\cdot, t) \in \mathcal{Q}$ such that

$$\langle \rho \ddot{u}, v \rangle + a(u, v) + b(v, \mu) = \langle l, v \rangle \quad \forall v \in \mathcal{V}, \tag{3.42a}$$

$$b(u, \vartheta) \qquad\qquad = \langle m, \vartheta \rangle \quad \forall \vartheta \in \mathcal{Q}. \tag{3.42b}$$

The initial values $u(\cdot, t_0)$ and $\dot{u}(\cdot, t_0)$ are given in a weak sense like in (3.32). In [Sim00], the transient saddle point problem (3.42a), (3.42b) is called *weak descriptor form*, a name that stresses the analogy to the equations of constrained rigid body motion. For vanishing acceleration, $\ddot{u} \equiv 0$, the formulation reduces to a standard saddle point problem.

3.3 Mathematical Analysis

The transient saddle point problem (3.42a), (3.42b) represents an infinite-dimensional analogue of the differential-algebraic equations of rigid body motion (2.9a), (2.9b). In this section, we are concerned with an analysis of (3.42a), (3.42b) from this perspective. Which results can be carried over, and where are the differences to systems of rigid bodies? Additionally, an estimate on the influence of perturbations is given. The mathematical framework developed here sheds also new light on other model equations with constraints such as dynamic contact or incompressibility that will be discussed in the subsequent section.

3.3.1 Existence of Solutions

We start our analysis with a fundamental notion in the theory of saddle point problems.

A continuous bilinear form b on $\mathcal{V} \times \mathcal{Q}$ satisfies the *inf-sup condition* iff there exists a $\beta \in \mathbb{R}$ such that

$$\inf_{\vartheta \in \mathcal{Q}} \sup_{v \in \mathcal{V}} \frac{b(v, \vartheta)}{\|v\|_{\mathcal{V}} \|\vartheta\|_{\mathcal{Q}}} \geq \beta > 0. \tag{3.43}$$

Often, (3.43) is also called the LBB-condition due to Ladyshenskaja, Babuška, and Brezzi [BF91]. Looking back to the minimax characterization (2.2) of the finite-dimensional case, we see that (3.43) is a generalized regularity condition on the weak constraint equations. In standard saddle point theory, (3.43) guarantees, along with ellipticity of the bilinear form a, that the Lagrange multiplier is unique.

Since the trace γ is a continuous mapping, the bilinear form b of (3.41) is also continuous and thus satisfies the prerequisite for the inf-sup condition. In case of an exclusive Dirichlet boundary $\Gamma_0 = \partial\Omega$ and $\Gamma_1 = \emptyset$, the inf-sup condition itself can be shown by means of the properties of the trace operator (3.36), see [DK01]. In fact, appending Dirichlet boundary conditions by Lagrange multipliers has become fairly established since the original work of Babuška [Bab73]. The situation where $\Gamma_0 \subset \Omega$ and $\Gamma_1 \neq \emptyset$, however, requires a somewhat more involved discussion in order to prove the inf-sup condition for b.

Lemma 3.1 *If the boundary $\partial\Omega$ is sufficiently smooth, the inf-sup condition (3.43) for the bilinear form b of (3.41) is satisfied with a constant $\beta > 0$ that depends solely on the geometry.*

Proof It suffices to consider a scalar function $u \in H^1(\Omega) =: V$ with the boundary condition $u = u_0$ on Γ_0 in the sense of traces. As described above, the boundary conditions are now appended by Lagrange multipliers.

The restriction of the trace space $H^{1/2}(\partial\Omega)$ to Γ_0 defines $H^{1/2}(\Gamma_0)$ with its dual space $Q := H^{1/2}(\Gamma_0)'$, cf. Sect. 3.1.4. Next, we introduce the bilinear form $b: V \times Q \to \mathbb{R}$ where

$$b(v, \vartheta) := \langle \gamma v_{|\Gamma_0}, \vartheta \rangle.$$

On Q we have the norm

$$\|\vartheta\|_Q = \sup_{v \in H^{1/2}(\Gamma_0)} \frac{\langle v, \vartheta \rangle}{\|v\|_{H^{1/2}(\Gamma_0)}}. \tag{3.44}$$

For each $\vartheta \in Q$ there is a function $v_\vartheta \in H^{1/2}(\Gamma_0)$ with

$$\|\vartheta\|_Q \leq 2 \frac{\langle v_\vartheta, \vartheta \rangle}{\|v_\vartheta\|_{H^{1/2}(\Gamma_0)}}. \tag{3.45}$$

This v_ϑ is simply selected by taking a function which is way below the supremum in (3.44). For v_ϑ in turn there exists a continuous extension $\tilde{v}_\vartheta \in H^{1/2}(\partial\Omega)$ such that $v_\vartheta = \tilde{v}_\vartheta$ on Γ_0, see Stein [Ste70]. The norm of $\tilde{v}_\vartheta$ satisfies the estimate

$$\|\tilde{v}_\vartheta\|_{H^{1/2}(\partial\Omega)} \leq c_1 \|v_\vartheta\|_{H^{1/2}(\Gamma_0)} \tag{3.46}$$

with a positive real constant c_1 that depends on the shape and the smoothness of $\partial\Omega$. Moreover, by the trace theorem $\tilde{v}_\vartheta$ has an extension $\hat{v}_\vartheta \in H^1(\Omega)$ such that $\gamma \hat{v}_\vartheta = \tilde{v}_\vartheta$ and

$$\|\hat{v}_\vartheta\|_{H^1(\Omega)} \leq c_2 \|\tilde{v}_\vartheta\|_{H^{1/2}(\partial\Omega)} \tag{3.47}$$

with a second constant c_2. Combining (3.45), (3.46), and (3.47), we obtain

$$\sup_{v \in H^1(\Omega)} \frac{b(v,\vartheta)}{\|v\|_{H^1(\Omega)}} \geq \frac{b(\hat{v}_\vartheta, \vartheta)}{\|\hat{v}_\vartheta\|_{H^1(\Omega)}} \geq \frac{1}{2c_1 c_2} \|\vartheta\|_Q,$$

which concludes the proof with $\beta = 1/(2c_1 c_2)$. $\square$

Though the term "sufficiently smooth" in Lemma 3.1 can be made more precise, we omit these technical details here and just remark that boundaries defined by functions that are continuously differentiable fall into this class, see Fig. 3.2.

Associated with the weak constraints, we next define the operator

$$B : \mathcal{V} \to \mathcal{Q}', \quad \langle Bv, \vartheta \rangle = b(v, \vartheta) \quad \forall v \in \mathcal{V}, \vartheta \in \mathcal{Q}. \tag{3.48}$$

Thus, the operator B maps each $v \in \mathcal{V}$ to a linear functional $Bv \in \mathcal{Q}'$. It can be shown that the inf-sup condition (3.43) is equivalent with a property of the operator B. To this end, let $\mathcal{V}_0^\perp$ be the orthogonal complement of $\mathcal{V}_0$ with respect to the standard scalar product in $\mathcal{V}$. If B defines an isomorphism between the spaces $\mathcal{V}_0^\perp$ and $\mathcal{Q}'$ and satisfies

$$\|Bv\|_{\mathcal{Q}'} \geq \beta \|v\|_{\mathcal{V}} \quad \forall v \in \mathcal{V}_0^\perp, \tag{3.49}$$

the inf-sup condition is satisfied [Bra01, §III.4]. Conversely, the inf-sup condition implies (3.49) and the isomorphism property.

After these preparations, we look at the weak constraint equations in more detail. We have the equivalence

$$u|_{\Gamma_0} = u_0 \text{ in the sense of traces} \Leftrightarrow b(u, \vartheta) = \langle m, \vartheta \rangle \quad \forall \vartheta \in \mathcal{Q}. \tag{3.50}$$

The implication from left to right follows from

$$0 = b(u, \vartheta) - \langle m, \vartheta \rangle = \langle u - u_0, \vartheta \rangle \leq \|u - u_0\|_{\mathcal{Q}'} \|\vartheta\|_{\mathcal{Q}},$$

and the reverse direction is a consequence of the duality argument

$$\|u - u_0\|_{\mathcal{Q}'} = \sup_{\vartheta \in \mathcal{Q}} \frac{\langle u - u_0, \vartheta \rangle}{\|\vartheta\|_{\mathcal{Q}}}.$$

3.3 Mathematical Analysis

Based on the equivalence (3.50), we can characterize the subspace $\mathcal{V}_0$ of $\mathcal{V} = H^1(\Omega)^3$ defined in (3.26) by

$$\mathcal{V}_0 = \{v \in \mathcal{V} : b(v, \vartheta) = 0 \ \forall \vartheta \in \mathcal{Q}\} = \text{Ker}(B). \tag{3.51}$$

We are now in a position to show the relation between the constrained system (3.42a), (3.42b) and the unconstrained weak form (3.31). By restricting the test functions v in the transient saddle point formulation (3.42a), (3.42b) to the subspace $\mathcal{V}_0$ and exploiting that $b(v, \mu) = 0$ for $v \in \mathcal{V}_0$, we obtain the following variational problem: Find $u(\cdot, t) \in \mathcal{V}$ such that

$$\langle \rho \ddot{u}, v \rangle + a(u, v) = \langle l, v \rangle \qquad \forall v \in \mathcal{V}_0, \tag{3.52a}$$

$$b(u, \vartheta) = \langle m, \vartheta \rangle \qquad \forall \vartheta \in \mathcal{Q}. \tag{3.52b}$$

Obviously, the equivalence (3.50) implies that the constraint (3.52b) is nothing else than the Dirichlet boundary condition $u|_{\Gamma_0} = u_0$ in the sense of traces. In other words, the restricted problem (3.52a), (3.52b) and the weak form (3.31) are also equivalent. If we start from a solution of the saddle point formulation (3.42a), (3.42b), the projection of (3.42a) onto the space $\mathcal{V}_0 = \text{Ker}(B)$ brings us back to the unconstrained case.

In the following, we analyze the existence of solutions in the unconstrained case and turn then again our attention to the saddle point formulation.

Theory of Duvaut and Lions

For the unconstrained weak form (3.31), Duvaut/Lions [DL75, §III.4] showed existence and uniqueness of the solution u under certain conditions. We briefly summarize here the main results for later use in combination with the saddle point formulation.

It is convenient to restate the equations of motion (3.31). They read

$$\langle \rho \ddot{u}, v \rangle + a(u, v) = \langle l, v \rangle \qquad \forall v \in \mathcal{V}_0,$$

with boundary condition $u(\cdot, t)|_{\Gamma_0} = u_0(\cdot, t)$ and initial values $u_{t_0}, \dot{u}_{t_0}$. We have the following assumptions:

$$\text{Bilinear form } a \text{ is continuous and } \mathcal{V}_0 - \text{elliptic;} \tag{3.53a}$$

$$\text{Initial values } u_{t_0}|_{\Gamma_0} = u_0(\cdot, t_0), \ \dot{u}_{t_0} \in L_2(\Omega)^3; \tag{3.53b}$$

$$\text{Volume force } \beta, \dot{\beta} \in L_2(\Omega \times (t_0, t_1))^3; \tag{3.53c}$$

$$\text{Surface traction } \tau, \dot{\tau} \in L_2(\Gamma_1 \times (t_0, t_1))^3; \tag{3.53d}$$

$$\text{Boundary displacement } u_0 = \bar{u}_0|_{\Gamma_0}$$

where $\partial^i \bar{u}_0(\cdot, t)/\partial t^i \in H^{1/2}(\partial\Omega)^3$ (3.53e)

and $\left(\int_{t_0}^{t_1} \|\partial^i \bar{u}_0/\partial t^i\|^2_{H^{1/2}(\partial\Omega)^3} \, dt \right)^{1/2} < \infty, \quad i = 0, \ldots, 3.$

Note that the last requirement (3.53e) for the extension $\bar{u}_0$ of u_0 defined on the complete boundary $\partial\Omega$ postulates the existence of temporal derivatives of the boundary data in a weak sense. This will turn out to be similar to the smoothness assumption when determining the index of a differential-algebraic equation by differentiation of the constraints.

In order to show existence and uniqueness of the displacement field, one applies the usual splitting

$$u = w + r, \qquad w|_{\Gamma_0} = 0 \text{ and } r|_{\Gamma_0} = u_0 \qquad (3.54)$$

where r is assumed to be known. Thus we can transform (3.31) to a problem in the unknown variable w with homogeneous boundary conditions: Find $w(\cdot, t) \in \mathcal{V}_0$ such that

$$\langle \rho \ddot{w}, v \rangle + a(w, v) = \langle \psi, v \rangle \qquad \forall v \in \mathcal{V}_0. \qquad (3.55)$$

The linear functional $\psi \in \mathcal{V}'_0$ is defined by

$$\langle \psi, v \rangle := \langle l, v \rangle - \langle \rho \ddot{r}, v \rangle - a(r, v),$$

and the inhomogeneous part r satisfies

$$\frac{\partial^i r(\cdot, t)}{\partial t^i} \in \mathcal{V} = H^1(\Omega)^3 \quad \text{for } i = 0, \ldots, 3, \qquad (3.56)$$

which can be shown to be a consequence of assumption (3.53e).

As main result, one has the following theorem [DL75, The. 4.1, §III.4]:

Theorem 3.2 *If the assumptions* (3.53a)–(3.53e) *are satisfied, the weak form* (3.55) *possesses a unique solution w. Furthermore, the unconstrained equations of motion* (3.31) *possess a unique solution $u = w + r$ with initial values $u_{t_0}, \dot{u}_{t_0}$. On finite time intervals, w, $\dot{w}$ and $\ddot{w}$ are bounded with*

$$w(\cdot, t) \in \mathcal{V}_0, \quad \dot{w}(\cdot, t) \in L_2(\Omega)^3, \quad \ddot{w}(\cdot, t) \in \mathcal{V}'_0. \qquad (3.57)$$

There are several important aspects to this result. First, the splitting (3.54) follows the same reasoning as in the stationary case. However, we require here obviously additional smoothness of u_0 and r with respect to time. Secondly, the initial value $\dot{u}_{t_0}$ for the velocity field is an element of $L_2(\Omega)^3$ and hence not affected by the boundary condition since the trace is not defined in this setting. And finally, the velocity $\dot{w}$ and the acceleration $\ddot{w}$ are less smooth with respect to the spatial variable. In fact, the smoothness in space decreases with each differentiation in time, and the

3.3 Mathematical Analysis

same holds for the corresponding derivatives $\dot{u}(\cdot, t) \in L_2(\Omega)^3$ and $\ddot{u}(\cdot, t) \in \mathcal{V}_0'$ of the displacement field.

Theorem 3.2 is also the key to analyze the projected equations of motion (3.52a), (3.52b). Since we showed in (3.50) that boundary conditions in the sense of traces and weak constraints are equivalent, any solution of the unconstrained weak form (3.31) also satisfies the constrained formulation (3.52a), (3.52b). For completeness, we mention briefly the stationary case $\dot{u} = \ddot{u} \equiv 0$ where the problem reduces to

$$a(u, v) = \langle l, v \rangle \qquad \forall v \in \mathcal{V}_0. \tag{3.58}$$

This weak form is equivalent to minimizing the potential energy,

$$U(u) = \frac{1}{2} a(u, u) - \langle l, u \rangle \to \min ! \tag{3.59}$$

By the Lemma of Lax–Milgram [Bra01, §II.2], a unique solution $u \in \mathcal{V}$ exists if the bilinear form a is $\mathcal{V}_0$-elliptic.

Existence of Lagrange Multiplier

We have seen above in (3.38c) that the Lagrange multiplier μ in the saddle point formulation (3.42a), (3.42b) can be viewed as the negative normal stress induced by the constraint $u = u_0$ on the boundary segment Γ_0. Instead of using Green's formula, we now seek to solve for μ directly from the weak form (3.42a), (3.42b). More precisely, we take the displacement field u as a solution of the projected formulation (3.52a), (3.52b) and look for a multiplier μ such that (u, μ) satisfies (3.42a), (3.42b). The following steps proceed as in the stationary case described in [BF91, Chap. 2]. Besides the assumptions (3.53a)–(3.53e), the inf-sup condition (3.43) for the continuous bilinear form b comes into play here.

First, the space $\mathcal{V}$ is decomposed into $\mathcal{V} = \mathcal{V}_0 \oplus \mathcal{V}_0^\perp$ where orthogonality is again to be understood with respect to the inner product in $\mathcal{V}$. Projected onto $\mathcal{V}_0^\perp$, the dynamic equation (3.42a) yields

$$b(v, \mu) = \langle l, v \rangle - a(u, v) - \langle \rho \ddot{u}, v \rangle \qquad \forall v \in \mathcal{V}_0^\perp. \tag{3.60}$$

For any $w \in \mathcal{V}$, the operator B from (3.48) defines a linear functional $Bw \in \mathcal{Q}'$, which we can write as inner product in $\mathcal{Q}$ due to the Riesz theorem. Accordingly, the mapping $\mu \mapsto \langle Bw, \mu \rangle = b(w, \mu)$ possesses a representation $(jBw, \mu)_\mathcal{Q}$ where j stands for the corresponding Riesz operator. The bilinear form

$$d(v, w) := (jBv, jBw)_\mathcal{Q}$$

defined on $\mathcal{V} \times \mathcal{V}$ is hence continuous and symmetric. Additionally, d satisfies the estimate

$$d(v, v) = \|jBv\|_\mathcal{Q}^2 = \|Bv\|_{\mathcal{Q}'}^2 \geq \beta^2 \|v\|_\mathcal{V}^2 \qquad \forall v \in \mathcal{V}_0^\perp, \tag{3.61}$$

due to the inf-sup condition (3.43) and its variant (3.49).

The estimate (3.61) implies that d is $\mathcal{V}_0^\perp$-elliptic. As a consequence, the variational problem

$$d(v, w) = \langle l, v \rangle - a(u, v) - \langle \rho \ddot{u}, v \rangle \qquad \forall v \in \mathcal{V}_0^\perp \qquad (3.62)$$

has a unique solution $w \in \mathcal{V}_0^\perp$ if the right hand side is well-defined, i.e., if it can be viewed as linear functional in $(\mathcal{V}_0^\perp)'$. For this reasoning, we require additionally that $\rho \ddot{u}$ represents a linear functional in $\mathcal{V}' \subset \mathcal{V}_0'$.

Finally, one sets $\mu(\cdot, t) := jBw \in \mathcal{Q}$ to obtain

$$(jBv, \mu)_\mathcal{Q} = \langle Bv, \mu \rangle = b(v, \mu) = \langle l, v \rangle - a(u, v) - \langle \rho \ddot{u}, v \rangle$$

for all $v \in \mathcal{V}_0^\perp$. The multiplier μ and the displacement field u solve thus the transient saddle point problem (3.42a), (3.42b). Since the operator B is surjective, μ is unique. Summarizing, we have

Theorem 3.3 *Let $u(\cdot, t) \in \mathcal{V}$ be a solution of the projected variational problem (3.52a), (3.52b) with $\rho \ddot{u}(\cdot, t) \in \mathcal{V}'$, and assume the inf-sup condition (3.43) to hold. Then the transient saddle point problem (3.42a), (3.42b) possesses a unique solution $(u(\cdot, t), \mu(\cdot, t)) \in \mathcal{V} \times \mathcal{Q}$.*

3.3.2 Influence of Perturbations

Finally, we derive an estimate on the influence of perturbations for the transient saddle point problem (3.42a), (3.42b). In the rigid body case, the corresponding estimate (2.76) revealed a particular sensitivity of the Lagrange multiplier with respect to perturbations of the constraints. Is there a similar property of the infinite-dimensional constraints encountered here?

Three constants play an important role in the following analysis:

$$|a(u, v)| \leq \eta \|u\|_\mathcal{V} \|v\|_\mathcal{V} \quad \forall u, v \in \mathcal{V} \quad \text{(continuity)}, \qquad (3.63a)$$

$$a(u, u) \geq \alpha \|u\|_\mathcal{V}^2 \quad \forall u \in \mathcal{V}_0 \quad \text{(ellipticity)}, \qquad (3.63b)$$

$$\|Bv\|_{\mathcal{Q}'} \geq \beta \|v\|_\mathcal{V} \quad \forall v \in \mathcal{V}_0^\perp \quad \text{(inf-sup condition)}. \qquad (3.63c)$$

Let (u, μ) be a solution of the saddle point problem (3.42a), (3.42b). Starting with (3.60), we get the estimate

$$|b(v, \mu)| \leq |\langle l, v \rangle| + |a(u, v)| + |\langle \rho \ddot{u}, v \rangle|$$
$$\leq \|l\|_{\mathcal{V}'} \|v\|_\mathcal{V} + \eta \|u\|_\mathcal{V} \|v\|_\mathcal{V} + \|\rho \ddot{u}\|_{\mathcal{V}'} \|v\|_\mathcal{V} \qquad \forall v \in \mathcal{V}.$$

The inf-sup condition yields

$$\beta \|\mu\|_\mathcal{Q} \leq \sup_{v \in \mathcal{V}} \frac{|b(v, \mu)|}{\|v\|_\mathcal{V}} \leq \|l\|_{\mathcal{V}'} + \eta \|u\|_\mathcal{V} + \|\rho \ddot{u}\|_{\mathcal{V}'}$$

3.3 Mathematical Analysis

and thus

$$\|\mu\|_Q \le \frac{1}{\beta}(\|l\|_{\mathcal{V}'} + \eta\|u\|_{\mathcal{V}} + \|\rho\ddot{u}\|_{\mathcal{V}'}). \tag{3.64}$$

We proceed as in the stationary case and replace $\|u\|_{\mathcal{V}}$ in (3.64), treating $\|\rho\ddot{u}\|_{\mathcal{V}'}$ for the moment as known quantity.

Following the splitting in (3.54), we set $u = w + r$ and insert $v = w$ in (3.42a) or (3.55), respectively, to obtain

$$a(w, w) = \langle l, w \rangle - a(r, w) - \langle \rho\ddot{u}, w \rangle.$$

Due to $\mathcal{V}_0$-ellipticity and continuity of a, one concludes

$$\|w\|_{\mathcal{V}} \le \frac{1}{\alpha}(\|l\|_{\mathcal{V}'} + \eta\|r\|_{\mathcal{V}} + \|\rho\ddot{u}\|_{\mathcal{V}'}).$$

Therefore, we have

$$\|u\|_{\mathcal{V}} \le \|w\|_{\mathcal{V}} + \|r\|_{\mathcal{V}} \le \frac{1}{\alpha}(\|l\|_{\mathcal{V}'} + \|\rho\ddot{u}\|_{\mathcal{V}'}) + (1 + \frac{\eta}{\alpha})\|r\|_{\mathcal{V}},$$

and from

$$\|m\|_{Q'} = \|Br\|_{Q'} \ge \beta\|r\|_{\mathcal{V}}$$

we get

$$\|u\|_{\mathcal{V}} \le \frac{1}{\alpha}(\|l\|_{\mathcal{V}'} + \|\rho\ddot{u}\|_{\mathcal{V}'}) + \frac{1}{\beta}(1 + \frac{\eta}{\alpha})\|m\|_{Q'}.$$

Upon insertion of this relation into (3.64), it follows

$$\|\mu\|_Q \le (1 + \frac{\eta}{\alpha})\left(\frac{1}{\beta}\|l\|_{\mathcal{V}'} + \frac{1}{\beta}\|\rho\ddot{u}\|_{\mathcal{V}'} + \frac{\eta}{\beta^2}\|m\|_{Q'}\right). \tag{3.65}$$

In the stationary case $\ddot{u} \equiv 0$, the estimate (3.65) is a fundamental result on the influence of the data l, m on the Lagrange multiplier [BF91, §II.1]. In the dynamic case, however, the acceleration term $\|\rho\ddot{u}\|_{\mathcal{V}'}$ on the right hand side is still intruding. Using the weak constraint equations, it is possible to provide a more detailed estimate.

Theorem 3.4 *Assume the conditions (3.63a)–(3.63c) and let $(u(\cdot, t), \mu(\cdot, t)) \in \mathcal{V} \times Q$ be a solution of the transient saddle point problem (3.42a), (3.42b) with $\rho\ddot{u} \in \mathcal{V}'$ and constant mass density ρ. Then we have*

$$\|\mu\|_Q \le (1 + \frac{\eta}{\alpha})\left(\frac{1}{\beta}\|l\|_{\mathcal{V}'} + \frac{\eta}{\beta^2}\|m\|_{Q'} + \frac{\rho}{\beta}\|\ddot{w}\|_{\mathcal{V}'} + \frac{\rho}{\beta^2}\|\ddot{m}\|_{Q'}\right). \tag{3.66}$$

Here, $w = u - r$ stands for the homogeneous part of the solution from the splitting (3.54).

Proof The inhomogeneous part r satisfies the constraints

$$b(r, \vartheta) - \langle m, \vartheta \rangle = 0 \qquad \forall \vartheta \in \mathcal{Q}$$

and can be differentiated with respect to time since $\partial^i r(\cdot, t)/\partial t^i \in \mathcal{V}$ for $i = 0, \ldots, 3$ from (3.56). Furthermore, the assumption (3.53e) on the boundary displacement u_0 guarantees the existence of $\dot{m}, \ddot{m} \in \mathcal{Q}'$. Thus, we may differentiate the constraints as in the DAE case to obtain

$$\frac{d}{dt}(b(r, \vartheta) - \langle m, \vartheta \rangle) = b(\dot{r}, \vartheta) - \langle \dot{m}, \vartheta \rangle = 0, \tag{3.67}$$

$$\frac{d^2}{dt^2}(b(r, \vartheta) - \langle m, \vartheta \rangle) = b(\ddot{r}, \vartheta) - \langle \ddot{m}, \vartheta \rangle = 0. \tag{3.68}$$

If $\ddot{r} \neq 0$ on the boundary, we can assume $\ddot{r} \in \mathcal{V}_0^\perp$ without loss of generality. The inf-sup condition (3.49) implies then

$$\|\ddot{m}\|_{\mathcal{Q}'} = \|B\ddot{r}\|_{\mathcal{Q}'} \geq \beta \|\ddot{r}\|_{\mathcal{V}}$$

and for constant mass density ρ

$$\|\rho\ddot{u}\|_{\mathcal{V}'} \leq \|\rho\ddot{w}\|_{\mathcal{V}'} + \|\rho\ddot{r}\|_{\mathcal{V}'} \leq \rho\|\ddot{w}\|_{\mathcal{V}'} + \frac{\rho}{\beta}\|\ddot{m}\|_{\mathcal{Q}'}.$$

Insertion of this estimate into (3.65) completes the proof. $\square$

Though still containing an acceleration term $\ddot{w}$, the estimate (3.66) allows several conclusions about the influence of the data on the Lagrange multiplier. In order to better compare this result with the rigid body estimate (2.76), we introduce a perturbed system

$$\langle \rho\ddot{\hat{u}}, v \rangle + a(\hat{u}, v) + b(v, \hat{\mu}) = \langle l, v \rangle + \langle \delta, v \rangle \qquad \forall v \in \mathcal{V},$$

$$b(\hat{u}, \vartheta) \qquad\qquad = \langle m, \vartheta \rangle + \langle \theta, \vartheta \rangle \qquad \forall \vartheta \in \mathcal{Q}$$

with perturbations $\delta(\cdot, t) \in \mathcal{V}', \theta(\cdot, t) \in \mathcal{Q}'$ and corresponding solution $(\hat{u}, \hat{\mu})$. Subtracting the unperturbed saddle point problem (3.42a), (3.42b) from the perturbed one, we get

$$\langle \rho(\ddot{\hat{u}} - \ddot{u}), v \rangle + a(\hat{u} - u, v) + b(v, \hat{\mu} - \mu) = \langle \delta, v \rangle \qquad \forall v \in \mathcal{V},$$

$$b(\hat{u} - u, \vartheta) \qquad\qquad = \langle \theta, \vartheta \rangle \qquad \forall \vartheta \in \mathcal{Q}.$$

The decomposition $\hat{u} - u = w + r$ and Theorem 3.4 immediately yield the estimate

$$\|\hat{\mu} - \mu\|_{\mathcal{Q}} \leq \left(1 + \frac{\eta}{\alpha}\right)\left(\frac{1}{\beta}\|\delta\|_{\mathcal{V}'} + \frac{\eta}{\beta^2}\|\theta\|_{\mathcal{Q}'} + \frac{\rho}{\beta}\|\ddot{w}\|_{\mathcal{V}'} + \frac{\rho}{\beta^2}\|\ddot{\theta}\|_{\mathcal{Q}'}\right). \tag{3.69}$$

3.3 Mathematical Analysis

In the stationary case, the time derivatives on the right hand side vanish, and (3.69) reduces again to the standard result for saddle point problems. We observe that the perturbation θ of the constraints is amplified by the factor $1/\beta^2$, and for this reason it is important that the constant β is bounded below from zero. In the non-stationary case, time derivatives come additionally into play, and the second time derivative $\ddot{\theta}$ shows the connection to the estimate (2.76) for the rigid body equations where the second time derivative of the perturbation in the constraint also appeared and lead to the conclusion that the index is 3. Accordingly, we may say that the perturbation index of the transient saddle point problem (3.42a), (3.42b) with respect to the temporal behavior is also 3.

3.3.3 Remarks

At the end of this section, several remarks should be made on the foregoing analysis. The perturbation result (3.69) holds for any abstract time-dependent saddle point problem that has the form (3.42a), (3.42b) and satisfies the assumptions on the bilinear forms a and b. In other words, the analysis is not restricted to linear elasticity and can be generalized to other applications such as domain decomposition or incompressibility, see the next section.

A further remark concerns the notation. Alternatively, the transient saddle point problem (3.42a), (3.42b) is written as operator equation

$$\rho \ddot{u} + Au + B'\mu = l, \qquad (3.70a)$$

$$Bu = m \qquad (3.70b)$$

with operators $A : \mathcal{V} \to \mathcal{V}'$, $\langle Au, v \rangle = a(u, v)$ and $B' : \mathcal{Q} \to \mathcal{V}'$, $\langle B'\mu, v \rangle = b(v, \mu)$ for all $v \in \mathcal{V}$. Note that B' is the dual operator of B defined in (3.48).

The representation (3.70a), (3.70b) stresses the analogy to the finite-dimensional case. In fact, we may view the operators in (3.70a), (3.70b) as matrices and redo the analysis in a finite-dimensional setting. But one has to be careful when working with different norms and corresponding estimates. For example, the weak form (3.55) for the homogeneous solution w reduces to

$$v^T \rho \ddot{w} = v^T l - v^T Au - v^T \rho \ddot{r} \qquad \forall v \in \text{Ker}(B)$$

in the finite-dimensional context. Here we may set $v = \ddot{w}$ in order to derive the estimate

$$\rho \|\ddot{w}\|_2 \leq \eta \|u\|_2 + \|l\|_2 + \rho \|\ddot{r}\|_2.$$

The right hand side in this estimate could be used to replace the term containing $\ddot{w}$ in (3.66), but unfortunately this simple elimination step is not feasible in the infinite-dimensional case. For one, $\ddot{w}$ is not smooth enough to be applied as test function, and furthermore the norms in $L_2(\Omega)$ and $H^1(\Omega)$ are not equivalent.

Fig. 3.4 Nonoverlapping domain decomposition

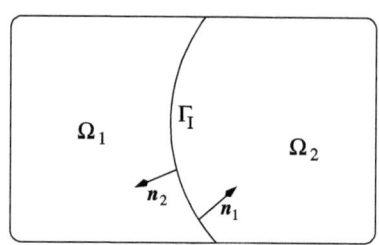

In the emerging field of partial differential-algebraic equations (PDAEs), transient saddle point problems like (3.70a), (3.70b) are definitely a rewarding and at the same time challenging subject where a number of open questions remain. We mention here the work of Higueras [Hig01] who studies the operator equation (3.70a), (3.70b) in the context of logarithmic norms and Günther [Gün01] who investigates coupled problems in electrical circuit simulation.

3.4 Related Mathematical Models

The abstract mathematical model of constrained elastic motion (3.42a), (3.42b) is closely related to approaches for nonoverlapping domain decomposition and contact mechanics. Even incompressible elastic motion can be cast in a similar framework. At the end of this chapter, we briefly point out these connections and give corresponding references.

3.4.1 Domain Decomposition

Consider the domain $\Omega \subset \mathbb{R}^3$ in Fig. 3.4 and its partition into two nonoverlapping subdomains Ω_1 and Ω_2,

$$\overline{\Omega} = \overline{\Omega_1 \cup \Omega_2}, \quad \Omega_1 \cap \Omega_2 = \emptyset.$$

The joint interface of the subdomains is defined by

$$\Gamma_I := \partial \Omega_1 \cap \partial \Omega_2$$

and, like the remaining boundaries, assumed to be sufficiently smooth. To demonstrate the basic idea, we start from the equations of linear elasticity (3.18a)–(3.18c) and state them as pure Neumann problem

$$\rho \ddot{u}(x,t) = \mathbf{div}\, \sigma(u(x,t)) + \boldsymbol{\beta}(x,t) \quad \text{in } \Omega, \qquad (3.71a)$$

$$\sigma(u(x,t))n(x) = \tau(x,t) \quad \text{on } \partial \Omega. \qquad (3.71b)$$

3.4 Related Mathematical Models

Under suitable regularity assumptions on the source term $\boldsymbol{\beta}$ and the surface load $\boldsymbol{\tau}$, this problem can be seen to be equivalent to the coupled problem

$$\rho \ddot{\boldsymbol{u}}_1(x,t) = \text{div}\,\sigma(\boldsymbol{u}_1(x,t)) + \boldsymbol{\beta}(x,t) \quad \text{in } \Omega_1, \tag{3.72a}$$

$$\sigma(\boldsymbol{u}_1(x,t))\boldsymbol{n}_1(x) = \boldsymbol{\tau}(x,t) \quad \text{on } \partial\Omega_1\backslash\Gamma_I, \tag{3.72b}$$

$$\rho \ddot{\boldsymbol{u}}_2(x,t) = \text{div}\,\sigma(\boldsymbol{u}_2(x,t)) + \boldsymbol{\beta}(x,t) \quad \text{in } \Omega_2, \tag{3.72c}$$

$$\sigma(\boldsymbol{u}_2(x,t))\boldsymbol{n}_2(x) = \boldsymbol{\tau}(x,t) \quad \text{on } \partial\Omega_2\backslash\Gamma_I, \tag{3.72d}$$

$$\boldsymbol{u}_1(x,t) = \boldsymbol{u}_2(x,t) \quad \text{on } \Gamma_I, \tag{3.72e}$$

$$\sigma(\boldsymbol{u}_1(x,t))\boldsymbol{n}_1(x) = -\sigma(\boldsymbol{u}_2(x,t))\boldsymbol{n}_2(x) \quad \text{on } \Gamma_I. \tag{3.72f}$$

We thus seek the solutions $\boldsymbol{u}_1$ in Ω_1 and $\boldsymbol{u}_2$ in Ω_2 subject to the *transmission conditions* (3.72e) and (3.72f) that ensure continuity of the displacements along the interface and of the stresses across the interface in normal direction.

For a weak formulation of the *nonoverlapping domain decomposition problem* (3.72a)–(3.72f), recall the definitions of Sect. 3.1.3 that we now extend by indices for the different subdomains. The bilinear form of stress and strain (3.27) leads to

$$a_i(\boldsymbol{u}_i, \boldsymbol{v}_i) := \int_{\Omega_i} \sigma(\boldsymbol{u}_i) : \boldsymbol{\varepsilon}(\boldsymbol{v}_i)\,dx, \quad i = 1, 2; \tag{3.73}$$

while body and surface load terms define the linear functionals

$$\langle l_i, \boldsymbol{v}_i \rangle := \int_{\Omega_i} \boldsymbol{v}_i^T \boldsymbol{\beta}\,dx + \int_{\partial\Omega_i\backslash\Gamma_I} \boldsymbol{v}_i^T \boldsymbol{\tau}\,ds, \quad i = 1, 2. \tag{3.74}$$

An appropriate function space for the displacements $\boldsymbol{u}_i$ and test functions $\boldsymbol{v}_i$ is given by $\mathcal{V}_i := H^1(\Omega_i)^3$. Inertia terms are treated in a similar fashion by setting

$$\langle \rho \ddot{\boldsymbol{u}}_i, \boldsymbol{v}_i \rangle := \int_{\Omega_i} \boldsymbol{v}_i^T \rho \ddot{\boldsymbol{u}}_i\,dx, \quad i = 1, 2. \tag{3.75}$$

As in the definition of the bilinear form (3.41), the interface condition (3.72e) is eventually expressed in weak form as

$$\int_{\Gamma_I} \boldsymbol{\vartheta}^T(\boldsymbol{u}_1 - \boldsymbol{u}_2)\,ds = 0 \quad \forall \boldsymbol{\vartheta} \in \mathcal{Q}$$

with $\mathcal{Q} = (H^{1/2}(\Gamma_I)^3)'$ being again the dual space of the trace space on the interface and the integral understood as duality pairing. Using bilinear forms b_1 and b_2, we recast this condition as

$$b_1(\boldsymbol{u}_1, \boldsymbol{\vartheta}) - b_2(\boldsymbol{u}_2, \boldsymbol{\vartheta}) = 0, \quad b_i(\boldsymbol{u}_i, \boldsymbol{\vartheta}) := \int_{\Gamma_I} \boldsymbol{\vartheta}^T \boldsymbol{u}_i\,ds \tag{3.76}$$

for all $\boldsymbol{\vartheta} \in \mathcal{Q}$ and $i = 1, 2$.

Fig. 3.5 Dynamic contact problem

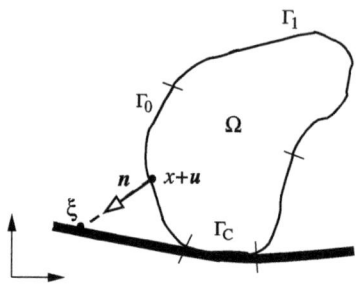

After these preparations, the weak form of (3.72a)–(3.72f) can be formulated as transient saddle point problem: At each point of time t, find displacement fields $(u_1(\cdot,t), u_2(\cdot,t)) \in \mathcal{V}_1 \times \mathcal{V}_2$ and a multiplier $\mu(\cdot,t) \in \mathcal{Q}$ such that

$$\langle \rho \ddot{u}_1, v_1 \rangle + a_1(u_1, v_1) + b_1(v_1, \mu) = \langle l_1, v_1 \rangle \quad \forall v_1 \in \mathcal{V}_1, \quad (3.77a)$$

$$\langle \rho \ddot{u}_2, v_2 \rangle + a_2(u_2, v_2) - b_2(v_2, \mu) = \langle l_2, v_2 \rangle \quad \forall v_2 \in \mathcal{V}_2, \quad (3.77b)$$

$$b_1(u_1, \vartheta) - b_2(u_2, \vartheta) = 0 \quad \forall \vartheta \in \mathcal{Q}. \quad (3.77c)$$

Obviously, this mathematical model features basically the same structure as the constrained equations of motion (3.42a), (3.42b), and properties such as the inf-sup condition for the bilinear forms b_1 and b_2 will play once more a crucial role in its analysis. The generalization to a decomposition into N subdomains and the additional treatment of mixed boundary conditions are straightforward to derive. For a more detailed exposition of domain decomposition methods and their relation to the iterative solution of large linear systems obtained from discretized PDEs, we refer to Toselli/Widlund [TW05] and Wohlmuth [Woh01].

3.4.2 Dynamic Contact

Dynamic contact problems are the second field closely related to the constrained model equations (3.42a), (3.42b). We consider here the extension of Signorini's problem to the dynamic case, i.e., the contact of a single elastic body on a rigid foundation, Fig. 3.5.

Under the assumption of linear elasticity, the outward normal vector $n(x)$ defines the surface point $\xi(x,t)$ on the rigid foundation for potential contact points $x + u$ on the boundary. Notice that the infinitesimal theory implies that we may use here $n(x)$ instead of the normal vector on the deformed boundary $n(x+u)$. The scalar *gap function*

$$g(x,t,u) := \Big(\xi(x,t) - (x + u(x,t))\Big)^T n(x) \quad (3.78)$$

measures the distance in normal direction between the boundary and the foundation. Based on the gap function, the impenetrability condition for the elastic body reads $g \geq 0$ on the boundary Γ_0, and if $g = 0$ the body is in contact on $\Gamma_c \subset \Gamma_0$.

3.4 Related Mathematical Models

On the other hand, for the scalar quantity of the *contact pressure* $p(x,t)$ in normal direction, one postulates $p \geq 0$ on Γ_0, and if $p(x,t) = 0$, there is no contact in the point x at time t. It follows that we have either contact with $g = 0$ and $p > 0$ or no contact with $g > 0$ and $p = 0$, which is conveniently expressed by the complementarity condition $g \cdot p = 0$ on Γ_0.

Summing up, the strong form of dynamic contact can be written as

$$\rho \ddot{u} = \text{div}\,\sigma(u) + \beta \qquad \text{in } \Omega, \tag{3.79a}$$

$$\sigma(u)n = \tau \qquad \text{on } \Gamma_1, \tag{3.79b}$$

$$g \geq 0 \qquad \text{on } \Gamma_0, \tag{3.79c}$$

$$p \geq 0 \qquad \text{on } \Gamma_0, \tag{3.79d}$$

$$g \cdot p = 0 \qquad \text{on } \Gamma_0. \tag{3.79e}$$

In order to pass to a weak formulation, we make use of the function space $\mathcal{V} = H^1(\Omega)^3$ for the displacements like above and set

$$\mathcal{Q} := \left\{ r \in H^{1/2}(\Gamma_0)' : \int_{\Gamma_0} rw\,ds \geq 0 \; \forall w \in H^{1/2}(\Gamma_0),\, w \geq 0 \right\}$$

as function space for the contact pressure, which represents a Lagrange multiplier. For $r \in \mathcal{Q}$, one has, due to the complementarity condition (3.79e), the inequality

$$0 \leq \int_{\Gamma_0} gr\,ds = \int_{\Gamma_0} g(r-p)\,ds = \int_{\Gamma_0} u^T n(p-r)\,ds - \int_{\Gamma_0} (\xi - x)^T n(p-r)\,ds.$$

The last two integrals give rise to the definitions of the bilinear form

$$b_c(v, p) := \int_{\Gamma_0} v^T n \cdot p\,ds$$

on $\mathcal{V} \times \mathcal{Q}$ and of the linear functional $m \in \mathcal{Q}'$ where

$$\langle m, p \rangle := \int_{\Gamma_0} (\xi - x)^T np\,ds.$$

After these preparations, we may state the dynamic contact problem in weak form as follows: At each point of time, find the displacement $u(\cdot, t) \in \mathcal{V}$ and the contact pressure $p(\cdot, t) \in \mathcal{Q}$ such that

$$\langle \rho \ddot{u}, v \rangle + a(u, v) + b_c(v, p) = \langle l, v \rangle \qquad \forall v \in \mathcal{V}, \tag{3.80a}$$

$$b_c(u, p-r) - \langle m, p-r \rangle \geq 0 \qquad \forall r \in \mathcal{Q}. \tag{3.80b}$$

The weak form is thus a time-dependent variational inequality with saddle point structure, and if we replace the inequality sign by an equality sign in (3.80b), the relation to the constrained model (3.42a), (3.42b) becomes apparent. In this case, a closer look at the equations reveals even more that the contact pressure p in (3.80a), (3.80b) and the Lagrange multiplier μ in (3.42a), (3.42b) satisfy

$$p = -(\sigma n)^T n = \mu^T n. \tag{3.81}$$

A more detailed discussion of contact problems is beyond the scope of this book, but in Part II we will occasionally come back to the treatment of contact in flexible multibody systems. The classical reference on this topic is the monograph by Kikuchi/Oden [KO88]. A more recent exposition of mathematical models and finite element discretizations can be found in [SP05, WK03, Wri02].

Finally, it should be remarked that in the weak form (3.80a), (3.80b), it is straightforward to eliminate the Lagrange multiplier p by restricting the displacement u and the test functions v to the cone

$$\mathcal{K} := \left\{ v \in H^1(\Omega)^3 : g(x, v) \geq 0 \text{ on } \Gamma_0 \right\},$$

where we have assumed that the position of the rigid foundation is independent of time t. This elimination step leads to the variational inequality

$$\langle \rho \ddot{u}, v - u \rangle + a(u, v - u) \geq \langle l, v - u \rangle \qquad \forall v \in \mathcal{K},$$

which reduces to a classical obstacle problem in the stationary case.

Furthermore, for dynamic contact quite often the setting of linear elasticity is not appropriate. Large displacements and rigid body modes can either be taken into account by the strain tensor

$$E = \frac{1}{2}(\nabla u + \nabla u^T + \nabla u^T \nabla u),$$

cf. (3.4), or by the method of floating reference frames that will be introduced in the next chapter.

3.4.3 Incompressible Elastic Body

The last mathematical model that we want to mention here refers to the dynamics of an incompressible elastic body. While in the previous models constraints were always formulated with respect to some part of the boundary, incompressibility concerns the domain Ω itself, and accordingly, the corresponding Lagrange multiplier or pressure variable requires a different treatment.

Let as before $u(x, t)$ denote the displacement field of the elastic body under the assumption of linear elasticity, and let $p(x, t)$ be the pressure. The equations of motion under the incompressibility constraint read

$$\rho \ddot{u} + \nabla p = \mathbf{div}\, \sigma_d(u) + \beta \qquad \text{in } \Omega, \tag{3.82a}$$

$$\text{div } u = 0 \qquad \text{in } \Omega. \tag{3.82b}$$

3.4 Related Mathematical Models

Here, $\boldsymbol{\sigma}_d$ stands for the deviatoric stress tensor [SSP09], which stems from decomposing the stress tensor $\boldsymbol{\sigma}$ into $\operatorname{\mathbf{div}} \boldsymbol{\sigma} = \operatorname{\mathbf{div}} \boldsymbol{\sigma}_d - \nabla p$.

Like in fluid dynamics, the displacement field is required to be divergence-free, and this condition can be formulated in weak form as

$$b(\boldsymbol{u}, q) = 0, \quad b(\boldsymbol{u}, q) := -\int_\Omega \operatorname{div} \boldsymbol{u} \cdot q \, dx$$

for all test functions q in the normalized L_2-space

$$\mathcal{M} := \left\{ q \in L_2(\Omega) : \int_\Omega q \, dx = 0 \right\}.$$

For simplicity, we consider zero Dirichlet boundary conditions and take

$$\mathcal{X} := H_0^1(\Omega)^3 = \left\{ \boldsymbol{v} \in H^1(\Omega)^3 : \boldsymbol{v} = \boldsymbol{0} \text{ on } \partial\Omega \right\}$$

as function space for the displacement field. In this case, we have by the Gauss theorem

$$-\int_\Omega \operatorname{div} \boldsymbol{v} \cdot q \, dx = \int_\Omega \boldsymbol{v}^T \nabla q \, dx \qquad \forall \boldsymbol{v} \in \mathcal{X}, q \in \mathcal{M}.$$

Using this relation, the weak form is easily derived from (3.82a), (3.82b) and reads: At each point of time, find the displacement $\boldsymbol{u}(\cdot, t) \in \mathcal{X}$ and the pressure $p(\cdot, t) \in \mathcal{M}$ such that

$$\langle \rho \ddot{\boldsymbol{u}}, \boldsymbol{v} \rangle + a(\boldsymbol{u}, \boldsymbol{v}) + b(\boldsymbol{v}, p) = \langle \boldsymbol{l}, \boldsymbol{v} \rangle \qquad \forall \boldsymbol{v} \in \mathcal{X}, \tag{3.83a}$$

$$b(\boldsymbol{u}, q) \qquad\qquad\qquad\quad = 0 \qquad \forall q \in \mathcal{M}. \tag{3.83b}$$

One clearly observes once again the structural equivalence with the constrained equations (3.42a), (3.42b). For a more elaborate treatment of incompressibility see [Bra01, Hug87].

Chapter 4
Flexible Multibody Dynamics

Flexible multibody systems contain both rigid and elastic components and aim at applications such as lightweight and high-precision mechanical systems where the elasticity of certain bodies needs to be taken into account. Since elastic bodies are governed by PDEs, as described in the previous chapter, and rigid bodies by ODEs or DAEs, the mathematical model of a flexible multibody system is heterogeneous by nature.

This chapter introduces the method of floating reference frames as standard approach for flexible multibody dynamics. The treatment of constraints is extended to this problem class, and issues related to the modeling of joints are discussed. Moreover, we cover special bodies such as beams and also touch shortly upon additional nonlinearities, the so-called geometric stiffening terms. Several examples conclude this first part of the book.

As pointed out before, the literature on flexible multibody systems is rich, and various simulation codes offer corresponding features. Standard references are the monographs of Bauchau [Bau10], Bremer [Bre08], Bremer and Pfeiffer [BP92], Géradin and Cardona [GC00], Schwertassek and Wallrapp [SW99], and Shabana [Sha98].

4.1 Floating Reference Frame

So far, we have introduced a standard model for the small deformation of an elastic body. In a multibody system, however, each body may perform large rotations and translations, which is in contradiction to the assumption of small deformation. To overcome this difficulty, we apply the *method of floating reference frames:* The body's motion is split into rigid motion variables and small deformation variables. The latter are conveniently expressed in a floating reference frame that moves with the body.

In the following, we once again use the least action principle and derive the equations of unconstrained motion for a single elastic body as part of a multibody sys-

Fig. 4.1 Floating reference frame

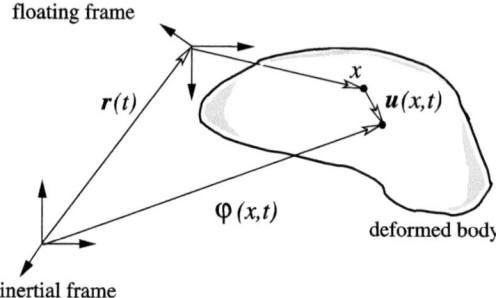

tem. More details on this standard approach can be found in the literature mentioned above.

4.1.1 Variational Formulation

Consider the elastic body in Fig. 4.1 and the two reference frames, an inertial reference frame and a body-fixed frame, which moves with the body and for which we postulate that there is no rigid body motion between the body and this frame. While in the previous chapter, both reference frames coincided, we now take up the model of Sect. 2.2 for the rigid body motion in three-dimensional space and let the translation vector $r(t) \in \mathbb{R}^3$ locate the origin of the floating frame. Furthermore, the transformation from the floating coordinate system to the inertial one is accomplished by a rotation matrix $A(\alpha(t)) \in SO(3)$ that depends on the angles $\alpha(t) \in \mathbb{R}^3$. For simplicity, we restrict the discussion here to Euler angles, but other representations such as Cardan angles or Euler parameters can be used instead.

Kinematics

Each material point of the deformed body is located by the deformation vector

$$\varphi(x,t) = r(t) + A(\alpha(t))(x + u(x,t)). \tag{4.1}$$

The kinematic equation (4.1) describes a rigid body motion if the displacement u vanishes. On the other hand, we have the setting of small deformation theory (3.1) if r is equal to zero and A is the identity matrix.

Principle of Least Action

Instead on the displacement u, all energy expressions are now based on the deformation φ as function of r, α, and u according to the kinematic equation (4.1).

4.1 Floating Reference Frame

The kinetic energy is given by

$$T = \frac{1}{2} \int_{\Omega} \rho \dot{\boldsymbol{\varphi}}^T \dot{\boldsymbol{\varphi}} \, dx, \tag{4.2}$$

compare the kinetic energy (3.11) of linear elasticity. Furthermore, the potential energy reads

$$U = \frac{1}{2} \int_{\Omega} \sigma(\boldsymbol{u}) : \varepsilon(\boldsymbol{u}) \, dx - \int_{\Omega} \boldsymbol{\varphi}^T A \boldsymbol{\beta} \, dx - \int_{\Gamma_1} \boldsymbol{\varphi}^T A \boldsymbol{\tau} \, ds. \tag{4.3}$$

Since the strain energy is invariant under rigid body motion, the first expression in (4.3) remains the same as in linear elasticity. The density of volume forces $\boldsymbol{\beta}(x,t)$ and the surface traction $\boldsymbol{\tau}(x,t)$, on the other hand, need to be transformed to the inertial reference frame by a premultiplication with rotation matrix A.

Though the next steps are straightforward, one point merits special attention. The motion $\boldsymbol{\varphi} = \boldsymbol{\varphi}(r, \alpha, u)$ is composed of translation r, rotation α, and displacement u. A perturbed motion takes therefore the form

$$\boldsymbol{\varphi}(r + \theta z, \alpha + \theta \zeta, u + \theta v) = r + \theta z + A(\alpha + \theta \zeta)(x + u + \theta v). \tag{4.4}$$

Here, θ is a real parameter while $z(t) \in \mathbb{R}^3$ and $\zeta(t) \in \mathbb{R}^3$ are perturbations of translation and rotation. In the literature, the *variations* $\theta z = \delta r$ and $\theta \zeta = \delta \alpha$ are referred to as *virtual translation* and *virtual rotation*, respectively. Both variations depend on time t and satisfy $z(t_0) = z(t_1) = \mathbf{0}$ and $\zeta(t_0) = \zeta(t_1) = \mathbf{0}$, i.e., they vanish at the starting point and end point of the motion. The variation of the displacement $\theta v = \delta u$ has the same properties as above in Sect. 3.1.1, i.e., $v = v(x,t)$ is a function of space and time with $v(\cdot, t) = \mathbf{0}$ on Γ_0 and $v(\cdot, t_0) = v(\cdot, t_1) = \mathbf{0}$ in Ω.

The variation of the deformation $\boldsymbol{\varphi}$ is defined to be the directional derivative

$$\delta \boldsymbol{\varphi} = \theta \left(\frac{d}{d\theta} \boldsymbol{\varphi}(r + \theta z, \alpha + \theta \zeta, u + \theta v) \Big|_{\theta=0} \right) \tag{4.5}$$

$$= \theta \left(z + \frac{d}{d\theta} A(\alpha + \theta \zeta) \Big|_{\theta=0} (x + u) + A(\alpha) v \right).$$

Due to orthogonality, $A^T A = I$, we have

$$\left(\frac{d}{d\theta} A^T \right) A + A^T \left(\frac{d}{d\theta} A \right) = 0 \quad \Rightarrow \quad A^T \left(\frac{d}{d\theta} A \right) \text{ is skew symmetric,}$$

compare the definition of the angular velocity (2.23). Hence there exists a vector $\gamma \in \mathbb{R}^3$ such that

$$A(\alpha)^T \left(\frac{d}{d\theta} A(\alpha + \theta \zeta) \Big|_{\theta=0} \right) = \begin{pmatrix} 0 & -\gamma_3 & \gamma_2 \\ \gamma_3 & 0 & -\gamma_1 \\ -\gamma_2 & \gamma_1 & 0 \end{pmatrix} = \tilde{\gamma}. \tag{4.6}$$

Like in Sect. 2.2, we make here use of the compact "tilde" notation $\tilde{a}$ for a skew symmetric matrix based on the vector $a \in \mathbb{R}^3$.

We finally premultiply (4.6) by A, insert the result in (4.5), and get with $\tilde{\gamma}(x+u) = -(\tilde{x}+\tilde{u})\gamma$ the variation

$$\delta\varphi = \theta \cdot \eta, \quad \eta := z - A(\alpha)(\tilde{x}+\tilde{u})\gamma + A(\alpha)v. \tag{4.7}$$

Summing up, admissible variations show the special structure (4.7), which may be viewed as tangential to a given motion φ. We prefer the notation $\theta\eta = \theta\eta(x,t)$ with admissible test function η instead of $\delta\varphi$ since we want to stress the analogy to the weak formulation of linear elasticity, see below.

One more time, the stationarity condition (3.13) of Hamilton's principle is now exploited, with kinetic energy (4.2), potential energy (4.3), and perturbed motion (4.4). This leads to

$$0 = \int_{t_0}^{t_1} \left(\int_{\Omega} (\rho \dot{\eta}^T \dot{\varphi} - \sigma(u):\varepsilon(v) + \eta^T A \beta)\, dx + \int_{\Gamma_1} \eta^T A \tau\, ds \right) dt.$$

Partial integration of the inertia term with respect to time yields in the same way as in (3.15)

$$0 = \int_{t_0}^{t_1} \left(\int_{\Omega} \left(\rho \eta^T \ddot{\varphi} + \sigma(u):\varepsilon(v) - \eta^T A \beta \right) dx - \int_{\Gamma_1} \eta^T A \tau\, ds \right) dt, \tag{4.8}$$

which must hold for all admissible η.

4.1.2 Equations of Unconstrained Motion

We next insert the special structure (4.7) of the test function η into (4.8), apply Green's formula (3.16), and rearrange the terms as

$$0 = \int_{t_0}^{t_1} z^T \left(\int_{\Omega} (\rho \ddot{\varphi} - A\beta)\, dx - \int_{\Gamma_1} A\tau\, ds \right) dt \tag{4.9a}$$

$$+ \int_{t_0}^{t_1} \gamma^T \left(\int_{\Omega} (\tilde{x}+\tilde{u})(A^T \rho \ddot{\varphi} - \beta)\, dx - \int_{\Gamma_1} (\tilde{x}+\tilde{u})\tau\, ds \right) dt \tag{4.9b}$$

$$+ \int_{t_0}^{t_1} \left(\int_{\Omega} v^T (\rho \ddot{\varphi} - \operatorname{div} \sigma(u) - \beta)\, dx + \int_{\Gamma_1} v^T (\sigma n - \tau)\, ds \right) dt. \tag{4.9c}$$

This formulation holds for arbitrary vectors $z(t)$ and $\gamma(t)$ where $t_0 \le t \le t_1$. The test function $v = v(x,t)$ is independent of z and γ but still satisfies $v = 0$ on Γ_0.

4.1 Floating Reference Frame

Strong Form of the Equations of Motion

The characterization (4.9a)–(4.9c) features a splitting of the action integral into translation (4.9a), rotation (4.9b), and displacement part (4.9c). Since z, γ, and v may be varied independently, each of the three terms must vanish identically, and consequently the fundamental lemma of the calculus of variations can be applied to each of them. In this way, we eventually obtain the strong form

$$\int_\Omega \rho \ddot{\varphi}\, \mathrm{d}x = \int_\Omega A\beta\, \mathrm{d}x + \int_{\Gamma_1} A\tau\, \mathrm{d}s, \tag{4.10a}$$

$$\int_\Omega (\tilde{x}+\tilde{u}) A^T \rho \ddot{\varphi}\, \mathrm{d}x = \int_\Omega (\tilde{x}+\tilde{u})\beta\, \mathrm{d}x + \int_{\Gamma_1} (\tilde{x}+\tilde{u})\tau\, \mathrm{d}s, \tag{4.10b}$$

$$\rho \ddot{\varphi} = \mathbf{div}\, \sigma(u) + \beta \quad \text{in } \Omega. \tag{4.10c}$$

The boundary conditions are still missing here. This means that the Dirichlet boundary condition $u(\cdot,t) = u_0(\cdot,t)$ on Γ_0 must still hold but could be equivalently expressed in terms of the deformation φ. On the other hand, the boundary integral in (4.9c) yields, as in linear elasticity, the natural or Neumann boundary condition $\sigma(u(\cdot,t))n = \tau(\cdot,t)$ on Γ_1.

Compared to the strong form of linear elasticity (3.18a)–(3.18c), Eqs. (4.10a)–(4.10c) feature additionally the balance of momentum (4.10a) and the balance of angular momentum (4.10b) for the rigid motion. This becomes clearer if one evaluates the integrals on the left hand side, which results in mass and inertia matrix expressions, see the end of this subsection.

Finally, we compute the acceleration $\ddot{\varphi}$. Total differentiation of φ with respect to time yields

$$\frac{\mathrm{d}}{\mathrm{d}t}\varphi = \dot{r} + \left(\frac{\mathrm{d}}{\mathrm{d}t}A\right)(x+u) + A\dot{u} = \dot{r} + A\tilde{\omega}(x+u) + A\dot{u}$$

where the angular velocity $\omega(t) \in \mathbb{R}^3$ with respect to the floating frame is given by

$$A(\alpha)^T \frac{\mathrm{d}}{\mathrm{d}t} A(\alpha) = \tilde{\omega}. \tag{4.11}$$

Conversely, with respect to the inertial reference frame the angular velocity has the representation $\omega_I = (\mathrm{d}A/\mathrm{d}t) A^T$.

Making use of the relation (4.11) and of the rules for the tilde operator after a second differentiation with respect to time, one obtains the acceleration vector in the form

$$\frac{\mathrm{d}^2}{\mathrm{d}t^2}\varphi = \ddot{r} + A(\dot{\tilde{\omega}} + \tilde{\omega}\tilde{\omega})(x+u) + 2A\tilde{\omega}\dot{u} + A\ddot{u}. \tag{4.12}$$

Inserting (4.12) in the strong form (4.10a)–(4.10c) is straightforward but omitted here.

Weak Form of the Equations of Motion

As discussed in the previous chapter, the weak formulation stands in the middle of strong form and variational principle. We regard η now as admissible and time-independent test function, i.e., η has the special structure given in (4.7), which we restate for convenience as

$$\eta = \eta(z, \gamma, v; \alpha, u) = z - A(\alpha)(\tilde{x} + \tilde{u})\gamma + A(\alpha)v. \tag{4.13}$$

Both z and γ are constant but otherwise arbitrary vectors, whereas the test function with respect to the displacement $v = v(x) \in \mathcal{V}_0$ satisfies the homogeneous boundary condition $v = 0$ on Γ_0.

We multiply next (4.10a) by z from the left, (4.10b) by γ, and (4.10c) by v. The last equation is integrated over Ω, Green's formula (3.16) is applied in reverse, and all three work expressions are finally summed up and rearranged as in (4.8). As result, we get the weak form or principle of virtual work

$$\int_\Omega \rho \eta^T \ddot{\varphi}\, dx + \int_\Omega \sigma(u) : \varepsilon(v)\, dx = \int_\Omega \eta^T A \beta\, dx + \int_{\Gamma_1} \eta^T A \tau\, ds, \tag{4.14}$$

which must hold for all admissible η from (4.13).

The equations of motion (4.14) are a generalization of the weak form of linear elasticity (3.19). Due to the rotation matrix A, they are nonlinear; a fact that is hidden behind the definitions of φ and η. Like the strong form (4.10a)–(4.10c), the weak form (4.14) may be used to state the equations partitioned into momentum, angular momentum and displacement parts. For this purpose, we insert the definition (4.13) of the test function η into (4.14) and sort the terms like in (4.9a)–(4.9c). This yields

$$z^T \int_\Omega \rho \ddot{\varphi}\, dx = z^T \int_\Omega A\beta\, dx + z^T \int_{\Gamma_1} A\tau\, ds, \tag{4.15a}$$

$$\gamma^T \int_\Omega (\tilde{x}+\tilde{u})A^T \rho \ddot{\varphi}\, dx = \gamma^T \int_\Omega (\tilde{x}+\tilde{u})\beta\, dx + \gamma^T \int_{\Gamma_1} (\tilde{x}+\tilde{u})\tau\, ds, \tag{4.15b}$$

$$\langle \rho \ddot{\varphi}, v \rangle + a(u, v) = \langle l, v \rangle \quad \forall v \in \mathcal{V}_0. \tag{4.15c}$$

Momentum and angular momentum equations (4.15a) and (4.15b) coincide with the strong form because the prefactors z (virtual translation) and γ (virtual rotation) can be simply omitted. Moreover, we used the abstract notation of Sect. 3.1.3 for the displacement equation (4.15c).

We point out that the partitioned equations (4.15a)–(4.15c) will change if the definition (4.1) of the deformation φ changes. The principle (4.14), on the other hand, is independent of a specific form of φ. It can also be extended to dissipative or discontinuous right hand sides, and for these reasons we refer to (4.14) as the basic equation of motion.

4.1 Floating Reference Frame

For completeness, we finally list the left hand sides of (4.10a)–(4.10c) and (4.15a)–(4.15c), respectively, with $\ddot{\boldsymbol{\varphi}}$ from (4.12) and evaluated rigid motion integrals. We have

$$\int_\Omega \rho \ddot{\boldsymbol{\varphi}} \, \mathrm{d}x = m\ddot{\boldsymbol{r}} + A\tilde{\mathbf{s}}^T \dot{\boldsymbol{\omega}} + A\tilde{\boldsymbol{\omega}}\tilde{\boldsymbol{\omega}}\mathbf{s} + A \int_\Omega \rho(2\tilde{\boldsymbol{\omega}}\dot{\boldsymbol{u}} + \ddot{\boldsymbol{u}}) \, \mathrm{d}x \quad (4.16)$$

for the translational part,

$$\int_\Omega (\tilde{\boldsymbol{x}} + \tilde{\boldsymbol{u}}) A^T \rho \ddot{\boldsymbol{\varphi}} \, \mathrm{d}x = \tilde{\mathbf{s}} A^T \ddot{\boldsymbol{r}} + \mathbf{J}\dot{\boldsymbol{\omega}} + \tilde{\boldsymbol{\omega}}\mathbf{J}\boldsymbol{\omega} + \int_\Omega \rho((\tilde{\boldsymbol{x}} + \tilde{\boldsymbol{u}})\ddot{\boldsymbol{u}} + 2(\tilde{\boldsymbol{x}} + \tilde{\boldsymbol{u}})\tilde{\boldsymbol{\omega}}\dot{\boldsymbol{u}}) \, \mathrm{d}x \quad (4.17)$$

for the rotational part, and

$$\langle \rho \ddot{\boldsymbol{\varphi}}, \boldsymbol{v} \rangle = \langle \rho A^T \ddot{\boldsymbol{r}}, \boldsymbol{v} \rangle + \langle \rho(\dot{\tilde{\boldsymbol{\omega}}} + \tilde{\boldsymbol{\omega}}\tilde{\boldsymbol{\omega}})(\boldsymbol{x} + \boldsymbol{u}), \boldsymbol{v} \rangle + \langle \rho(2\tilde{\boldsymbol{\omega}}\dot{\boldsymbol{u}} + \ddot{\boldsymbol{u}}), \boldsymbol{v} \rangle \quad (4.18)$$

for the displacement. Here, $m = \int \rho \, \mathrm{d}x$ denotes the mass and

$$\mathbf{J} = \mathbf{J}(\boldsymbol{u}) := \int_\Omega \rho(\tilde{\boldsymbol{x}} + \tilde{\boldsymbol{u}})^T (\tilde{\boldsymbol{x}} + \tilde{\boldsymbol{u}}) \, \mathrm{d}x$$

the 3×3 inertia matrix of the deformed body. A further integral leads to the definition of the vector

$$\mathbf{s} = \mathbf{s}(\boldsymbol{u}) := \int_\Omega \rho(\boldsymbol{x} + \boldsymbol{u}) \, \mathrm{d}x,$$

which is related to the center of gravity (centroid) of the deformed body given by $\mathbf{s}/m$. If the origin of the body-fixed reference frame coincides with the centroid, we have $\mathbf{s} = \mathbf{0}$.

4.1.3 Extensions and Special Cases

Though the rigid body motion is taken into account, the equations of motion (4.14) require additional explanation in order to make them suitable for use in a multibody system.

Dirichlet Boundary Condition

The model (4.14) is based on the assumption that the motion $\boldsymbol{u}_0$ on the boundary Γ_0 is known a priori. In a multibody system, however, $\boldsymbol{u}_0$ may depend on the unknown motion of adjacent bodies, in particular if Γ_0 is the interface to a joint. The next section introduces the Lagrange multiplier technique to handle this case.

The situation simplifies substantially if the boundary condition expresses a rigid motion, e.g., induced by a joint, and if the origin of the floating reference frame is placed *in the boundary* Γ_0. Then, the choice $\boldsymbol{u}_0 = \boldsymbol{0}$ ensures that Γ_0 is not deformed, and the coupling via translation $\boldsymbol{r}$ and rotation $\boldsymbol{\alpha}$ can be modeled like in the rigid body case. At the same time, this approach guarantees that the floating reference frame moves with the body.

Generalized Coordinates

In practice, the rigid body motion is often not expressed in terms of three translation and three rotation coordinates. E.g., a joint may be attached to the elastic body, and the corresponding constraint is resolved by a set of minimum coordinates $\boldsymbol{y}(t)$. As a consequence, translation $\boldsymbol{r}$ and rotation $\boldsymbol{\alpha}$ depend on $\boldsymbol{y}$ in this case. Moreover, in case of beam models, the displacement $\boldsymbol{u}$ depends on the position of the neutral fiber. To simplify notation, we write here also $\boldsymbol{u}(\boldsymbol{y})$ and assume that $\boldsymbol{y} = \boldsymbol{y}(x, t)$ stands for the independent or so-called *generalized coordinates* of both the rigid and the elastic motion.

When evaluating the equations of motion (4.14), one needs to take this dependence into account. More specifically, deformation and admissible test function are given by

$$\boldsymbol{\varphi}(\boldsymbol{y}) = \boldsymbol{r}(\boldsymbol{y}) + \boldsymbol{A}(\boldsymbol{\alpha}(\boldsymbol{y}))(x + \boldsymbol{u}(\boldsymbol{y})), \tag{4.19a}$$

$$\boldsymbol{\eta} = \left.\frac{\mathrm{d}}{\mathrm{d}\theta}\boldsymbol{\varphi}(\boldsymbol{y} + \theta \boldsymbol{w})\right|_{\theta=0} \tag{4.19b}$$

in the weak form (4.14). Note that (4.19b) requires the test function $\boldsymbol{\eta}$ to be tangential to the motion $\boldsymbol{\varphi}(\boldsymbol{y})$.

Force Elements

Force elements between bodies are a typical feature of multibody systems. They usually act on certain points of the boundary but not on a whole surface. The surface integral $\int_{\Gamma_1} \boldsymbol{\eta}^T \boldsymbol{A}\boldsymbol{\tau} \, ds$ in the equations of motion (4.14) is not appropriate here, and for this reason we look at a standard force element in more detail.

Figure 4.2 shows the elastic body and the force element with translational spring, damper, and actuator. The element is attached to the point x_f on the boundary Γ_1 and to the point $\boldsymbol{r}_0$, which is assumed to be fixed with respect to the inertial reference frame. The vector

$$\boldsymbol{c}(\boldsymbol{\varphi}(x_f, t)) := \boldsymbol{r}_0 - \boldsymbol{\varphi}(x_f, t) = \boldsymbol{r}_0 - \boldsymbol{r}(t) - \boldsymbol{A}(\boldsymbol{\alpha}(t))(x_f + \boldsymbol{u}(x_f, t))$$

connects both points, and the corresponding distance is measured by ζ with $\zeta^2 := \|\boldsymbol{c}\|_2^2 = \boldsymbol{c}^T \boldsymbol{c}$.

Depending on ζ, the magnitude of the acting force is

$$f(\zeta, \dot{\zeta}, t) = k(\zeta - \zeta_0) + d\dot{\zeta} + h(\zeta, \dot{\zeta}, t) \tag{4.20}$$

4.1 Floating Reference Frame

Fig. 4.2 Elastic body and force element

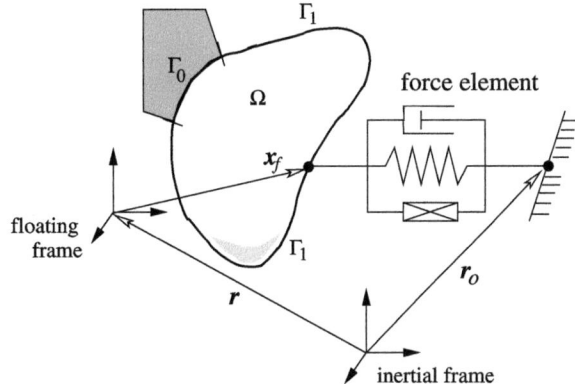

where k denotes the spring constant, ζ_0 the nominal spring length, d the damping coefficient, and h the actuator law. The force element exerts the force $\boldsymbol{F} = f\boldsymbol{c}/\zeta$, and this force leads to an inner product term instead of a surface integral in the weak form (4.14). Formally, we can distribute the force uniformly on some boundary $\Gamma_2 \subset \Gamma_1$ around x_f by setting $\boldsymbol{A}\boldsymbol{\tau} := \boldsymbol{F}/(\text{area } \Gamma_2)$ as surface traction. In the limit we obtain

$$\int_{\Gamma_2} \boldsymbol{\eta}^T \boldsymbol{A}\boldsymbol{\tau}\, ds = \int_{\Gamma_2} \boldsymbol{\eta}^T \frac{\boldsymbol{F}}{\text{area } \Gamma_2}\, ds \quad \overset{\text{area } \Gamma_2 \to 0}{\longrightarrow} \quad \boldsymbol{\eta}^T \boldsymbol{F} = \boldsymbol{\eta}^T f \frac{\boldsymbol{c}}{\zeta} \qquad (4.21)$$

as contribution in the weak form.

The point force $\boldsymbol{F}$ may be viewed as a Dirac δ-function acting in x_f, and this of course means a singularity. Such force elements, on the other hand, are essential in multibody dynamics and cannot be avoided. The discrete structure of rigid body motion collides here with the continuous nature of the elastic body. A way out of this dilemma is to actually distribute the point force over a small boundary segment Γ_2 as indicated in (4.21) on the left.

Newton–Euler Equations

If we assume rigid body motion without elastic displacement, both the strong form (4.10a)–(4.10c) and the weak form (4.14) reduce to three equations (4.15a) for the translation and three equations (4.15b) for the rotation. Making use of the relations (4.16) and (4.17) for the acceleration terms involving $\ddot{\boldsymbol{\varphi}}$ and placing the origin of the body-fixed frame into the centroid, we arrive at the Newton–Euler equations

$$m\ddot{\boldsymbol{r}} = \boldsymbol{f}_f, \qquad \boldsymbol{f}_f := \boldsymbol{A} \int_\Omega \boldsymbol{\beta}\, dx + \boldsymbol{A} \int_{\Gamma_1} \boldsymbol{\tau}\, ds;$$
$$\boldsymbol{J}\dot{\boldsymbol{\omega}} + \tilde{\boldsymbol{\omega}}\boldsymbol{J}\boldsymbol{\omega} = \boldsymbol{f}_m, \qquad \boldsymbol{f}_m := \int_\Omega \tilde{\boldsymbol{x}}\boldsymbol{\beta}\, dx + \int_{\Gamma_1} \tilde{\boldsymbol{x}}\boldsymbol{\tau}\, ds \qquad (4.22)$$

with applied forces and torques $\boldsymbol{f}_f$ and $\boldsymbol{f}_m$, respectively. Compare with the representation (2.25) in Sect. 2.2.

Planar Motion

Rigid body motion in the plane is much easier to describe. As a consequence, the method of floating reference frames simplifies considerably, compare the elastic pendulum example of Chap. 1.

More specifically, assume that the displacement field $u = (u_1, u_2)^T$ is two-dimensional, as it will be described below in Sect. 4.4 for the cases of plane strain, plane stress, and a planar beam model. We restrict accordingly the rigid body motion to the (x_1, x_2)-plane and express the deformation of a material point as

$$\varphi(x,t) = r(t) + A(\alpha(t))(x + u(x,t)), \quad A(\alpha) := \begin{pmatrix} \cos\alpha & -\sin\alpha \\ \sin\alpha & \cos\alpha \end{pmatrix}, \quad (4.23)$$

with translation $r(t) \in \mathbb{R}^2$, rotation matrix $A(\alpha) \in SO(2)$, and scalar angle $\alpha(t)$. An admissible test function has the form

$$\eta = z + A'(\alpha)(x + u)\gamma + A(\alpha)v \quad (4.24)$$

where $A' = \partial A/\partial \alpha$.

The equations of planar motion follow then from inserting (4.23) and (4.24) into the weak form (4.14), with an added trivial zero component. Note that integration with respect to the third coordinate x_3 results in a common constant factor that should not be canceled here since it enters mass and inertia parameters.

Abstract Notation

For subsequent use, the abstract notation introduced in Sect. 3.1.3 for the weak form of linear elasticity is now extended to the current setting of a floating reference frame. We write the equations of motion (4.14) in compact form as

$$\langle \rho\ddot{\varphi}, \eta \rangle + a(\varphi, \eta) = \langle l(\varphi), \eta \rangle \quad \forall \eta \text{ admissible due to (4.13)}. \quad (4.25)$$

The expressions on the left hand side are given by

$$\langle \rho\ddot{\varphi}, \eta \rangle := \int_\Omega \rho \eta^T \ddot{\varphi}\, dx, \quad a(\varphi, \eta) := \int_\Omega \sigma(u) : \varepsilon(v)\, dx,$$

whereas the right hand side reads

$$\langle l(\varphi), \eta \rangle := \int_\Omega \eta^T A\beta\, dx + \int_{\Gamma_1} \eta^T A\tau\, ds.$$

Notice that the right hand side is nonlinear in general due to the rotation matrix A. Moreover, force elements as described in (4.21) can also be subsumed under this notation. Though the functional l may depend explicitly on time, e.g., if the surface load τ varies with time, we skip this dependence here to keep the notation simple.

4.2 Constraints

Based on the framework developed in Sect. 3.2 for the equations of constrained motion (3.42a), (3.42b), it is straightforward to extend the coupling technique with Lagrange multipliers to the method of floating reference frames.

4.2.1 Weak Form of Equations of Constrained Motion

We take the deformation vector $\boldsymbol{\varphi}$ as in (4.1) and replace the boundary condition $\boldsymbol{u} = \boldsymbol{u}_0$ by the equivalent condition $\boldsymbol{\varphi} = \boldsymbol{\varphi}_0$ on Γ_0. Modifying the least action integral (3.37) correspondingly and proceeding in the same fashion as in Sect. 3.2, we obtain the equations of constrained motion in weak form

$$\int_\Omega \boldsymbol{\eta}^T \rho \ddot{\boldsymbol{\varphi}}\, dx + \int_\Omega \boldsymbol{\sigma} : \boldsymbol{\varepsilon}\, dx + \int_{\Gamma_0} \boldsymbol{\eta}^T \boldsymbol{\mu}\, ds = \int_\Omega \boldsymbol{\eta}^T \boldsymbol{A}\boldsymbol{\beta}\, dx + \int_{\Gamma_1} \boldsymbol{\eta}^T \boldsymbol{A}\boldsymbol{\tau}\, ds, \quad (4.26a)$$

$$\int_{\Gamma_0} \boldsymbol{\vartheta}^T \boldsymbol{\varphi}\, ds = \int_{\Gamma_0} \boldsymbol{\vartheta}^T \boldsymbol{\varphi}_0\, ds, \quad (4.26b)$$

which must hold for all admissible $\boldsymbol{\eta}$ and $\boldsymbol{\vartheta}$. While the latter may vary arbitrarily and can be defined in terms of the dual trace space $\mathcal{Q}$ as in (3.40), admissible test functions $\boldsymbol{\eta}$ have the form

$$\boldsymbol{\eta} = \boldsymbol{z} - \boldsymbol{A}(\boldsymbol{\alpha})(\tilde{\boldsymbol{x}} + \tilde{\boldsymbol{u}})\boldsymbol{\gamma} + \boldsymbol{A}(\boldsymbol{\alpha})\boldsymbol{v}. \quad (4.27)$$

This differs only in one point from the unconstrained case (4.13): The test function $\boldsymbol{v} \in \mathcal{V}$ is now arbitrary on Γ_0, while in (4.7) it had to vanish on Γ_0.

By inserting $\boldsymbol{\varphi} = \boldsymbol{r} + \boldsymbol{A}(\boldsymbol{x} + \boldsymbol{u})$ with corresponding acceleration $\ddot{\boldsymbol{\varphi}}$ and admissible test function $\boldsymbol{\eta}$, the analogue of the partitioned equations of motion (4.15a)–(4.15c) follows directly. Instead of redoing the foregoing rather tedious calculations, we skip the display of the partitioned equations and proceed with the extension of the abstract notation (4.25) to the constrained case, which will provide more insight.

For this purpose, the weak constraints (4.26b) are expressed as

$$b(\boldsymbol{\varphi}, \boldsymbol{\vartheta}) = \langle \boldsymbol{m}, \boldsymbol{\vartheta} \rangle \quad \forall \boldsymbol{\vartheta} \in \mathcal{Q} \quad (4.28)$$

where

$$b(\boldsymbol{\varphi}, \boldsymbol{\vartheta}) := \int_{\Gamma_0} \boldsymbol{\vartheta}^T \boldsymbol{\varphi}\, ds, \quad \langle \boldsymbol{m}, \boldsymbol{\vartheta} \rangle := \int_{\Gamma_0} \boldsymbol{\vartheta}^T \boldsymbol{\varphi}_0\, ds.$$

Observe that we do not distinguish here between integration with respect to inertial or body-fixed coordinates. Based on the bilinear form b and the notation introduced in (4.25), the equations of constrained elastic motion (4.26a), (4.26b) are then written as transient saddle point problem

$$\langle \rho\ddot{\boldsymbol{\varphi}}, \boldsymbol{\eta} \rangle + a(\boldsymbol{\varphi}, \boldsymbol{\eta}) + b(\boldsymbol{\eta}, \boldsymbol{\mu}) = \langle l(\boldsymbol{\varphi}), \boldsymbol{\eta} \rangle \quad \forall \boldsymbol{\eta} \text{ from (4.27)}, \quad (4.29a)$$

$$b(\boldsymbol{\varphi}, \boldsymbol{\vartheta}) = \langle \boldsymbol{m}, \boldsymbol{\vartheta} \rangle \quad \forall \boldsymbol{\vartheta} \in \mathcal{Q}. \quad (4.29b)$$

If one omits the rigid motion variables of the floating reference frame by setting $\varphi = x + u$, this representation of the elastic deformation is completely equivalent to the constrained equations (3.42a), (3.42b) discussed and analyzed in the previous chapter.

While the unconstrained equations of motion with floating reference frame (4.15a)–(4.15c) or (4.25), respectively, can be found in most of the above references of the engineering literature, the constrained counterpart (4.29a), (4.29b) breaks apparently new ground. The reason for this lies in the fact that in flexible multibody applications, modeling and discretization are usually intertwined. The elastic body is first discretized in space by techniques such as the finite element method and thereafter, the bonds due to joints are taken into account. In contrast to this, we advocate here a strict separation of modeling and discretization, which clearly requires an approach that includes constraints before discretization.

To point out one advantage of this approach, we consider the special case of a *point constraint* on the elastic motion. Let $x_p \in \Omega$ be a point where the motion is prescribed by the function $\varphi_0 = \varphi_0(t)$. We formulate the constraint in weak form by

$$0 = b(\varphi, \vartheta) - \langle m, \vartheta \rangle := \vartheta^T (r + A(\alpha)(x_p + u(x_p, t))) - \vartheta^T \varphi_0 \qquad (4.30)$$

where the test functions ϑ are actually vectors in $\mathcal{Q} := \mathbb{R}^3$. Correspondingly, the bilinear form b in (4.29a) can be replaced by

$$b(\eta, \mu) = \mu^T \left(z - A(\alpha)(\tilde{x}_p + \tilde{u}(x_p, t))\gamma + A(\alpha)v(x_p) \right)$$

with Lagrange multiplier $\mu(t) \in \mathbb{R}^3$. One could even more skip the elastic motion and go back to the rigid motion case given by the Newton–Euler equations (4.22) augmented by a point constraint, which can be found to read

$$\begin{aligned} m\ddot{r} &= f_f - \mu, \\ J\dot{\omega} + \tilde{\omega}J\omega &= f_m - \tilde{x}_p A(\alpha)^T \mu, \\ 0 &= r + A(\alpha)x_p - \varphi_0. \end{aligned} \qquad (4.31)$$

However, there is a flaw in this argumentation. While a point constraint is well-defined in the rigid body case, this does not hold true in general for elastic bodies. We mentioned above in Sect. 3.1.3 that for functions in $H^1(\Omega)$, a pointwise evaluation $u(x_p)$ makes no sense in case of two or three space dimensions, and for this reason the mathematical setting that we have derived in the previous chapter does not apply to (4.30). In other words, we lose the firm mathematical ground if we formulate constraints for elastic bodies in isolated points.

This example shows that a careful analysis of the mathematical model *before discretization* is valuable and gives hints on well-defined or questionable approaches. Note that this reasoning does not apply to beams, as will be explained below in Sect. 4.4.

4.2 Constraints

Fig. 4.3 Joint between two rigid bodies

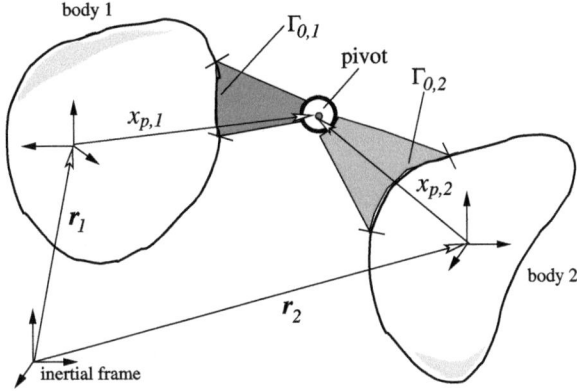

4.2.2 Modeling Joints

We turn now our attention to the modeling of joints in connection with elastic bodies, which will demonstrate the hybrid character of flexible multibody systems and, at the same time, will point out open questions. The discussion is exemplary and considers solely two bodies that are interconnected by a spherical joint in the spatial case or by a revolute joint in the planar case, Fig. 4.3. Recall that the standard modeling assumption on joints is that they are *massless* and have a *fixed geometry* that imposes geometric constraints on the motion.

If both bodies were rigid, the constraint expressing the joint would simply read

$$\varphi_1(x_{p,1}, t) - \varphi_2(x_{p,2}, t) = \mathbf{0} \qquad (4.32)$$

where the index $i = 1, 2$ stands for body i and $x_{p,i}$ denotes the position of the pivot with respect to the body-fixed reference frame of body i. In terms of the rigid motion variables r_i for translation and α_i for rotation, one has the equivalent condition

$$r_1(t) + A(\alpha_1(t))x_{p,1} - (r_2(t) + A(\alpha_2(t))x_{p,2}) = \mathbf{0}. \qquad (4.33)$$

A particular property of this constraint equation is that it plays no role where the body-fixed coordinate systems have been placed. A change of a floating frame leads automatically to an equivalent constraint formulation and does not alter the model.

In case of a joint between two elastic bodies, we distinguish between the following approaches:

1. Floating Frame Attached to Joint

A straightforward approach is to place the floating reference frames of bodies 1 and 2 somewhere on the boundary segment $\Gamma_{0,i}$ where the joint is attached, see Fig. 4.4 on top. If one postulates $u_i = 0$ on $\Gamma_{0,i}$ where $i = 1, 2$ for the corresponding elastic displacements, both boundaries perform a rigid motion, which is in accordance with

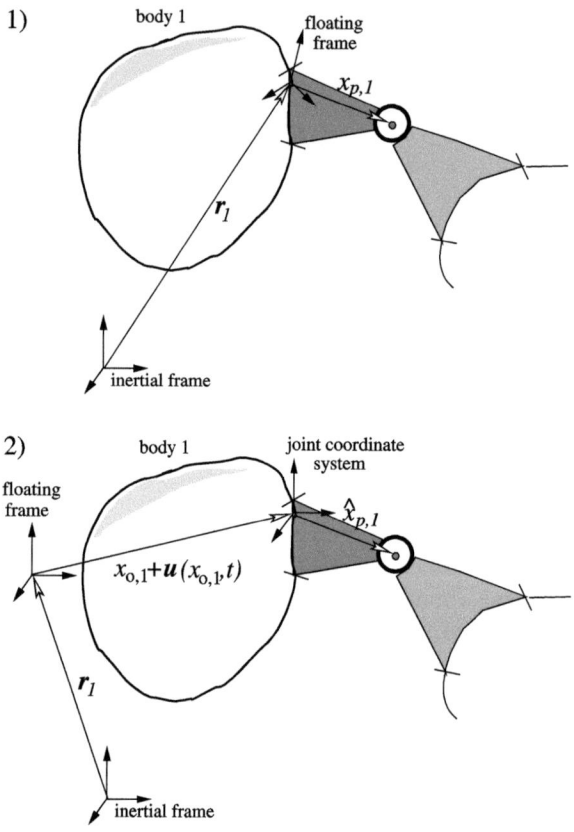

Fig. 4.4 Floating frame attached to joint (*top*) and intermediate joint coordinate system (*bottom*)

the requirement of a fixed joint geometry (one says that the joint is "rigidly attached" [Sha98]). Any joint is hence easily expressed in terms of the rigid motion variables only. For the spherical joint in our example, this means that the constraint equation (4.33) remains valid and does not depend on the displacement fields. In this way, it is also guaranteed that the floating frames move with the bodies.

In general, only a single joint of a particular body can be treated in this way. If there is a second joint, the body-fixed frame has already been specified and cannot be moved to the other joint. In such a situation, a different approach is required.

2. Intermediate Joint Coordinate Systems

Suppose the floating reference frames are not placed in the boundary segments $\Gamma_{0,1}$ and $\Gamma_{0,2}$ connected with the joint. The motion in the joint between the two bodies can be defined by the relative motion of two additional reference frames, so-called intermediate joint coordinate systems [Sha98, §5.4] whose origins we denote by $x_{o,1} \in \Gamma_{0,1}$ and $x_{o,2} \in \Gamma_{0,2}$, Fig. 4.4 at the bottom. With respect to the inertia frame,

4.2 Constraints

one defines the vectors

$$r_i(t) + A(\alpha_i(t))(x_{o,i} + u_i(x_{o,i}, t)) =: \hat{r}_i(t), \quad i = 1, 2;$$

that localize the intermediate joint coordinate systems. Let $\hat{A}_i$ denote the corresponding rotation matrices and $\hat{x}_{p,i}$ the pivot location with respect to the joint coordinate system, then the constraint can be expressed similar to (4.33) by

$$\hat{r}_1(t) + \hat{A}_1 \hat{x}_{p,1} - (\hat{r}_2(t) + \hat{A}_2 \hat{x}_{p,2}) = \mathbf{0}. \tag{4.34}$$

There are several concerns associated with this model. For one, the boundary segments $\Gamma_{0,i}$ are not required to perform a rigid motion. In practice, such a joint coordinate system is mostly placed in a single point without any further specification of a corresponding segment. The introduction of rotation matrices $\hat{A}_i$, however, should be based on some notion of rigid motion. One could use additional observer points in the boundary to specify the orientation of the intermediate coordinate system, and in combination with a finite element grid and corresponding specialized elements for joint modeling, this is a common approach. From a practical viewpoint, the requirement of a fixed geometry of the interface between the joint and the elastic body can thus be weakened. But if the boundary segments $\Gamma_{0,i}$ do not perform a rigid motion, the model (4.34) contains a point constraint since the vectors $\hat{r}_i$ depend on the elastic displacements $u(x_{o,i})$. As we have seen above in the discussion of the point constraint (4.30), this implies a singularity, and the setting based on the function space $H^1(\Omega)$ for the displacement fields is no more valid.

3. Weak Constraints

We sketch next models that avoid point constraints. The discussion, however, is restricted to situations where the second body is rigid, which means that the vector $\varphi_2(x_{p,2}, t)$ locates the joint pivot with respect to this body, cf. the rigid body constraint (4.32). Let on the other hand $\xi(x)$ for $x \in \Gamma_{0,1}$ denote the vector from a point on the boundary of the elastic body to the joint pivot. Such a parametrization of the boundary is a priori known due to the fixed joint geometry and is now employed to introduce the condition

$$\varphi_1(x, t) + \xi(x) - \varphi_2(x_{p,2}, t) = \mathbf{0} \quad \text{for } x \in \Gamma_{0,1}. \tag{4.35}$$

This constraint can also be formulated in weak form by postulating

$$\int_{\Gamma_{0,1}} \vartheta^T \left(\varphi_1(\cdot, t) + \xi - \varphi_2(x_{p,2}, t) \right) ds = 0 \quad \forall \vartheta \in \mathcal{Q}, \tag{4.36}$$

which is compatible with the framework of the transient saddle point problem (4.29a), (4.29b) with weak constraint (4.29b). The constraint (4.36) avoids a point evaluation of the displacement field.

As a special case of the weak constraint (4.36), we view ϑ as prolongation of the test function η defined on body 1 and set it to a variation of a rigid body motion, i.e., $\vartheta = z - A(\alpha_1)\tilde{x}y$ with arbitrary vectors $z, y \in \mathbb{R}^3$. Upon insertion of ϑ in (4.36) one obtains

$$0 = z^T \int_{\Gamma_{0,1}} \left(\varphi_1(\cdot, t) + \xi - \varphi_2(x_{p,2}, t)\right) ds$$

$$+ y^T \int_{\Gamma_{0,1}} \tilde{x} A(\alpha_1)^T \left(\varphi_1(\cdot, t) + \xi - \varphi_2(x_{p,2}, t)\right) ds$$

for all $z, y \in \mathbb{R}^3$, which reduces to six constraint equations

$$0 = \int_{\Gamma_{0,1}} \left(\varphi_1(\cdot, t) + \xi - \varphi_2(x_{p,2}, t)\right) ds, \tag{4.37a}$$

$$0 = \int_{\Gamma_{0,1}} \tilde{x} A(\alpha_1)^T \left(\varphi_1(\cdot, t) + \xi - \varphi_2(x_{p,2}, t)\right) ds \tag{4.37b}$$

for the coupled rigid motion variables and the displacement field u_1. These conditions do not guarantee a rigid motion of $\Gamma_{0,1}$, but they constitute an *averaging technique*.

Associated with the six constraint equations (4.37a), (4.37b), we have six Lagrange multipliers $\mu(t) \in \mathbb{R}^6$ where the first three can be interpreted as resultant force and the last three as resultant torque that are induced by the joint. In the planar case, this technique leads to three averaged constraints with corresponding multipliers.

4. Joints as Additional Rigid Bodies

While the above approaches are based on the assumption of a massless joint, it is also possible to treat the joint by way of two small additional rigid bodies. The part of the joint left to the pivot defines the first rigid body that has a common interface $\Gamma_{0,1}$ with the elastic body 1 on the left, and the part of the joint right to the pivot stands for the second rigid body with interface $\Gamma_{0,2}$. Both interface conditions can then be expressed in terms of weak constraint equations (4.29b), which is very similar to the domain decomposition technique (3.77a)–(3.77c), and the joint itself leads to the rigid body constraint (4.32).

An advantage of this approach is that we get well-defined constraint equations with rigid boundary segments. On the other hand, two additional rigid bodies mean 12 additional degrees of freedom in the spatial case, and the small dimensions of the joint result in small entries in the mass matrix of the rigid body equations, which can be seen as a singular perturbation and is known to be prone for instabilities in the numerical time integration.

5. Rigid Subdomains for Joint Coupling

In this approach, the part of the joint left to the pivot is, as above, viewed as a continuum with a certain mass. Now, however, it is assumed to form a rigid subdomain of the elastic body that can be used to prevent point forces acting from the joint. Let $\Omega_{0,1}$ denote this part of the joint, then the displacement field is only allowed to perform a rigid motion in $\Omega_{0,1}$,

$$\boldsymbol{u}_1(x_1,t) = \tilde{\boldsymbol{r}}(t) + \boldsymbol{A}(\tilde{\boldsymbol{\alpha}}(t))x_1 \quad \text{in } \Omega_{0,1} \tag{4.38}$$

with translation $\tilde{\boldsymbol{r}}$ and rotation $\boldsymbol{A}(\tilde{\boldsymbol{\alpha}})$. Since a fully nonlinear rotation contradicts the assumptions of linear elasticity, (4.38) needs to be linearized for small angles $\tilde{\boldsymbol{\alpha}}$. In combination with finite elements, this approach leads to *rigid body elements*, and the linearized rigid body motion is imposed in terms of additional linear constraint equations that are applied to reduce the degrees of freedom in such a way that the elements that discretize $\Omega_{0,1}$ may perform solely a small rigid body motion. In Sect. 5.1.2, an example will illustrate this technique, which is widespread in commercial finite element codes.

In conclusion, the notion of a joint in a multibody system is a discrete modeling approach that uses massless links to restrict the motion of pairs of bodies by some geometric relation. While for rigid bodies the joint's interface to a body plays no role, the interface becomes crucial for elastic bodies as it transports momentum and angular momentum to the continuum. Obviously, two modeling approaches collide at this point and one has to compromise in some way. Moreover, the placement of the floating reference frame determines whether the constraints depend on the elastic displacement or not. Different choices of coordinates for the same physical system do thus not necessarily lead to equivalent mathematical models, which stands in contrast to rigid body dynamics.

The above discussion should have illustrated this and showed how apparent contradictions can be relaxed. Nevertheless, we emphasize that many questions in this field are still open. The situation becomes much simpler for beam models. Since beams feature a rigid cross section area, it is natural to use, e.g., the end cross section as interface to a joint. This leads to point constraints for the displacements of the neutral fiber, see the slider crank example at the end of this chapter.

4.2.3 Remarks

Besides the method of floating reference frames, the so-called *absolute nodal coordinate formulation* [Sha98] has also become popular in flexible multibody dynamics. This approach builds upon the large deformation theory with nonlinear strain tensor (3.4) and leads to a much simpler acceleration expression. The nonlinear framework for the deformation, on the other hand, is more involved and has been so far mainly elaborated for beams, truss structures, and shells [BS09].

We stress that the modeling issues associated with joints for elastic bodies apply in a similar way to the absolute nodal coordinate formulation. Again, with the exception of one-dimensional continua, point constraints should be avoided in this context. This holds also for the recently developed method of *isogeometric analysis* where splines are used as shape functions in the finite element discretization [CHB09, GDSV13].

4.3 Flexible Multibody System

This section is devoted to developing a model for flexible multibody dynamics. Based on the weak form of constrained motion (4.29a), (4.29b), we extend our approach to mixed systems of rigid and elastic bodies. The resulting framework incorporates the case of solely rigid bodies, which has already been discussed in Chap. 2. We deliberately leave some loose ends in our presentation, mainly with respect to the various types of links that may occur, and concentrate instead on the essential problem structure.

4.3.1 Variational Formulation

We assume n_b bodies in the multibody system and number the bodies from 1 to n_b. Let $\mathcal{K}$ be a set of index pairs such that

$$(i, j) \in \mathcal{K} \quad \Rightarrow \quad \text{there is a force element between bodies } i \text{ and } j. \tag{4.39}$$

The index 0 is admitted and indicates a link to a point that is directly defined in the inertial reference frame (a so-called absolute link, in contrast to relative links between pairs of bodies). Note that two or more links between the same bodies i and j can be included by a subnumbering that we omit here for simplicity. By symmetry, $(i, j) \in \mathcal{K}$ is equivalent with $(j, i) \in \mathcal{K}$.

Analogously, let $\mathcal{J}$ be a set of index pairs such that

$$(k, l) \in \mathcal{J} \quad \Rightarrow \quad \text{there is a joint between bodies } k \text{ and } l. \tag{4.40}$$

Only those joints are taken into account for which a constraint needs to be formulated. Joints that have already been eliminated due to a special choice of coordinates are not element of $\mathcal{J}$.

After the system topology has been specified in terms of the sets $\mathcal{K}$ and $\mathcal{J}$, we next introduce the motion of body i via

$$\boldsymbol{\varphi}_i = \boldsymbol{r}_i + \boldsymbol{A}(\boldsymbol{\alpha}_i)(x_i + \boldsymbol{u}_i). \tag{4.41}$$

The floating reference frame of body i leads to this decomposed motion with translation $\boldsymbol{r}_i$, rotation $\boldsymbol{\alpha}_i$, and small displacement $\boldsymbol{u}_i$. It should be stressed that the motion

4.3 Flexible Multibody System

may depend on some generalized coordinates y as in the single body case (4.19a), (4.19b). We write y without index and assume that it comprises all generalized coordinates in the system. With this in mind, the test function η_i of body i is defined as in (4.19b), i.e.,

$$\eta_i = \frac{\mathrm{d}}{\mathrm{d}\theta} \varphi_i(y + \theta w)\Big|_{\theta=0}. \tag{4.42}$$

If r_i, α_i, and u_i are independent coordinates, this general definition leads to

$$\eta_i = z_i - A(\alpha_i)(\tilde{x}_i + \tilde{u}_i)\gamma_i + A(\alpha_i)v_i,$$

compare (4.13) in the single body case.

Our modeling approach builds on a compact notation that may be regarded as a generalization of the constrained variational problem (4.29a), (4.29b). For the inertia integrals we write the pairing

$$\langle \rho_i \ddot{\varphi}_i, \eta_i \rangle := \int_{\Omega_i} \rho_i \eta_i^T \ddot{\varphi}_i \, \mathrm{d}x$$

where ρ_i is the mass density of body i and $\mathrm{d}x = \mathrm{d}x_i$ for brevity. Thus the system's kinetic energy becomes

$$T = \sum_{i=1}^{n_b} \frac{1}{2} \langle \rho_i \dot{\varphi}_i, \dot{\varphi}_i \rangle.$$

Furthermore, the bilinear form

$$a_i(\varphi_i, \eta_i) := \int_{\Omega_i} \sigma_i(u_i) : \varepsilon(v_i) \, \mathrm{d}x$$

stands for the variation of the strain energy. If body i is rigid, we have $u_i \equiv 0$ and hence $a_i \equiv 0$. Otherwise, the notation σ_i indicates that material parameters in Hooke's law (3.8) may differ from body to body. Volume forces are finally denoted by

$$\langle l_i(\varphi_i), \eta_i \rangle := \int_{\Omega_i} \eta_i^T A(\alpha_i) \beta_i \, \mathrm{d}x.$$

Instead of surface tractions, we solely consider force elements between bodies. In a straightforward extension of the force law (4.20) for a translational spring-damper-actuator, the scalar force acting between bodies $(i, j) \in \mathcal{K}$ is calculated to be

$$f_{ij}(\zeta_{ij}, \dot{\zeta}_{ij}, t) = k_{ij}(\zeta_{ij} - \zeta_{0,ij}) + d_{ij}\dot{\zeta}_{ij} + h_{ij}(\zeta_{ij}, \dot{\zeta}_{ij}, t). \tag{4.43}$$

Here, $\zeta_{ij} = \|c_{ij}\|_2$ is the distance between the two attachment points $x_{f,i}$ and $x_{f,j}$ of the force element while $c_{ij} := \varphi_j(x_{f,j}, t) - \varphi_i(x_{f,i}, t)$ stands for the corresponding vector. For $d_{ij} = 0$ and $h \equiv 0$, we are in the conservative case and find the potential energy to be

$$U = \sum_{i=1}^{n_b} \left(\frac{1}{2} a_i(\boldsymbol{\varphi}_i, \boldsymbol{\varphi}_i) - \langle l_i(\boldsymbol{\varphi}_i), \boldsymbol{\varphi}_i \rangle \right) + \sum_{(i,j) \in \mathcal{K}} \frac{1}{2} \frac{f_{ij}(\boldsymbol{\varphi}_i, \boldsymbol{\varphi}_j)^2}{k_{ij}}.$$

Like force elements, joints involve nonlinear relations. We write a constraint that models a joint between bodies k and l in weak form as

$$0 = b_{kl}(\boldsymbol{\varphi}_k, \boldsymbol{\varphi}_l, \boldsymbol{\vartheta}_{kl}).$$

This quite general notation includes weak constraints such as (4.26b) and (4.36). Furthermore, nonlinear point constraints like (4.32) or (4.34) can also be included. To this end, the usual strong formulation $\boldsymbol{0} = \boldsymbol{g}_{kl}(\boldsymbol{\varphi}_k, \boldsymbol{\varphi}_l)$ is taken into account via $0 = \boldsymbol{\vartheta}_{kl}^T \boldsymbol{g}_{kl}(\boldsymbol{\varphi}_k, \boldsymbol{\varphi}_l)$ with some appropriate vector $\boldsymbol{\vartheta}_{kl}$. Though the constraints are nonlinear in general, we require that their weak form is linear with respect to the test function (or test vector) $\boldsymbol{\vartheta}_{kl}$.

Making use of these prerequisites and sticking for the moment to the conservative case, the least action principle can be invoked. It states that the system performs motion such that

$$\int_{t_0}^{t_1} \left(T - U - \sum_{(k,l) \in \mathcal{J}} b_{kl}(\boldsymbol{\varphi}_k, \boldsymbol{\varphi}_l, \boldsymbol{\mu}_{kl}) \right) dt \to \text{stationary !} \quad (4.44)$$

Unknowns are the motions $\boldsymbol{\varphi}_1, \ldots, \boldsymbol{\varphi}_{n_b}$ (or the generalized coordinates $\boldsymbol{y}$, respectively) and the multipliers $\boldsymbol{\mu}_{kl}$ for $(k, l) \in \mathcal{J}$.

4.3.2 Equations of Motion

Care has to be taken when evaluating the stationarity condition. With respect to both force elements and joints, nonlinearities arise that we have not yet dealt with. For a spring element, we have

$$\frac{d}{d\theta} \left(\frac{f_{ij}(\boldsymbol{\varphi}_i(\boldsymbol{y} + \theta \boldsymbol{w}), \boldsymbol{\varphi}_j(\boldsymbol{y} + \theta \boldsymbol{w}))^2}{2k_{ij}} \right) \bigg|_{\theta=0} = f_{ij}(\boldsymbol{\varphi}_i, \boldsymbol{\varphi}_j) \frac{\boldsymbol{c}_{ij}^T}{\zeta_{ij}} (\boldsymbol{\eta}_i - \boldsymbol{\eta}_j). \quad (4.45)$$

For a constraint term, the directional derivative can be written as

$$\frac{d}{d\theta} b_{kl}(\boldsymbol{\varphi}_k(\boldsymbol{y} + \theta \boldsymbol{w}), \boldsymbol{\varphi}_l(\boldsymbol{y} + \theta \boldsymbol{w}), \boldsymbol{\mu}_{kl} + \theta \boldsymbol{\vartheta}_{kl})$$
$$= b'_{kl}(\boldsymbol{\varphi}_k, \boldsymbol{\varphi}_l, \boldsymbol{\eta}_k, \boldsymbol{\eta}_l, \boldsymbol{\mu}_{kl}) + b_{kl}(\boldsymbol{\varphi}_k, \boldsymbol{\varphi}_l, \boldsymbol{\vartheta}_{kl}) \quad (4.46)$$

4.3 Flexible Multibody System

where b'_{kl} is a sum of the derivatives of b_{kl} with respect to the first and the second slot, respectively. Like the weak expression b_{kl}, the derivative b'_{kl} is linear in the multiplier μ_{kl}.

We skip here the derivation of a strong form of the equations of motion and tackle instead directly the weak form. In the following, the arguments φ_i, φ_j in the force term f_{ij} and φ_k, φ_l in the constraint derivative term b'_{kl} are dropped to present the formulas in a concise way.

From the principle (4.44) we conclude, using the relations (4.45) and (4.46), that the equations of motion are given by

$$\sum_{i=1}^{n_b} \left(\langle \rho_i \ddot{\varphi}_i, \eta_i \rangle + a_i(\varphi_i, \eta_i) \right) + \sum_{(k,l) \in \mathcal{J}} b'_{kl}(\eta_k, \eta_l, \mu_{kl}) = \ell_{\text{mbs}}, \quad (4.47a)$$

$$\sum_{(k,l) \in \mathcal{J}} b_{kl}(\varphi_k, \varphi_l, \vartheta_{kl}) = 0 \quad (4.47b)$$

with right hand side

$$\ell_{\text{mbs}} := \sum_{i=1}^{n_b} \langle l_i, \eta_i \rangle - \sum_{(i,j) \in \mathcal{K}} f_{ij} \frac{c_{ij}^T}{\zeta_{ij}} (\eta_i - \eta_j). \quad (4.48)$$

The weak form (4.47a), (4.47b) must hold for all admissible η_i from (4.42) and ϑ_{kl}. In order to get a better understanding of the methodology used for this formulation, we discuss next the equations of motion (4.47a), (4.47b) from several viewpoints.

Constraints

As the test functions (or vectors) ϑ_{kl} in (4.47b) are independent, the sum over all pairs of indices $(k, l) \in \mathcal{J}$ can be replaced by the separate conditions

$$b_{kl}(\varphi_k, \varphi_l, \vartheta_{kl}) = 0 \quad \text{for } (k, l) \in \mathcal{J}.$$

On the other hand, if generalized coordinates are used to eliminate all constraints a priori, the equations of motion simplify to

$$\sum_{i=1}^{n_b} \langle \rho_i \ddot{\varphi}_i, \eta_i \rangle + a_i(\varphi_i, \eta_i) = \ell_{\text{mbs}} \quad (4.49)$$

for all admissible η_i.

Absolute Coordinates

If each body's rigid motion is expressed in terms of independent translational coordinates r_i and rotational coordinates α_i, the sum over all bodies in (4.47a) can be dropped, resulting in the equations of motion for body i

$$\langle \rho_i \ddot{\varphi}_i, \eta_i \rangle + a_i(\varphi_i, \eta_i) + \sum_{(i,l) \in \mathcal{J}} b'_{il}(\eta_i, \cdot, \mu_{il}) = \ell_{\mathrm{mbs},i}, \quad (4.50\mathrm{a})$$

$$\sum_{(i,l) \in \mathcal{J}} b_{il}(\varphi_i, \varphi_l, \vartheta_{il}) = 0 \quad (4.50\mathrm{b})$$

with right hand side

$$\ell_{\mathrm{mbs},i} = \langle l_i, \eta_i \rangle - \sum_{(i,j) \in \mathcal{K}} f_{ij} \frac{c_{ij}^T}{\zeta_{ij}} (\eta_i - \eta_j). \quad (4.51)$$

Here, the index i is arbitrary but fixed, and the notation $b'_{il}(\eta_i, \cdot, \mu_{il})$ indicates that solely the derivatives of b_{il} with respect to the first slot are taken into account. The derivatives with respect to the second slot, i.e., with respect to φ_l, belong to the corresponding equation of motion for φ_l.

Nonconservative Forces

The least action principle (4.44) possesses an extension to nonconservative forces. Restricting the discussion to the unconstrained case, this extension is usually written as

$$\delta \int_{t_0}^{t_1} (T - U) \, dt + \int_{t_0}^{t_1} \delta W \, dt = 0. \quad (4.52)$$

The first term denotes the first variation of the action integral, and δW in the second integral stands for the *virtual work done by nonconservative forces* [Sha98, §3.6]. Assuming the force elements defined in (4.43) to be nonconservative and evaluating the corresponding virtual work expressions, it can be shown that

$$\delta W = \sum_{(i,j) \in \mathcal{K}} f_{ij} \frac{c_{ij}^T}{\zeta_{ij}} (\eta_i - \eta_j),$$

from which we conclude that the equations of motion (4.47a), (4.47b) even hold in the nonconservative case. One simply inserts the definition of f_{ij} from (4.43) with damper and actuator law in the right hand side (4.48).

The same observation holds for additional dissipative terms such as Rayleigh damping (3.20), which are readily included in the left hand side of (4.47a). Constraints involve no additional difficulties and are taken into account as before by means of the multiplier technique.

Rigid Multibody Dynamics

Last but not least, the model equations (4.47a), (4.47b) include the case of solely rigid bodies. We have thus also gained new insight into the structure of the equations of motion (2.9a), (2.9b) presented in Chap. 2. E.g., the kinetic energy has the two representations

$$T = \frac{1}{2}\dot{q}^T M(q)\dot{q} = \sum_{i=1}^{n_b} \frac{1}{2}\langle \rho_i \dot{\varphi}_i, \dot{\varphi}_i \rangle,$$

and the force vector $f(q, \dot{q}, t)$ in (2.9a), (2.9b) is computed by sorting and evaluating the right hand side entries of (4.48) plus Coriolis and centrifugal terms.

4.4 Special Bodies

So far, the elastic body was treated as general three-dimensional continuum. In various applications, however, lower dimensional models are sufficient. For this reason, we briefly mention here plane strain and plane stress and present also a planar beam model.

4.4.1 Plane Strain and Plane Stress

Under certain assumptions, the three-dimensional elastic body may be simplified to a two-dimensional continuum. An example is the plane strain assumption where only a cross section of the body, say the plane spanned by coordinates (x_1, x_2), is considered. In the third direction x_3, the body's geometry as well as the applied forces are required to stay constant. We thus have

$$u(x,t) = \begin{pmatrix} u_1(x_1, x_2, t) \\ u_2(x_1, x_2, t) \\ 0 \end{pmatrix} \quad (4.53)$$

as displacement field.

In the strain tensor definition (3.5), the components $\varepsilon_{33}, \varepsilon_{13}, \varepsilon_{23}$ vanish then due to $\partial u_i / \partial x_3 = 0$, and the strain tensor can be written as vector $\underline{\varepsilon} := (\varepsilon_{11}, \varepsilon_{22}, 2\varepsilon_{12})^T$. Hooke's law (3.8) and the representation (3.10) yield a standard stress-strain relation $\underline{\sigma} = C_{PA}\underline{\varepsilon}$ where the stress vector reads $\underline{\sigma} := (\sigma_{11}, \sigma_{22}, \sigma_{12})^T$ and

$$C_{PA} := \frac{E}{(1+\nu)(1-2\nu)} \begin{pmatrix} 1-\nu & \nu & 0 \\ \nu & 1-\nu & 0 \\ 0 & 0 & (1-2\nu)/2 \end{pmatrix}. \quad (4.54)$$

Note that $\sigma_{13} = \sigma_{23} = 0$ but $\sigma_{33} = \nu(\sigma_{11} + \sigma_{22}) \neq 0$ in general.

If one inserts the essentially two-dimensional displacement (4.53) and replaces the product of stress and strain by the above plane strain version of Hooke's law, the structure of the equations of motion (3.19) or (3.31), respectively, remains the same. Integration with respect to the third coordinate x_3 yields a constant factor for all integrals, which cancels out immediately.

Recalling the discussion in Sect. 4.2.2 on different joint models, we observe that the two-dimensional continuum of plane strain requires the Sobolev space $H^1(\Omega)^2$ as basic setting for the displacement field. This implies that point evaluations and thus point constraints are also questionable for this model while weak constraints are well-defined.

Plane Stress

This two-dimensional model is based on the assumption that stresses in x_3-direction vanish, i.e., $\sigma_{33} = \sigma_{13} = \sigma_{23} = 0$, in contrast to the plane strain case where the corresponding strains vanish. A thin disk is a typical example where plane stress applies, while the cross section of a long body usually leads to a plane strain problem.

As above, we have the displacement field (4.53) and $\underline{\sigma} = (\sigma_{11}, \sigma_{22}, \sigma_{12})^T$ and $\underline{\varepsilon} = (\varepsilon_{11}, \varepsilon_{22}, 2\varepsilon_{12})^T$ as stress and strain vectors. Hooke's law (3.8) takes now the form $\underline{\sigma} = C_{PS}\underline{\varepsilon}$ with matrix

$$C_{PS} := \frac{E}{(1-\nu)^2} \begin{pmatrix} 1 & \nu & 0 \\ \nu & 1 & 0 \\ 0 & 0 & (1-\nu)/2 \end{pmatrix}. \tag{4.55}$$

Up to a constant transformation, plane stress and plane strain are equivalent models.

4.4.2 Beam Model

Beams are long bodies with small cross sections. The basic assumption is that planar cross sections remain planar under deformation. Each cross section is attached to the *neutral fiber*, and changes in the neutral fiber fully determine the body's behavior. Additionally, in case of an Euler–Bernoulli beam cross-sections perpendicular to the neutral fiber remain perpendicular. Figure 4.5 shows a planar Euler–Bernoulli beam with neutral fiber and deformed state.

In the remainder of this section, we derive a model for this beam assuming a constant cross section. The displacement field u is again two-dimensional as in (4.53). In addition, it satisfies [SW99, §4.2]

$$\begin{aligned} u_1(x,t) &= w_1(x_1,t) - x_2 w_2'(x_1,t), \\ u_2(x,t) &= w_2(x_1,t), \end{aligned} \tag{4.56}$$

where w_1 denotes the longitudinal displacement of the neutral fiber and w_2 the lateral displacement. Compare also the beam model for the elastic pendulum (1.1).

4.4 Special Bodies

Fig. 4.5 Planar Euler–Bernoulli beam

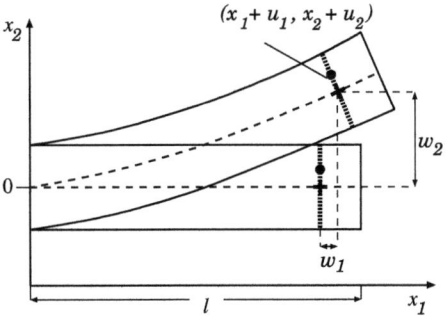

Differentiation with respect to x_1 is written as $w' := \partial w/\partial x_1$. Note that (4.56) holds solely for small displacements.

Because of the material law of St. Venant–Kirchhoff (3.6) and the definition (4.56), the strain energy W_{nl} from (3.7) depends merely on the (1, 1)-element of the strain tensor E,

$$W_{nl} = \frac{1}{2}E \int_\Omega E_{11}^2 \, dx$$

$$= \frac{1}{2}E \int_\Omega \left(\underbrace{w_1' - x_2 w_2''}_{=\varepsilon_{11}} + \underbrace{1/2\left((w_1' - x_2 w_2'')^2 + (w_2')^2\right)}_{\text{from } \nabla u^T \nabla u} \right)^2 dx \quad (4.57)$$

with modulus of elasticity E.

In linear elasticity we take only those terms into account that belong to ε_{11}. After integration with respect to x_2 and x_3, the strain energy in terms of $\boldsymbol{w} := (w_1, w_2)^T$ reads

$$W = \frac{1}{2}a(\boldsymbol{w}, \boldsymbol{w}) = \frac{1}{2}\text{EA} \int_0^l (w_1')^2 \, dx_1 + \frac{1}{2}\text{EI} \int_0^l (w_2'')^2 \, dx_1$$

with cross section area A, axial moment of inertia I, and length l. Correspondingly, we get instead of the bilinear form $a(\boldsymbol{u}, \boldsymbol{v})$ defined in (3.27) the sum

$$a(\boldsymbol{w}, \boldsymbol{v}) = \text{EA} \int_0^l w_1' v_1' \, dx_1 + \text{EI} \int_0^l w_2'' v_2'' \, dx_1. \quad (4.58)$$

The test functions $\boldsymbol{v} := (v_1, v_2)^T$ are element of $H^1((0, l)) \times H^2((0, l))$ and have to satisfy additional boundary conditions that depend on the specific choice of the floating reference frame.

Point evaluations are well-defined in both $H^1((0, l))$ and $H^2((0, l))$, which means that constraints such as (4.34) based on intermediate joint coordinate systems lead to sound models for beams. Moreover, the assumption that planar cross sections do not deform allows a straightforward attachment of the intermediate coordinate system to one of the beam tips, and there is no need to introduce an additional rotation between the floating and the intermediate frame.

Geometric Stiffening

Linear elasticity models are not always sufficient in flexible multibody dynamics. In particular slim bodies like beams may require additional *geometric stiffening terms* [Wal91]. E.g., in the planar beam discussed here, the term $(w_2')^2$ from the nonlinear part $\nabla \boldsymbol{u}^T \nabla \boldsymbol{u}$ of the strain energy (4.57) plays a key role in order to model coupling effects between longitudinal and lateral displacements.

Neglecting other second order terms in (4.57), we get as strain energy

$$W_{nl} \doteq \frac{1}{2} a(\boldsymbol{w}, \boldsymbol{w}) + \frac{1}{2} EA \int_0^l w_1'(w_2')^2 \, dx_1. \tag{4.59}$$

Consequently, the variation $a(\boldsymbol{w}, \boldsymbol{v})$ has to be replaced by the sum $a(\boldsymbol{w}, \boldsymbol{v}) + a_s(\boldsymbol{w}, \boldsymbol{v})$ with additional geometric or nonlinear stiffening term

$$a_s(\boldsymbol{w}, \boldsymbol{v}) := \frac{1}{2} EA \int_0^l \left(v_1'(w_2')^2 + 2 v_2' w_1' w_2' \right) dx_1. \tag{4.60}$$

The slider crank example below and the corresponding simulation results in Chap. 8 will illustrate the role of this stiffening term.

4.5 Examples

The chapter closes with three examples that cover different aspects of flexible multibody dynamics: the slider crank with elastic connecting rod, the truck with elastic load area, and the coupled system of pantograph and catenary.

4.5.1 Slider Crank

The planar slider crank shown in Fig. 4.6 consists of two rigid bodies, which are the crank and the sliding block, and one elastic body, the connecting rod. The mechanism has been widely used as simple benchmark problem for flexible multibody dynamics, see, e.g., Koppens [Kop89] and Jahnke et al. [JPD93].

Like in the rigid body model described in Sect. 2.2.1, three revolute joints and one translational joint connect the bodies and constrain their motion. Furthermore, we employ the same angles α_1, α_2 and the same translation r_3 to describe the rigid motion of the bodies. The new feature in this example is the connecting rod, which is now modeled by a planar Euler–Bernoulli beam and the method of floating reference frames.

As sketched in Fig. 4.6, the origin of the body-fixed frame for the crank (body 1) has been placed in the pivot of the first revolute joint, which means that

$$\boldsymbol{\varphi}_1(x_1, t) = \boldsymbol{A}(\alpha_1(t)) x_1$$

4.5 Examples

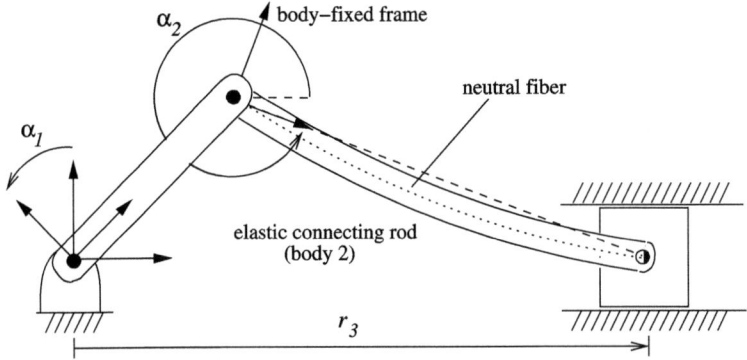

Fig. 4.6 Planar slider crank mechanism with elastic connecting rod

describes the motion of any point x_1 of the crank. Here, the matrix

$$A(\alpha) = \begin{pmatrix} \cos\alpha & -\sin\alpha \\ \sin\alpha & \cos\alpha \end{pmatrix}$$

from (4.23) expresses the planar rotation from the body-fixed frame to the inertial frame, which is also placed in the joint pivot. On the other hand, the motion of the sliding block (body 3) reads

$$\varphi_3(x_3, t) = \begin{pmatrix} r_3(t) \\ 0 \end{pmatrix} + x_3,$$

and by setting the vertical degree of freedom of body 3 to zero, the horizontal sliding constraint is automatically satisfied. Note that this body can even be substituted by a point mass with a single degree of freedom r_3.

While bodies 1 and 3 are fully specified by this choice of φ_1 and φ_3, the motion of the elastic connecting rod is treated by the method of floating reference frames. We set

$$\varphi_2(x_2, t) = A(\alpha_1(t)) \begin{pmatrix} l_1 \\ 0 \end{pmatrix} + A(\alpha_2(t))(x_2 + u_2(x_2, t)) \quad (4.61)$$

where l_1 is the distance between both pivots of body 1 and x_2 stands for the material points of body 2. Hoping not to create too much confusion with the notation, we drop the index 2 for body 2 in the following and write x instead of x_2. Moreover, $x = (\xi_1, \xi_2)^T$ stands now for the two coordinate directions in the plane with respect to the floating reference frame of body 2.

Similar to body 1, the floating reference frame of body 2 has been placed in the left pivot and is assumed to move with the body; a prerequisite that has to be ensured by additional boundary conditions that will be outlined in a moment. According to Euler–Bernoulli theory (4.56), the displacement field $u = u_2$ in (4.61) can be written as

$$u(x,t) = \begin{pmatrix} w_1(\xi_1,t) - \xi_2 w_2'(\xi_1,t) \\ w_2(\xi_1,t) \end{pmatrix}$$

with longitudinal and lateral displacements of the neutral fiber w_1 and w_2, respectively. By postulating the boundary conditions

$$w_1(0,t) = w_2(0,t) = w_2(l_2,t) = 0$$

where l_2 is the nominal distance between both pivots of the connecting rod, the ξ_1-axis of the floating reference frame represents a secant through the left and right pivots. Thus, the reference frame moves with the body, and no additional rigid motion can occur. The additional boundary conditions

$$w_1'(l_2,t) = 0, \qquad w_2''(0,t) = w_2''(l_2,t) = 0$$

are finally required to fully specify the strong form of the beam equations.

With respect to the weak formulation and the equations of motion (4.47a), (4.47b), the admissible test functions of body 2 are determined from the derivative

$$\begin{aligned}\eta &= \frac{d}{d\theta}\varphi(\alpha_1+\theta\beta_1,\alpha_2+\theta\beta_2,w_1+\theta v_1,w_2+\theta v_2)\Big|_{\theta=0} \\ &= A'(\alpha_1)\begin{pmatrix}l_1\\0\end{pmatrix}\beta_1 + A'(\alpha_2)(x+u)\beta_2 + A(\alpha_2)\begin{pmatrix}v_1-\xi_2 v_2'\\v_2\end{pmatrix}\end{aligned} \quad (4.62)$$

with the matrix

$$A'(\alpha) = \frac{\partial A(\alpha)}{\partial \alpha} = \begin{pmatrix} -\sin\alpha & -\cos\alpha \\ \cos\alpha & -\sin\alpha \end{pmatrix}.$$

In (4.62), the test functions $v := (v_1, v_2)^T$ are element of the space $H^1((0,l_2)) \times H^2((0,l_2))$ and satisfy the Dirichlet boundary conditions $v_1(0,t) = v_2(0,t) = v_2(l_2,t) = 0$. Furthermore, the arbitrary scalars β_1 and β_2 stand for variations of the angles α_1 and α_2.

For the bilinear form expressing the variation of the strain energy, we have, compare (4.58),

$$a(w,v) = EA\int_0^{l_2} w_1' v_1' d\xi_1 + EI\int_0^{l_2} w_2'' v_2'' d\xi_1.$$

If required, the geometric stiffening term (4.60) can be added to include the nonlinear coupling between longitudinal and lateral displacement.

The volume force term of body 2 is found to be

$$\langle l, \eta \rangle = \int_{\Omega_2} \eta^T \begin{pmatrix} 0 \\ \rho\gamma \end{pmatrix} dx$$

with mass density ρ and gravitation constant γ. Since there is no force element involved, the index set $\mathcal{K}$ from (4.39) is empty. However, the set $\mathcal{J} = \{(2,3)\}$ contains the joint between connecting rod and sliding block that still needs to be addressed. The other two revolute joints and the sliding joint have already been included in the model by the choice of α_1, α_2, r_3 as rigid motion coordinates.

In strong form, we formulate the constraint for the revolute joint between body 2 and body 3 as

$$\begin{pmatrix} r_3 \\ 0 \end{pmatrix} = A(\alpha_1) \begin{pmatrix} l_1 \\ 0 \end{pmatrix} + A(\alpha_2) \begin{pmatrix} l_2 + w_1(l_2, t) \\ 0 \end{pmatrix}.$$

Not that the displacement w_2 vanishes at the right pivot due to the boundary condition $w_2(l_2, t) = 0$, which implies that the constraint equation does solely depend on the longitudinal displacement w_1.

For the weak form of the constraint, we define

$$b_{23}(\varphi_2, \varphi_3, \vartheta_{23}) := \vartheta_{23}^T \left(\begin{pmatrix} r_3 \\ 0 \end{pmatrix} - A(\alpha_1) \begin{pmatrix} l_1 \\ 0 \end{pmatrix} - A(\alpha_2) \begin{pmatrix} l_2 + w_1(l_2, t) \\ 0 \end{pmatrix} \right)$$

with arbitrary vector $\vartheta_{23} \in \mathbb{R}^2$. The constraint reads then

$$b_{23}(\varphi_2, \varphi_3, \vartheta_{23}) = 0 \quad \forall \vartheta_{23}.$$

After all main ingredients of the mathematical model have been outlined, one can now set up the equations of motion based on the formalism (4.47a), (4.47b). It is important to realize that these equations coincide with the rigid body model given in Sect. 2.2.1 for vanishing elastic motion. In other words, by applying the method of floating reference frames, the elastic displacement can be viewed as a *perturbation* of the rigid body motion.

Clearly, other coordinates could have been chosen to model the slider crank mechanism, and even the specification of the boundary conditions to attach the floating frame to the connecting rod is not unique and leaves extra freedom for alternative approaches such as, e.g., a tangential body-fixed coordinate system. A pitfall in this context is the fact that two models with different body-fixed coordinate systems will not be equivalent in general since the origin usually represents a linearization point for the elastic displacement. This major difference to the dynamics of rigid bodies renders the modeling process difficult and requires great care.

4.5.2 Truck Model

The truck model sketched in Fig. 4.7 is an example for a flexible multibody system in vehicle dynamics. Though still fairly simple due to its planar structure, this model is already adequate for preliminary analyses on ride comfort and dynamic wheel loads. It has been derived as an extension of the rigid multibody system given in [SGFR94], which is commonly used for benchmarking numerical methods.

Like in [SGFR94], we assume bodies 1 to 6, i.e., the front wheels, the frame, the engine block, the cabin, and the driver seat, to be rigid. The long loading area

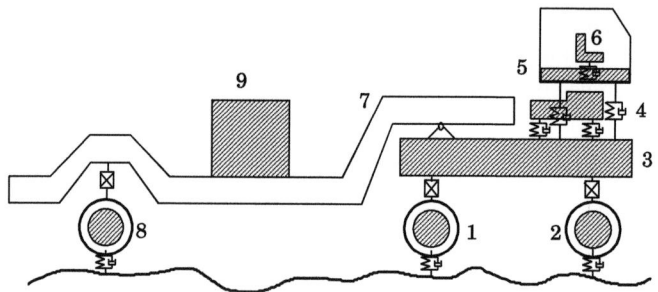

Fig. 4.7 Truck model with eight rigid bodies and elastic loading area

(body 7) and the extra wheel (body 8) expand the original truck to an articulate truck, also called semitrailer truck. We treat the loading area as planar elastic body that features three interconnection elements. On the one hand, it is linked to the frame in front via a revolute joint, and on the other hand, a nonlinear pneumatic spring element attaches the rear wheel. Finally, the load is modeled as rigid body 9 and leads to a rigid-elastic interface between bodies 7 and 9.

Making use of the choice of coordinates given in [SGFR94], we have nine variables for the motion of bodies 1 to 6 where solely vertical displacements with respect to a nominal configuration are taken into account plus additional rotations for bodies 3 to 5. Proceeding in a similar fashion with the extra bodies 7 to 9 of the articulate truck, the motion of the rear wheel is described by a single vertical displacement while loading area and load are associated with a full set of three rigid motion coordinates each. In total, we thus obtain $n_q = 16$ variables for the rigid motion:

q_1 vertical motion of rear tire
q_2 vertical motion of front tire
q_3 vertical motion of truck chassis
q_4 rotation of truck chassis
q_5 vertical motion of engine
q_6 rotation of engine
q_7 vertical motion of driver cabin
q_8 rotation of driver cabin
q_9 vertical motion of driver seat
q_{10} vertical motion of tire for loading area
q_{11} horizontal motion of load
q_{12} vertical motion of load
q_{13} rotation of load
q_{14} horizontal motion of loading area
q_{15} vertical motion of loading area
q_{16} rotation of loading area.

The loading area, body 7, is elastic and satisfies the assumptions of plane stress as described in Sect. 4.4. Hence, the pneumatic spring as interconnection to the wheel

4.5 Examples

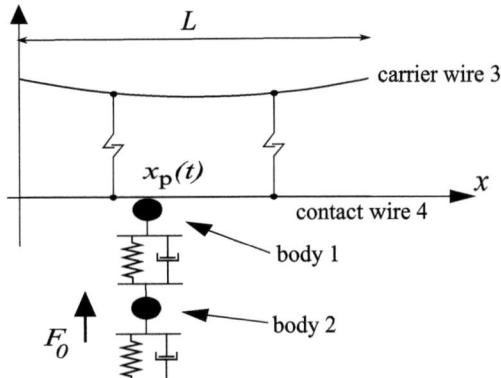

Fig. 4.8 Pantograph and catenary: simplified benchmark problem with two mass points modeling the pantograph

results in a force term of the form (4.43) whereas the revolute joint and the interface to body 9 can be modeled by one of the techniques outlined in Sect. 4.2.2. What is an adept placement of the body-fixed frame in this example? One option is to place the origin on that part of the boundary that serves as interface to the revolute joint. But equivalently, one could select the interface to the load or the attachment point of the force element. As remarked above at the slider crank example, this selection is not arbitrary and requires care since the resulting model equations are not invariant under a change of the body-fixed coordinate system.

4.5.3 Pantograph and Catenary

One of the most sensitive points in today's high speed trains is the transmission of electrical energy to the moving vehicle, which is typically accomplished by so-called *pantographs* that collect the current from an overhead contact wire. This contact wire and its suspension system form the *catenary*, a complex arrangement of cables and wires. Moving with high speed, the pantograph head induces traveling waves in the contact wire such that the contact force between pantograph and catenary varies strongly and the contact may even get interrupted for short moments, see Poetsch et al. [PEM+97].

In real life applications, the dynamical simulation of the interaction between pantograph and catenary is a challenging numerical problem. We sketch here a simplified planar benchmark problem that has been developed in [AS00] in order to discuss the basic steps of discretization in space and time.

The pantograph is a lumped mass model with two point masses representing the pantograph head (body 1) and the remaining components of the pantograph (body 2). There are spring–damper elements between bodies 1 and 2 and between body 2 and the ground, see Fig. 4.8. Both bodies move with fixed speed in x-direction along the simplified catenary, which consists of the carrier and contact wires (bodies 3 and 4) and two massless droppers, i.e., vertical wires that interconnect the contact wire and the carrier wire on top.

While bodies 1 and 2 form a multibody system with rigid members, bodies 3 and 4 are elastic one-dimensional structures. For the carrier wire, a standard string model suffices, but for the contact wire, the bending stiffness needs to be taken into account, which means that an Euler–Bernoulli beam model is an appropriate choice. In contrast to the above examples, the motion of the elastic catenary is not superimposed by translations or rotations, and therefore we do not require any floating reference frame here.

The position of the pantograph head at time t is denoted by $x_p(t)$. Since the independent variable is $x = x_1$, we omit in the following the corresponding subscript $_1$ and denote the vertical displacements of the pantograph bodies by r_1 and r_2. Furthermore, the displacements of carrier and contact wire are written as w_3 and w_4.

The system of pantograph and catenary illustrates once again the above introduced differential-algebraic methodology for setting up the equations of motion. In the first step, we neglect the contact conditions and consider the equations of unconstrained motion, which yields

$$m_1 \ddot{r}_1 = -d_1(\dot{r}_1 - \dot{r}_2) - c_1(r_1 - r_2), \tag{4.63}$$

$$m_2 \ddot{r}_2 = -d_2 \dot{r}_2 + d_1(\dot{r}_1 - \dot{r}_2) - c_2 r_2 + c_1(r_1 - r_2) + F_0, \tag{4.64}$$

$$\rho_3 A_3 \ddot{w}_3 = -\beta_3 \dot{w}_3 + T_3 w_3'' - \rho_3 A_3 \gamma, \tag{4.65}$$

$$\rho_4 A_4 \ddot{w}_4 = -\beta_4 \dot{w}_4 + T_4 w_4'' - E_4 I_4 w_4'''' - \rho_4 A_4 \gamma. \tag{4.66}$$

Here, the first two equations describe the pantograph motion with damper and spring constants d_1, d_2, c_1, c_2 and a constant force F_0 which includes the influence of the gravity. The carrier is expressed by the equation of a vibrating string with tensile force T_3 and viscous damping factor β_3. Finally, the beam equation for the contact wire includes both a pre-stress term due to the tensile force T_4 as well as a bending stiffness term with factor $E_4 I_4$. The notation of the parameters is the same as in Sect. 4.4.2; e.g., A stands for the cross section area and γ is the gravity constant.

The next step integrates the coupling conditions. For the moment, we assume bilateral contact to simplify the discussion. Contact wire and carrier are interconnected by two massless droppers with relative distances l_1 and l_2 and positions $x_{p,1}$ and $x_{p,2}$. The third constraint results from the coupling of contact wire and body 1. In strong form, we require thus

$$w_3(x_{p,1}, t) - w_4(x_{p,1}, t) + l_1 = 0,$$

$$w_3(x_{p,2}, t) - w_4(x_{p,2}, t) + l_2 = 0,$$

$$w_4(x_p(t), t) - r_1(t) = 0. \tag{4.67}$$

For the transition to constraints in weak form, we multiply each of these equations by constants $\vartheta_1, \vartheta_2, \vartheta_3 \in \mathbb{R}$ and write the constraints as

$$(w_3(x_{p,1}, t) - w_4(x_{p,1}, t) + l_1)\vartheta_1 =: b_1(w_3, w_4, \vartheta_1) = 0 \quad \forall \vartheta_1 \in \mathbb{R}, \tag{4.68}$$

$$(w_3(x_{p,2}, t) - w_4(x_{p,2}, t) + l_2)\vartheta_2 =: b_2(w_3, w_4, \vartheta_2) = 0 \quad \forall \vartheta_2 \in \mathbb{R}, \tag{4.69}$$

$$(w_4(x_p(t), t) - r_1(t))\vartheta_3 =: b_3(t; r_1, w_4, \vartheta_3) = 0 \quad \forall \vartheta_3 \in \mathbb{R}. \tag{4.70}$$

4.5 Examples

Notice that the weak constraint b_3 depends explicitly on time t due to the moving contact point.

For the last step, we have to couple constraints and equations of motion by Lagrange multipliers. We introduce the notation $\langle c, e \rangle = \int_0^L ce\,dx$, where L is the length of the catenary, and the symmetric bilinear forms

$$a_3(w, v) := \int_0^L T_3 w'v'\,dx, \quad a_4(w, v) := \int_0^L (T_4 w'v' + E_4 I_4 w''v'')\,dx.$$

Calculating the derivatives b'_i, $i = 1, 2, 3$, as given in (4.46), we get the coupling terms for the Lagrange multipliers and arrive at the weak equations of motion for bodies 3 and 4

$$\langle \rho_3 A_3 \ddot{w}_3, v_3 \rangle = -\langle \beta_3 \dot{w}_3, v_3 \rangle - a_3(w_3, v_3) - \langle \rho_3 A_3 \gamma, v_3 \rangle \qquad (4.71)$$
$$- v_3(x_{p,1})\mu_1 - v_3(x_{p,2})\mu_2,$$

$$\langle \rho_4 A_4 \ddot{w}_4, v_4 \rangle = -\langle \beta_4 \dot{w}_4, v_4 \rangle - a_4(w_4, v_4) - \langle \rho_4 A_4 \gamma, v_4 \rangle \qquad (4.72)$$
$$+ v_4(x_{p,1})\mu_1 + v_4(x_{p,2})\mu_2 - v_4(x_p(t))\mu_3.$$

From a mechanical viewpoint, the discrete Lagrange multipliers $\mu_i(t) \in \mathbb{R}$, $i = 1, 2, 3$, represent the constraint or contact forces that act on both carrier and contact wire. Furthermore, the multiplier μ_3 also enters the pantograph equation (4.63), which leads to

$$m_1 \ddot{r}_1 = -d_1(\dot{r}_1 - \dot{r}_2) - c_1(r_1 - r_2) + \mu_3$$

in the presence of the contact constraint. On the other hand, (4.64) remains unaltered since body 2 is not subject to constraints.

For completeness, we mention that we assumed zero boundary conditions when introducing the bilinear forms a_3 and a_4. The test functions v_3 for the string equation are element of $H_0^1((0, L))$ while for the beam equation one requires $v_4 \in H_0^2((0, L))$.

At the end of this example, we give two further remarks. First of all, if we admit unilateral contact of the pantograph head, we have to deal with the inequality constraint

$$w_4(x_p(t), t) - r_1(t) \geq 0.$$

The contact pressure (multiplier) μ_3 now must satisfy the complementarity condition

$$(w_4(x_p(t), t) - r_1(t))\mu_3(t) = 0,$$

which means that either contact takes place or the multiplier must vanish. The rest of the mathematical model remains unaffected.

Secondly, if we try to realize the coupling of the constraint equations in strong form, which is possible in this 1D setting, we have to use δ-distributions to represent the forces acting in the discrete points $x_{p,1}, x_{p,2}$ and $x_p(t)$, see [AS00] for further details.

4.6 Summary of Key Formulas in Part I

For easy referencing while reading Part II of the book, we summarize in the following the most important mathematical models and formulas of Part I.

Chapter 2

The motion of rigid multibody systems is described by the Lagrange equations of the first kind (2.9a), (2.9b)

$$M(q)\ddot{q} = f(q,\dot{q},t) - G(q)^T \lambda,$$
$$0 = g(q)$$

with coordinate vector $q(t) \in \mathbb{R}^{n_q}$, Lagrange multiplier $\lambda(t) \in \mathbb{R}^{n_\lambda}$, mass matrix $M(q) \in \mathbb{R}^{n_q \times n_q}$, force vector $f(q,\dot{q},t) \in \mathbb{R}^{n_q}$, holonomic constraints $g(q) \in \mathbb{R}^{n_\lambda}$, and constraint Jacobian $G(q) = \partial g(q)/\partial q \in \mathbb{R}^{n_\lambda \times n_q}$.

For the mathematical analysis, the main regularity assumption (2.70) reads

$$\begin{pmatrix} M(q) & G(q)^T \\ G(q) & 0 \end{pmatrix} \quad \text{is invertible.}$$

A necessary condition for (2.70) is the full rank of the constraint Jacobian, which can be expressed by the minimax characterization (2.74)

$$\sigma_{\min}(G) = \min_{\lambda} \max_{v} \frac{\lambda^T G v}{\|v\|_2 \|\lambda\|_2} > 0.$$

Chapter 3

In the setting of linear elasticity, the weak form or principle of virtual work (3.19) is given by

$$\int_\Omega \rho v^T \ddot{u}\, dx + \int_\Omega \sigma(u):\varepsilon(v)\, dx = \int_\Omega v^T \beta\, dx + \int_{\Gamma_1} v^T \tau\, ds$$

with displacement field $u(x,t) \in \mathbb{R}^3$ and test functions $v(x) \in \mathbb{R}^3$. Furthermore, the strain tensor reads $\varepsilon(u) = \frac{1}{2}(\nabla u + \nabla u^T)$, and the stress tensor is $\sigma(u) = \Lambda_1(\text{trace}\, \varepsilon(u))I + 2\Lambda_2 \varepsilon(u)$.

Using an abstract notation, we can formulate the corresponding variational problem (3.31): At each instant of time t, find a displacement field $u(\cdot,t) \in \mathcal{V}$ such that $u(\cdot,t) = u_0(\cdot,t)$ on Γ_0 and

$$\langle \rho \ddot{u}, v \rangle + a(u,v) = \langle l, v \rangle \qquad \forall v \in \mathcal{V}_0$$

where $\mathcal{V} = H^1(\Omega)^3$ and $\mathcal{V}_0 = \{v \in \mathcal{V}: v = 0 \text{ on } \Gamma_0\}$.

4.6 Summary of Key Formulas in Part I

The equations of constrained dynamic elasticity (3.42a), (3.42b) follow as

$$\langle \rho \ddot{u}, v \rangle + a(u, v) + b(v, \mu) = \langle l, v \rangle \quad \forall v \in \mathcal{V},$$
$$b(u, \vartheta) \qquad\qquad = \langle m, \vartheta \rangle \quad \forall \vartheta \in \mathcal{Q} := (H^{1/2}(\Gamma_0)^3)'$$

with unknown multiplier $\mu(\cdot, t) \in \mathcal{Q}$. The inf-sup condition (3.43) requires the existence of $\beta \in \mathbb{R}$ such that

$$\inf_{\vartheta \in \mathcal{Q}} \sup_{v \in \mathcal{V}} \frac{b(v, \vartheta)}{\|v\|_{\mathcal{V}} \|\vartheta\|_{\mathcal{Q}}} \geq \beta > 0.$$

Chapter 4

The weak form or principle of virtual work (4.14) for the deformation $\varphi(x, t) = r(t) + A(\alpha(t))(x + u(x, t))$ reads

$$\int_\Omega \rho \eta^T \ddot{\varphi} \, dx + \int_\Omega \sigma(u) : \varepsilon(v) \, dx = \int_\Omega \eta^T A\beta \, dx + \int_{\Gamma_1} \eta^T A\tau \, ds,$$

for all admissible $\eta = z - A(\alpha)(\tilde{x} + \tilde{u})\gamma + A(\alpha)v$.
In abstract notation, we write this in (4.25) as

$$\langle \rho \ddot{\varphi}, \eta \rangle + a(\varphi, \eta) = \langle l(\varphi), \eta \rangle \quad \forall \eta \text{ admissible due to (4.13).}$$

The corresponding transient saddle point formulation (4.29a), (4.29b) is

$$\langle \rho \ddot{\varphi}, \eta \rangle + a(\varphi, \eta) + b(\eta, \mu) = \langle l(\varphi), \eta \rangle \quad \forall \eta \text{ from (4.27),}$$
$$b(\varphi, \vartheta) \qquad\qquad = \langle m, \vartheta \rangle \quad \forall \vartheta \in \mathcal{Q}.$$

The unified differential-algebraic methodology (4.47a), (4.47b) for setting up the equations of motion of a flexible multibody system yields

$$\sum_{i=1}^{n_b} \left(\langle \rho_i \ddot{\varphi}_i, \eta_i \rangle + a_i(\varphi_i, \eta_i) \right) + \sum_{(k,l) \in \mathcal{J}} b'_{kl}(\eta_k, \eta_l, \mu_{kl}) = \ell_{\text{mbs}},$$

$$\sum_{(k,l) \in \mathcal{J}} b_{kl}(\varphi_k, \varphi_l, \vartheta_{kl}) = 0$$

with right hand side

$$\ell_{\text{mbs}} := \sum_{i=1}^{n_b} \langle l_i, \eta_i \rangle - \sum_{(i,j) \in \mathcal{K}} f_{ij} \frac{c_{ij}^T}{\zeta_{ij}} (\eta_i - \eta_j).$$

Part II
Numerical Methods

Chapter 5
Spatial Discretization

The model equations for elastic bodies developed in Chaps. 3 and 4 are now discretized with respect to the spatial variable. In case of the unconstrained equations of motion, this *Galerkin projection* is a widespread approach and usually applied with finite elements or eigenfunctions as spatial approximations. It leads in a natural way to a system of second order ODEs, which is then coupled with other components in the assembly to express the corresponding interactions.

As argued in Part I, the most general model equations include the constraints already before discretization, which raises the question how to discretize the transient saddle point models from above. In this chapter, we generalize the Galerkin projection to such models and investigate the properties of the resulting differential-algebraic system. For this purpose, we employ some of the results that have become standard for the classes of mixed and hybrid finite element methods.

Performing first the discretization in space—as opposed to starting with the time discretization—is mainly motivated by the requirements of commercial simulation software where elastic bodies are treated as *discrete substructures*. Even more, standard interfaces between finite element and multibody codes allow an easy model set-up by the finite element method followed by the export of data such as mass and stiffness matrices to the multibody formalism. This approach, however, requires a careful selection of the ansatz functions for the Galerkin projection, in particular if additional *model reduction techniques* are applied.

5.1 Finite Element Approximation of Elastic Body

This section considers, in analogy to Chap. 3, the equations of motion for a single elastic body under the presence of boundary constraints (3.42a), (3.42b) and introduces the corresponding Galerkin projection. Floating reference frames and additional rigid motion variables will be treated in the subsequent section. To get started, we review briefly the typical spatial discretization for the equations of unconstrained motion (3.31).

5.1.1 Unconstrained Equations

The Galerkin projection for the dynamic equations of motion (3.31) stems from a straightforward generalization of the stationary case [Hug87, Chap. 7]. Let $u_h(x,t)$ denote the approximation to the displacement field with the subscript h standing for the spatial grid. We decompose the approximation into

$$u_h = w_h + r, \qquad w_h|_{\Gamma_0} = \mathbf{0} \text{ and } r|_{\Gamma_0} = u_0, \qquad (5.1)$$

where the inhomogeneous part r is assumed to be known, cf. Sect. 3.3.1.

Next, let $\mathcal{V}_{0,h} \subset \mathcal{V}_0$ be a finite-dimensional subspace of $\mathcal{V}_0$. By restriction onto $\mathcal{V}_{0,h}$, the variational problem (3.31) or (3.55), respectively, is transformed into

$$\langle \rho \ddot{w}_h, v \rangle + a(w_h, v) = \langle \psi, v \rangle \qquad \forall v \in \mathcal{V}_{0,h}. \qquad (5.2)$$

We are seeking the approximation $w_h(\cdot, t) \in \mathcal{V}_{0,h}$ of the homogeneous solution part for the right hand side functional ψ, which is defined by

$$\langle \psi, v \rangle := \langle l, v \rangle - \langle \rho \ddot{r}, v \rangle - a(r, v).$$

At each point of time, w_h can be written as linear combination of basis functions $v_i \in \mathcal{V}_{0,h}$, and thus we have

$$w_h(x,t) = \sum_{i=1}^{n_d} v_i(x) d_i(t) = N_w(x) d(t), \qquad n_d = \dim \mathcal{V}_{0,h}, \qquad (5.3)$$

with unknown coefficient vector $d(t) \in \mathbb{R}^{n_d}$ and matrix $N_w(x) \in \mathbb{R}^{3 \times n_d}$.

Equations of Structural Dynamics

Upon insertion of (5.3) into the variational equation (5.2) and application of the test functions, we obtain the equations of *structural dynamics*

$$M_h \ddot{d}(t) + A_h d(t) = \psi_h(t). \qquad (5.4)$$

The mass matrix $M_h \in \mathbb{R}^{n_d \times n_d}$ and stiffness matrix $A_h \in \mathbb{R}^{n_d \times n_d}$ are symmetric positive definite or semi-definite, respectively, and given by

$$M_h := \Big(\langle \rho v_i, v_j \rangle \Big)_{i,j=1,\ldots,n_d}, \qquad A_h := \Big(a(v_i, v_j) \Big)_{i,j=1,\ldots,n_d}.$$

The load vector $\psi_h(t) \in \mathbb{R}^{n_d}$ defined by

$$\psi_h := \Big(\langle \psi, v_i \rangle \Big)_{i=1,\ldots,n_d}$$

5.1 Finite Element Approximation of Elastic Body

depends on body and surface loads and furthermore on the inhomogeneous solution part.

How well does w_h approximate the true solution w? The answer goes back to Strang/Fix [SF73] and yields immediately a corresponding statement for the approximation u_h of the displacement field u. We split the error into

$$w - w_h = \underbrace{w - Pw}_{=:\epsilon} + \underbrace{Pw - w_h}_{=:\zeta} = \epsilon + \zeta. \tag{5.5}$$

Here, $Pw(\cdot, t) \in \mathcal{V}_{0,h}$ is the *Ritz projection* of w onto $\mathcal{V}_{0,h}$ defined by

$$a(w - Pw, v) = 0 \quad \forall v \in \mathcal{V}_{0,h}. \tag{5.6}$$

From Céa's lemma [BS91, Chap. 2] we have for the error term ϵ the estimate

$$\|\epsilon\|_{\mathcal{V}} = \|w - Pw\|_{\mathcal{V}} \le \frac{\eta}{\alpha} \inf_{v \in \mathcal{V}_{0,h}} \|w - v\|_{\mathcal{V}} \tag{5.7}$$

with positive constants α and η characterizing the $\mathcal{V}_0$-ellipticity (3.28) and the continuity of the bilinear form a, respectively. If we consider spatial discretizations that converge in the stationary case—tacitly assuming that the exact solution w exists and is smooth enough, we have additionally the approximation property

$$\lim_{n_d \to \infty} \inf_{v \in \mathcal{V}_{0,h}} \|w - v\|_{\mathcal{V}} = 0. \tag{5.8}$$

In this way, the error term ϵ in the splitting (5.5) can be bounded.

In order to derive a bound for the error term ζ, we require the extra smoothness assumptions $\dot{w}(\cdot, t) \in \mathcal{V}_0$ and $\ddot{w}(\cdot, t) \in \mathcal{V}_0$. They imply that the projections $P\dot{w}$ and $P\ddot{w}$ are well-defined. Moreover, the bound (5.7) and the property (5.8) are then also valid for the time derivatives $\dot{\epsilon}$ and $\ddot{\epsilon}$.

The reasoning for an estimate for ζ starts by comparing the variational equation (5.2) for w_h with the one for w, which reads

$$\langle \rho \ddot{w}, v \rangle + a(w, v) = \langle \psi, v \rangle \quad \forall v \in \mathcal{V}_{0,h}.$$

We subtract both equations from each other, make use of (5.6), and arrive at

$$\langle \rho(P\ddot{w} - \ddot{w}_h), v \rangle + a(Pw - w_h, v) = \langle \rho(P\ddot{w} - \ddot{w}), v \rangle$$

for all $v \in \mathcal{V}_{0,h}$. In particular, the test function $v = P\dot{w} - \dot{w}_h = \dot{\zeta}$ is admissible and implies the differential equation

$$\langle \rho \ddot{\zeta}, \dot{\zeta} \rangle + a(\zeta, \dot{\zeta}) = \langle -\rho \ddot{\epsilon}, \dot{\zeta} \rangle \tag{5.9}$$

for the error ζ. The *total energy* of ζ at time t is defined as

$$E(\zeta, \dot{\zeta}) := \frac{1}{2}\Big(\langle \rho \dot{\zeta}, \dot{\zeta} \rangle + a(\zeta, \zeta)\Big). \tag{5.10}$$

Its square root $\sqrt{E}$ induces a norm on $\mathcal{V}_0 \times L_2(\Omega)^3$, which we write shortly as *energy norm* $\|\boldsymbol{\zeta}\|_E := \sqrt{E(\boldsymbol{\zeta}, \dot{\boldsymbol{\zeta}})}$. Based on this definition, the differential equation (5.9) yields

$$\frac{\mathrm{d}}{\mathrm{d}t} \|\boldsymbol{\zeta}\|_E^2 = \langle -\rho \ddot{\boldsymbol{\epsilon}}, \dot{\boldsymbol{\zeta}} \rangle.$$

Direct integration of this equation is feasible if we assume the mass density ρ to be constant. For the right hand side we get

$$|\langle \rho \ddot{\boldsymbol{\epsilon}}, \dot{\boldsymbol{\zeta}} \rangle| \leq \sqrt{2\rho}\, \|\ddot{\boldsymbol{\epsilon}}\|_{L_2(\Omega)^3} \|\boldsymbol{\zeta}\|_E,$$

from which we conclude

$$\frac{\mathrm{d}}{\mathrm{d}t} \|\boldsymbol{\zeta}\|_E \leq \sqrt{\rho/2}\, \|\ddot{\boldsymbol{\epsilon}}\|_{L_2(\Omega)^3}.$$

By integration it follows

$$\|\boldsymbol{\zeta}\|_E \leq \|\boldsymbol{\zeta}_0\|_E + \sqrt{\rho/2} \int_{t_0}^{t} \|\ddot{\boldsymbol{\epsilon}}\|_{L_2(\Omega)^3} \,\mathrm{d}\tau. \tag{5.11}$$

The bound for the error term $\boldsymbol{\zeta}$ depends thus on the error in the initial value $\boldsymbol{\zeta}_0 := P\boldsymbol{w}(\cdot, t_0) - \boldsymbol{w}_h(\cdot, t_0)$ and on the second time derivative $\ddot{\boldsymbol{\epsilon}}$ of the projection error. Application of the triangle inequality leads finally to the estimate

$$\|\boldsymbol{w} - \boldsymbol{w}_h\|_E = \|\boldsymbol{\epsilon} + \boldsymbol{\zeta}\|_E \leq \|\boldsymbol{\epsilon}\|_E + \|\boldsymbol{\zeta}\|_E,$$

which implies the convergence [SF73, p. 254]:

Theorem 5.1 *Let $\boldsymbol{w}(\cdot, t) \in \mathcal{V}_0$ be the solution of the weak from (3.55) with homogeneous Dirichlet boundary conditions, and let $\dot{\boldsymbol{w}}(\cdot, t), \ddot{\boldsymbol{w}}(\cdot, t) \in \mathcal{V}_0$. Suppose the subspace $\mathcal{V}_{0,h} \subset \mathcal{V}_0$ satisfies the approximation property (5.8), and the initial error $\boldsymbol{\zeta}_0 = P\boldsymbol{w}(\cdot, t_0) - \boldsymbol{w}_h(\cdot, t_0)$ fulfills $\lim_{n_d \to \infty} \|\boldsymbol{\zeta}_0\|_E = 0$.*

Then the projected solution $\boldsymbol{w}_h(\cdot, t)$ of (5.2) converges for $n_d \to \infty$ to the true solution $\boldsymbol{w}(\cdot, t)$. Furthermore, we have

$$\|\boldsymbol{w} - \boldsymbol{w}_h\|_E \leq \|\boldsymbol{\zeta}_0\|_E + \|\boldsymbol{\epsilon}\|_E + \sqrt{\rho/2} \int_{t_0}^{t} \|\ddot{\boldsymbol{\epsilon}}\|_{L_2(\Omega)^3} \,\mathrm{d}\tau. \tag{5.12}$$

The usual finite element discretizations satisfy the assumptions of Theorem 5.1, at least in case of uniform or quasi-uniform grids with gridsize h. For Lagrange elements, e.g., it can be shown that

$$\|\boldsymbol{\zeta}_0\|_{L_2(\Omega)^3} \leq c h^2 \|\dot{\boldsymbol{w}}(\cdot, t_0)\|_{\mathcal{V}}, \quad \|\ddot{\boldsymbol{\epsilon}}\|_{L_2(\Omega)^3} \leq c h^2 \|\dddot{\boldsymbol{w}}\|_{\mathcal{V}}.$$

Here, the initial velocity $\boldsymbol{w}_h(\cdot, t_0)$ is computed by L_2-projection of $\dot{\boldsymbol{w}}(\cdot, t_0)$ or by interpolation in $\mathcal{V}_{0,h}$. An analogous estimate holds for the norm of the second derivative $\|\ddot{\boldsymbol{\epsilon}}\|_{L_2(\Omega)^3}$. Treating the error terms $\boldsymbol{\epsilon}$ and $\boldsymbol{\zeta}_0$, however, in terms of the *stationary*

5.1 Finite Element Approximation of Elastic Body

energy norm, we lose one power of h,

$$\sqrt{a(\zeta_0, \zeta_0)} \leq ch\|w(\cdot, t_0)\|_V, \quad \sqrt{a(\epsilon, \epsilon)} \leq ch\|w\|_V.$$

In summary, the following error estimate can be derived for Lagrange elements, see [Hug87, p. 456]:

$$\|w - w_h\|_E \leq c_1 h \Big(\|w\|_V + \|w(\cdot, t_0)\|_V\Big)$$
$$+ c_2 h^2 \Big(\|\dot{w}\|_V + \|\dot{w}(\cdot, t_0)\|_V + \int_{t_0}^{t} \|\ddot{w}\|_V \, d\tau\Big).$$

Using a duality argument, it is furthermore possible to obtain an improved L_2-estimate, cf. [Dup73].

5.1.2 Constrained Equations of Motion

For the Galerkin projection of the constrained equations of motion (3.42a), (3.42b) we will employ arguments and discretizations that are standard in the numerical analysis of stationary saddle point problems. In a straightforward way, this will result in a differential-algebraic equation whose properties depend on a discrete analogue of the inf-sup condition (3.43). Basic references on the numerical treatment of saddle point problems are, among others, Braess [Bra01] and Brezzi/Fortin [BF91].

As above in the unconstrained case we select finite-dimensional subspaces $\mathcal{V}_h \subset \mathcal{V}$ and $\mathcal{Q}_h \subset \mathcal{Q}$ and seek approximations $u_h(\cdot, t) \in \mathcal{V}_h$ and $\mu_h(\cdot, t) \in \mathcal{Q}_h$ such that

$$\langle \rho \ddot{u}_h, v \rangle + a(u_h, v) + b(v, \mu_h) = \langle l, v \rangle \quad \forall v \in \mathcal{V}_h, \tag{5.13a}$$
$$b(u_h, \vartheta) = \langle m, \vartheta \rangle \quad \forall \vartheta \in \mathcal{Q}_h. \tag{5.13b}$$

The approximations u_h and μ_h are linear combinations of basis functions in $\mathcal{V}_h$ or $\mathcal{Q}_h$, respectively. Hence they can be written as

$$u_h(x, t) = \sum_{i=1}^{n_d} v_i(x) d_i(t) = N_u(x) d(t), \quad n_d = \dim \mathcal{V}_h; \tag{5.14}$$

$$\mu_h(x, t) = \sum_{j=1}^{n_\lambda} \vartheta_j(x) \lambda_j(t) = N_\mu(x) \lambda(t), \quad n_\lambda = \dim \mathcal{Q}_h \tag{5.15}$$

where $d(t) \in \mathbb{R}^{n_d}$ is the coefficient vector for the displacement, $\lambda(t) \in \mathbb{R}^{n_\lambda}$ is the vector of discrete Lagrange multipliers, and the matrices $N_u(x) \in \mathbb{R}^{3 \times n_d}$ and $N_\mu(x) \in \mathbb{R}^{3 \times n_\lambda}$ comprise the basis functions.

Structural Dynamics—Constrained Version

Inserting the basis functions into the saddle point formulation (5.13a), (5.13b) leads us directly to the differential-algebraic system

$$M_h \ddot{d}(t) + A_h d(t) + B_h^T \lambda(t) = l_h(t), \quad (5.16a)$$

$$B_h d(t) = m_h(t) \quad (5.16b)$$

in the unknown displacement vector d and Lagrange multiplier λ. The mass matrix $M_h \in \mathbb{R}^{n_d \times n_d}$ and the stiffness matrix $A_h \in \mathbb{R}^{n_d \times n_d}$ are, as above, symmetric positive definite and semi-definite, respectively, and given by

$$M_h := \Big(\langle \rho v_i, v_j \rangle\Big)_{i,j=1,\ldots,n_d}, \qquad A_h := \Big(a(v_i, v_j)\Big)_{i,j=1,\ldots,n_d}.$$

Additionally, the constraint matrix $B_h \in \mathbb{R}^{n_\lambda \times n_d}$ comes now into play. It combines both discretizations (or subspaces) via

$$B_h := \Big(b(v_i, \vartheta_j)\Big)_{\substack{j=1,\ldots,n_\lambda \\ i=1,\ldots,n_d}}.$$

The vectors $l_h(t) \in \mathbb{R}^{n_d}$ and $m_h(t) \in \mathbb{R}^{n_\lambda}$ on the right hand sides, finally, are defined by

$$l_h := \Big(\langle l, v_i \rangle\Big)_{i=1,\ldots,n_d} \quad \text{and} \quad m_h := \Big(\langle m, \vartheta_j \rangle\Big)_{j=1,\ldots,n_\lambda}.$$

Index of Semi-Discrete Equations

Obviously, the semi-discrete equations of motion (5.16a), (5.16b) possess the same structure as the equations of constrained mechanical motion (2.9a), (2.9b) discussed in Chap. 2 for rigid bodies. The same arguments apply in turn, and the index is found to be 3 if the full rank criterion (2.2) for the constraint matrix B_h holds.

We now resume the discussion about the minimax characterization (2.74), which was seen to be equivalent to the full rank criterion. Going back to Raviart [Rav74], the *compatibility* for the discretizations of the displacement field and the Lagrange multiplier is expressed in terms of the condition

$$\text{if there exists } \vartheta \in \mathcal{Q}_h \text{ with } b(v, \vartheta) = 0 \quad \forall v \in \mathcal{V}_h \quad \Rightarrow \quad \vartheta = 0. \quad (5.17)$$

This condition is necessary and sufficient for the full rank of B_h, as is easily verified by considering $b(v, \vartheta) = \bar{\vartheta}^T B_h \bar{v}$ for $v = N_u \bar{v} \in \mathcal{V}_h$, $\vartheta = N_\mu \bar{\vartheta} \in \mathcal{Q}_h$.

Theorem 5.2 *Assume that the subspaces $\mathcal{V}_h \subset \mathcal{V}$ and $\mathcal{Q}_h \subset \mathcal{Q}$ satisfy the compatibility condition (5.17). Then the differential-algebraic system (5.16a), (5.16b) resulting from the Galerkin projection (5.13a), (5.13b) is of index 3.*

5.1 Finite Element Approximation of Elastic Body

Theorem 5.2 implies immediately the existence and uniqueness of the semi-discrete solution $(\boldsymbol{d}, \boldsymbol{\lambda})$, if the right hand sides are sufficiently smooth and the initial values consistent, cf. Sect. 2.4. However, this reasoning does not include any statements about the approximation properties of $(\boldsymbol{u}_h, \boldsymbol{\mu}_h)$ with respect to the true solution $(\boldsymbol{u}, \boldsymbol{\mu})$. We will come back to this point in a moment.

The condition (5.17) is equivalent to [Rav74]

$$\sup_{v \in \mathcal{V}_h} \frac{b(v, \vartheta)}{\|v\|_{\mathcal{V}}} \geq \beta_h \|\vartheta\|_{\mathcal{Q}} \quad \forall \vartheta \in \mathcal{Q}_h, \quad \beta_h > 0. \tag{5.18}$$

From the numerical analysis of mixed and hybrid finite element methods it is well-known that (5.18) is prone for stability problems due to the dependence of the constant β_h on the gridsize [Bra01]. Therefore, one requires a stronger condition, the *discrete inf-sup condition*, which postulates that the *family of finite element spaces* $\mathcal{V}_h, \mathcal{Q}_h$ possesses a grid-independent constant $\beta > 0$ such that [BF91]

$$\inf_{\vartheta \in \mathcal{Q}_h} \sup_{v \in \mathcal{V}_h} \frac{b(v, \vartheta)}{\|\vartheta\|_{\mathcal{Q}} \|v\|_{\mathcal{V}}} \geq \beta. \tag{5.19}$$

One could also say that (5.19) guarantees a well-balanced combination of the subspaces $\mathcal{V}_h$ and $\mathcal{Q}_h$. Since the condition (5.19) includes the compatibility criterion (5.17), it also implies the full rank of the constraint matrix $\boldsymbol{B}_h$.

Convergence Analysis

The semi-discrete equations of motion (5.16a), (5.16b) are obviously expected to approximate the true solution of the transient saddle point problem (3.42a), (3.42b), and thus the question arises whether the convergence analysis outlined above for the unconstrained case can be generalized appropriately. To this end, we require, besides the discrete inf-sup condition (5.19), that the bilinear form a is $\mathcal{W}_h$-elliptic, i.e.,

$$a(v, v) \geq \alpha \|v\|_{\mathcal{V}}^2 \quad \forall v \in \mathcal{W}_h := \{w \in \mathcal{V}_h : b(w, \vartheta) = 0 \;\forall \vartheta \in \mathcal{Q}_h\}. \tag{5.20}$$

This property of a is essential in the stationary case. Notice that $\mathcal{W}_h \not\subset \mathcal{V}_0$ in general, which means that a non-conformal discretization is admitted.

For the analysis of the instationary case, let thus (5.19) be satisfied and let the bilinear form a be $\mathcal{W}_h$-elliptic according to (5.20). These prerequisites allow a convergence proof similar to the approach of Collino et al. [CJM97], who consider the wave equation in combination with zero boundary conditions and the *fictitious domain method*.

We first define the Ritz projection $\Pi(\boldsymbol{u}(\cdot, t), \boldsymbol{\mu}(\cdot, t)) \in \mathcal{V}_h \times \mathcal{Q}_h$ of the exact solution. With the splitting $\Pi(\boldsymbol{u}, \boldsymbol{\mu}) = (\Pi_u \boldsymbol{u}, \Pi_\mu \boldsymbol{\mu})$, this projection is given as the unique solution of the stationary saddle point problem [BF91]

$$a(\boldsymbol{u} - \Pi_u \boldsymbol{u}, v) + b(v, \boldsymbol{\mu} - \Pi_\mu \boldsymbol{\mu}) = 0 \quad \forall v \in \mathcal{V}_h, \tag{5.21a}$$

$$b(\boldsymbol{u} - \Pi_u \boldsymbol{u}, \vartheta) \qquad\qquad\qquad\qquad = 0 \quad \forall \vartheta \in \mathcal{Q}_h. \tag{5.21b}$$

Furthermore, we have the estimate

$$\|u - \Pi_u u\|_\mathcal{V} + \|\mu - \Pi_\mu \mu\|_\mathcal{Q} \le c \left(\inf_{v \in \mathcal{V}_h} \|u - v\|_\mathcal{V} + \inf_{\vartheta \in \mathcal{Q}_h} \|\mu - \vartheta\|_\mathcal{Q} \right), \quad (5.22)$$

which is the analogue of Cea's lemma (5.7) in the saddle point framework.

The discretizations for the displacement and the Lagrange multiplier are assumed to satisfy

$$\lim_{n_d \to \infty} \inf_{v \in \mathcal{V}_h} \|u - v\|_\mathcal{V} = 0, \quad \lim_{n_\lambda \to \infty} \inf_{\vartheta \in \mathcal{Q}_h} \|\mu - \vartheta\|_\mathcal{Q} = 0. \quad (5.23)$$

This basic approximation property implies with (5.22) the convergence of the Ritz projection (5.21a), (5.21b). Like in the unconstrained case, additional smoothness of the time derivatives of the exact solution is finally required,

$$(\dot{u}(\cdot, t), \dot{\mu}(\cdot, t)), \ (\ddot{u}(\cdot, t), \ddot{\mu}(\cdot, t)) \in \mathcal{V} \times \mathcal{Q}. \quad (5.24)$$

In this way, the projections $\Pi(\dot{u}, \dot{\mu})$ and $\Pi(\ddot{u}, \ddot{\mu})$ are well-defined by means of (5.21a), (5.21b) and possess the same approximation property as $\Pi(u, \mu)$.

After this formal set-up, we start the analysis with the error splitting

$$u - u_h = u - \Pi_u u + \Pi_u u - u_h =: \epsilon + \zeta, \quad (5.25)$$

$$\mu - \mu_h = \mu - \Pi_\mu \mu + \Pi_\mu \mu - \mu_h =: \theta + \gamma. \quad (5.26)$$

For the error terms ϵ, θ and the corresponding time derivatives, the estimate (5.22) and the limit (5.23) yield already a bound. We thus need to derive estimates for the terms ζ and γ.

The exact solution satisfies

$$\langle \rho \ddot{u}, v \rangle + a(u, v) + b(v, \mu) = \langle l, v \rangle \quad \forall v \in \mathcal{V}_h,$$

$$b(u, \vartheta) \hspace{5em} = \langle m, \vartheta \rangle \quad \forall \vartheta \in \mathcal{Q}_h.$$

By computing the difference with the semi-discrete system (5.13a), (5.13b), one gets

$$\langle \rho(\ddot{\epsilon} + \ddot{\zeta}), v \rangle + a(\epsilon + \zeta, v) + b(v, \theta + \gamma) = 0 \quad \forall v \in \mathcal{V}_h,$$

$$b(\epsilon + \zeta, \vartheta) \hspace{6em} = 0 \quad \forall \vartheta \in \mathcal{Q}_h.$$

The Ritz projection (5.21a), (5.21b) immediately yields $a(\epsilon, v) + b(v, \theta) = 0$ and $b(\epsilon, \vartheta) = 0$. Hence we obtain a saddle point problem for the error ζ,

$$\langle \rho \ddot{\zeta}, v \rangle + a(\zeta, v) + b(v, \gamma) = \langle -\rho \ddot{\epsilon}, v \rangle \quad \forall v \in \mathcal{V}_h, \quad (5.27a)$$

$$b(\zeta, \vartheta) = 0 \quad \forall \vartheta \in \mathcal{Q}_h. \quad (5.27b)$$

5.1 Finite Element Approximation of Elastic Body

Using the smoothness assumption (5.24), we differentiate the constraints (5.27b) with respect to time and arrive at

$$b(\dot{\zeta}, \vartheta) = 0 \quad \text{and} \quad b(\ddot{\zeta}, \vartheta) = 0 \quad \forall \vartheta \in \mathcal{Q}_h.$$

The last term on the left hand side of (5.27a) consequently cancels upon insertion of $v = \dot{\zeta}$, and by using the definition (5.10) of the energy norm $\|\cdot\|_E$ we deduce from (5.27a) the relation

$$\frac{d}{dt} \|\zeta\|_E^2 = \langle -\rho \ddot{\epsilon}, \dot{\zeta} \rangle.$$

An estimate analogously to the unconstrained case yields, cf. (5.11),

$$\|\zeta\|_E \leq \|\zeta_0\|_E + \sqrt{\rho/2} \int_{t_0}^{t} \|\ddot{\epsilon}\|_{L_2(\Omega)^3} \, d\tau. \tag{5.28}$$

We have thus found the desired bound for the error term ζ of the displacement field. In particular, the right hand side of (5.28) is a bound both for the L_2-norm of the velocity $\|\dot{\zeta}\|_{L_2(\Omega)^3}$ and the H^1-norm of the displacement $\|\zeta\|_V$.

Estimate for γ

The error term γ of the Lagrange multiplier μ requires a treatment similar to the analysis presented in Sect. 3.3. From (5.27a) it follows

$$|b(v, \gamma)| \leq |\langle \rho(\ddot{\epsilon} + \ddot{\zeta}), v \rangle| + |a(\zeta, v)|$$

$$\leq \|\rho(\ddot{\epsilon} + \ddot{\zeta})\|_{L_2(\Omega)^3} \|v\|_{L_2(\Omega)^3} + \eta \|\zeta\|_V \|v\|_V$$

$$\leq \left(\|\rho(\ddot{\epsilon} + \ddot{\zeta})\|_{L_2(\Omega)^3} + \eta \|\zeta\|_V \right) \|v\|_V$$

with the continuity constant η of the bilinear form a. Hence the discrete inf-sup condition (5.19) yields due to $\gamma = \Pi_\mu \mu - \mu_h \in \mathcal{Q}_h$ the estimate

$$\|\gamma\|_\mathcal{Q} \leq \frac{1}{\beta} \left(\eta \|\zeta\|_V + \|\rho(\ddot{\epsilon} + \ddot{\zeta})\|_{L_2(\Omega)^3} \right). \tag{5.29}$$

While we have already provided bounds for the contributions of ζ and $\ddot{\epsilon}$ on the right hand side, the term $\ddot{\zeta}$ is still troublesome. Though (5.27a) gives for $v = \ddot{\zeta}$ the relation

$$\|\ddot{\zeta}\|_{L_2(\Omega)^3}^2 \leq \|\ddot{\epsilon}\|_{L_2(\Omega)^3} \|\ddot{\zeta}\|_{L_2(\Omega)^3} + \eta/\rho \, \|\zeta\|_V \|\ddot{\zeta}\|_V,$$

the norms $\|\ddot{\zeta}\|_{L_2(\Omega)^3}$ and $\|\ddot{\zeta}\|_V$ are not equivalent and thus we may not cancel them.

In order to obtain, analogously to (5.28), an estimate for $\|\dot{\zeta}\|_E$ and accordingly for $\|\ddot{\zeta}\|_{L_2(\Omega)^3}$, we have to introduce the additional smoothness assumption [CJM97]

$$d^3/dt^3 \left(u(\cdot, t), \mu(\cdot, t) \right) \in \mathcal{V} \times \mathcal{Q}. \tag{5.30}$$

In this way, the third time derivative $\epsilon^{(3)}$ can be bounded by (5.22), and (5.23) and (5.27a) may be differentiated with respect to time,

$$\langle \rho \zeta^{(3)}, v\rangle + a(\dot{\zeta}, v) + b(v, \dot{\gamma}) = \langle -\rho \epsilon^{(3)}, v\rangle \qquad \forall v \in \mathcal{V}_h.$$

We set $v = \ddot{\zeta}$ to get once again $b(\ddot{\zeta}, \dot{\gamma}) = 0$, which yields

$$\frac{\mathrm{d}}{\mathrm{d}t} \|\dot{\zeta}\|_E^2 = \langle -\rho \epsilon^{(3)}, \ddot{\zeta}\rangle.$$

This implies, similarly to above,

$$\|\dot{\zeta}\|_E \leq \|\dot{\zeta}_0\|_E + \sqrt{\rho/2} \int_{t_0}^{t} \|\epsilon^{(3)}\|_{L_2(\Omega)^3} \,\mathrm{d}\tau.$$

Due to

$$\|\zeta\|_V \leq \frac{1}{\sqrt{\alpha}} \sqrt{a(\zeta, \zeta)} \leq \sqrt{2/\alpha} \|\zeta\|_E, \qquad \|\rho \ddot{\zeta}\|_{L_2(\Omega)^3} \leq \sqrt{2\rho} \|\dot{\zeta}\|_E,$$

we can transform (5.29) further into

$$\|\gamma\|_Q \leq \frac{\eta}{\beta} \sqrt{2/\alpha} \left(\|\zeta_0\|_E + \sqrt{\rho/2} \int_{t_0}^{t} \|\ddot{\epsilon}\|_{L_2(\Omega)^3} \,\mathrm{d}\tau \right) + \frac{\rho}{\beta} \|\ddot{\epsilon}\|_{L_2(\Omega)^3}$$

$$+ \frac{1}{\beta} \sqrt{2\rho} \left(\|\dot{\zeta}_0\|_E + \sqrt{\rho/2} \int_{t_0}^{t} \|\epsilon^{(3)}\|_{L_2(\Omega)^3} \,\mathrm{d}\tau \right). \tag{5.31}$$

Summarizing, we have shown the following result on the convergence of the semi-discrete system (5.13a), (5.13b).

Theorem 5.3 *Let (u, μ) be the solution of the time-dependent saddle point problem (3.42a), (3.42b) with the additional smoothness assumption (5.24). For the space discretization $\mathcal{V}_h \subset \mathcal{V}$ and $\mathcal{Q}_h \subset \mathcal{Q}$ let the discrete inf-sup condition (5.19) hold and let the bilinear form a be $\mathcal{W}_h$-elliptic. Furthermore, assume the approximation property (5.23) and additionally $\lim_{n_d, n_\lambda \to \infty} \|\zeta_0\|_E = 0$ for the initial error $\zeta_0 = \Pi_u u(\cdot, t_0) - u_h(\cdot, t_0)$.*

Then the displacement field $u_h(\cdot, t)$ of the semi-discrete system (5.13a), (5.13b) converges for $n_d \to \infty$ and $n_\lambda \to \infty$ to $u(\cdot, t)$. It holds the estimate

$$\|u - u_h\|_E \leq \|\zeta_0\|_E + \|\epsilon\|_E + \sqrt{\rho/2} \int_{t_0}^{t} \|\dot{\epsilon}\|_{L_2(\Omega)^3} \,\mathrm{d}\tau. \tag{5.32}$$

If the additional assumption (5.30) is satisfied and $\lim_{n_d, n_\lambda \to \infty} \|\dot{\zeta}_0\|_E = 0$, also the Lagrange multiplier μ_h converges to μ, and we have the estimate

$$\|\mu - \mu_h\|_Q \leq \|\theta\|_Q + \|\gamma\|_Q \tag{5.33}$$

where $\|\gamma\|_Q$ is bounded by (5.31).

5.1 Finite Element Approximation of Elastic Body

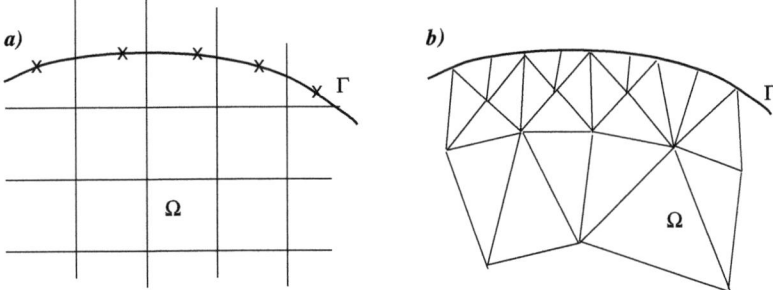

Fig. 5.1 Discretization of boundary Γ_0

As the factor $1/\beta$ is hidden in the constant c from (5.22), the estimate (5.31) for the error in the Lagrange multiplier features actually a factor of $1/\beta^2$. In this respect, we have here an analogy to the results of Theorem 3.4 on the influence of perturbations.

Discretization of Weak Constraints

There are several approaches for the discretization of the constrained equations (3.42a), (3.42b). Let us first assume that the discretized domain Ω_h resolves the boundary segment Γ_0 poorly, Fig. 5.1(a).

As introduced by Babuška [Bab73], the boundary can then be discretized by a second grid, which forms the basis for the subspace Q_h. Clearly, both subspaces V_h and Q_h should be well-balanced and satisfy the inf-sup condition (5.19). In essence, this means a requirement $h_\mu \geq c h_u$ for the gridsizes h_μ on the boundary and h_u in the interior, see, e.g., Pitkaranta [Pit79] and Dahmen/Kunoth [DK01] for details.

The discretization of Dirichlet boundary conditions formulated as weak constraints is thus feasible even for a coarse underlying grid of V_h.

The second approach plays a role in certain domain decomposition methods [Dor89] and is also connected to contact mechanics [KO88]. Assume that Γ_0 is well-resolved by the grid or triangulation for V_h, Fig. 5.1(b). Thus, the grid points on the boundary can be used for constructing discretizations of the constraint equations.

As example, we consider linear or quadratic Lagrange elements for V_h and the trace of these on the boundary as natural function space for Q_h. To analyze this approach in more detail, we write $\boldsymbol{u}_h = \boldsymbol{N}_u \boldsymbol{d}$ and $\boldsymbol{\mu}_h = \boldsymbol{N}_\mu \boldsymbol{\lambda}$ to express the weak constraints as

$$\int_{\Gamma_0} \boldsymbol{N}_\mu^T(x(s))\boldsymbol{N}_u(x(s))\mathrm{d}s\, \boldsymbol{d}(t) = \int_{\Gamma_0} \boldsymbol{N}_\mu^T(x(s))\boldsymbol{u}_0(x(s),t)\mathrm{d}s.$$

If $\boldsymbol{N}_\mu$ is defined as the trace of $\boldsymbol{N}_u$, it follows $\boldsymbol{N}_\mu(x(s)) = \boldsymbol{N}_u(x(s))$ on Γ_0, and solely those basis functions contribute to the integral that do not vanish on Γ_0. Let

$d_{\Gamma_0}(t) \in \mathbb{R}^{n_\lambda}$ denote the corresponding nodal variables. We rewrite the left hand side as

$$\int_{\Gamma_0} N_u^T(x(s)) N_u(x(s)) \, ds \, d(t) = M_{\Gamma_0} d_{\Gamma_0}(t),$$

with a positive definite mass matrix $M_{\Gamma_0} \in \mathbb{R}^{n_\lambda \times n_\lambda}$. Assume that the integral on the right hand side is computed using an interpolant of u_0, which is solely evaluated in the given discrete grid points. The resulting function values u_{Γ_0} and the functions N_u define an approximation

$$u_0(x(s), t) \doteq N_u(x(s)) u_{\Gamma_0}(t).$$

The right hand side of the constraints is hence transformed into

$$\int_{\Gamma_0} N_u^T(x(s)) N_u(x(s)) \, ds \, u_{\Gamma_0}(t) = M_{\Gamma_0} u_{\Gamma_0}(t),$$

and one concludes that $d_{\Gamma_0} = u_{\Gamma_0}$: The displacements coincide with the prescribed boundary conditions in the grid points.

A further consequence of using the trace for constructing $\mathcal{Q}_h$ is the formal equivalence of the discretized equations in the constrained and unconstrained case. Let (u_h, μ_h) be a solution of the constrained problem (5.13a), (5.13b). Choose

$$r|_{\Gamma_0} = N_u(x(s)) u_{\Gamma_0}(t)$$

as inhomogeneous solution part for the unconstrained problem, which means that r equals the interpolant of the boundary motion u_0 and hence also equals u_h. The difference $w_h := u_h - r$ from (5.13a) satisfies

$$\langle \rho \ddot{w}_h, v \rangle + a(w_h, v) + b(v, \mu_h) = \langle l, v \rangle - \langle \rho \ddot{r}, v \rangle - a(r, v)$$

for all $v \in \mathcal{V}_h$. In particular for $v \in \mathcal{V}_{0,h} = \{v \in \mathcal{V}_h : v|_{\Gamma_0} = 0\}$, the term $b(v, \mu_h)$ vanishes on the left hand side, and we obtain the unconstrained problem (5.2) for the homogeneous solution part w_h.

In view of the foregoing discussion, it might appear that the effort spent on the problem formulation with weak constraints and Lagrange multipliers is not worthwhile. As argued in Part I, however, the multibody applications lead to situations where the boundary motion u_0 is unknown a priori, and this precludes the transformation to the unconstrained model.

On the other hand, in most technical applications a fine discretization of the constraints is not required. Even more, certain interfaces between rigid and elastic bodies are qualified for *smearing*. This holds especially for the class of reduced models with few elastic degrees of freedom, which clearly demand for a coarse discretization of the interface.

5.1 Finite Element Approximation of Elastic Body

Fig. 5.2 Example of a rigid body element

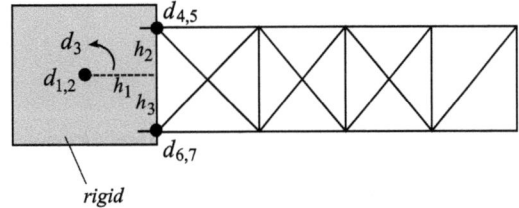

Rigid Body Elements

We now take up the discussion of Sect. 4.2.2 on the modeling of joints. The introduction of weak constraints and the corresponding discretization (5.16a), (5.16b) represents one way to avoid point constraints acting on the elastic body. Another approach are *rigid body elements* that may perform solely a rigid body motion and can be used for attaching joints. We study an example, which is inspired by [CK06, Ex. 14.6], to illustrate this technique.

Figure 5.2 displays a planar component that is modeled under the assumption of plane stress and discretized by triangular elements. At the left end, a rigid plate is connected to the structure, and its center point acts as a distinguished node in the finite element mesh. Three coordinates are assigned to the plate,

d_1, d_2 : horizontal/vertical displacement center node, d_3 : rotation.

Two nodes of the finite element mesh with displacements d_4, d_5 and d_6, d_7 are connected to the right boundary of the plate, which is expressed as

$$\boldsymbol{A}(d_3)\begin{pmatrix} h_1 \\ h_2 \end{pmatrix} + \begin{pmatrix} d_1 \\ d_2 \end{pmatrix} = \begin{pmatrix} d_4 \\ d_5 \end{pmatrix} + \begin{pmatrix} h_1 \\ h_2 \end{pmatrix}, \tag{5.34a}$$

$$\boldsymbol{A}(d_3)\begin{pmatrix} h_1 \\ -h_3 \end{pmatrix} + \begin{pmatrix} d_1 \\ d_2 \end{pmatrix} = \begin{pmatrix} d_6 \\ d_7 \end{pmatrix} + \begin{pmatrix} h_1 \\ -h_3 \end{pmatrix}, \tag{5.34b}$$

with h_1, h_2, h_3 denoting the distances between the nodes in the undeformed state as indicated in Fig. 5.2. In linear elasticity theory, only a small angle d_3 is admitted, and thus we may linearize the rotation matrix $\boldsymbol{A}$ to obtain

$$\boldsymbol{A}(d_3) = \begin{pmatrix} \cos d_3 & -\sin d_3 \\ \sin d_3 & \cos d_3 \end{pmatrix} \doteq \begin{pmatrix} 1 & -d_3 \\ d_3 & 1 \end{pmatrix}.$$

The interconnection conditions (5.34a), (5.34b) hence lead to the following set of linear constraint equations:

$$\begin{aligned} -h_2 d_3 + d_1 - d_4 &= 0, \\ h_1 d_3 + d_2 - d_5 &= 0, \\ h_3 d_3 + d_1 - d_6 &= 0, \\ h_1 d_3 + d_2 - d_7 &= 0. \end{aligned}$$

These equations can be written as

$$B_h d = 0 \qquad (5.35)$$

and obviously, they fit into the general framework developed so far. Note that the center node with displacements d_1, d_2 is an appropriate candidate for acting as, e.g., a pivot in a revolute joint. The resulting forces are then distributed via the rigid body element to the elastic component.

The conclusion from this example is that rigid body elements yield linear constraints of the form (5.35) that can either be treated explicitly in terms of the differential-algebraic equation (5.16a), (5.16b) or eliminated in the finite element assembly. The latter is common in many finite element packages to reduce the number of degrees of freedom [CK06].

5.1.3 Remarks

(i) In case of a conservative system, it is straightforward to show that the energy conservation property is preserved by the Galerkin projection, i.e., the semi-discrete equations (5.16a), (5.16b) also conserve the total energy.
(ii) As mentioned in Sect. 3.1.2, dissipation is important in practical applications. Rayleigh damping as introduced in (3.20) leads to a damping matrix $D_h := \zeta_1 A_h + \zeta_2 M_h$ with real parameters ζ_1 and ζ_2 for frequency-dependent and viscous damping. We then obtain an extra term $D_h \dot{d}$ in the semi-discrete equations, which in the constrained case results in

$$M_h \ddot{d}(t) + D_h \dot{d}(t) + A_h d(t) + B_h^T \lambda(t) = l_h(t), \qquad (5.36a)$$
$$B_h d(t) \phantom{+ D_h \dot{d}(t) + A_h d(t) + B_h^T \lambda(t)} = m_h(t). \qquad (5.36b)$$

The index and the convergence properties from above directly carry over to this model extension.
(iii) Nonlinear stiffness terms such as the geometric stiffening of beams discussed in Sect. 4.4.2 are not covered by the analysis presented above. A nonlinear strain tensor in combination with a hyperelastic material law (3.7) is typical for such models, compare (4.59) in case of a planar beam. In the weak form, this leads to a modified term $a_s(u, v)$ for the variation of the strain energy $W_{nl}(u_h)$ instead of $a(u, v)$. Notice that a_s is not a bilinear form but nonlinear with respect to the displacement field u, as illustrated in (4.60). For the Galerkin projection $u \mapsto u_h = N_u d$, one has to replace the linear stiffness term $A_h d$ by the gradient

$$\nabla W_h(d) := \frac{\partial}{\partial d} W_{nl}(N_u d) \qquad (5.37)$$

of the discretized strain energy.

5.2 Spatial Discretization of Flexible Multibody System

In Chap. 4, we introduced the method of floating reference frames to model elastic members in a multibody system. This approach is now combined with the Galerkin projection for the displacement field. We obtain thus a finite-dimensional description that represents a substructure for later integration into the multibody formalisms.

We solely address here the constrained equations of motion (4.26a), (4.26b) or (4.29a), (4.29b), respectively, if the abstract notation is preferred. The unconstrained case is discussed in great detail in [SW99, Sha98].

5.2.1 Galerkin Projection for Floating Reference Frame

Taking the rigid body motion as in (4.1) into account, we substitute the displacement field by its Galerkin approximation $\boldsymbol{u}_h(x,t) = \boldsymbol{N}_u(x)\boldsymbol{d}(t)$ *with respect to the floating or body-fixed reference frame*. Each material point of the deformed body is then located by the discretized deformation vector

$$\boldsymbol{\varphi}_h(x,t) := \boldsymbol{r}(t) + \boldsymbol{A}(\boldsymbol{\alpha}(t))(x + \boldsymbol{u}_h(x,t)). \tag{5.38}$$

Admissible variations or test functions $\boldsymbol{\eta}$ have the same structure as in (4.27), i.e.,

$$\boldsymbol{\eta} = \boldsymbol{z} - \boldsymbol{A}(\boldsymbol{\alpha})(\tilde{x} + \tilde{\boldsymbol{u}}_h)\boldsymbol{\gamma} + \boldsymbol{A}(\boldsymbol{\alpha})\boldsymbol{v} \tag{5.39}$$

with $\boldsymbol{v} \in \mathcal{V}_h$ being a test function of the Galerkin projection and $\boldsymbol{z}$ and $\boldsymbol{\gamma}$ as variations of translation $\boldsymbol{r}$ and rotation $\boldsymbol{\alpha}$, respectively.

Upon insertion of (5.38) into the equations of motion (4.29a), (4.29b), we obtain the semi-discrete equations

$$\langle \rho \ddot{\boldsymbol{\varphi}}_h, \boldsymbol{\eta} \rangle + a(\boldsymbol{\varphi}_h, \boldsymbol{\eta}) + b(\boldsymbol{\eta}, \boldsymbol{\mu}_h) = \langle \boldsymbol{l}, \boldsymbol{\eta} \rangle \quad \forall \boldsymbol{\eta} \text{ from (5.39)}, \tag{5.40a}$$

$$b(\boldsymbol{\varphi}_h, \boldsymbol{\vartheta}) \qquad\qquad = \langle \boldsymbol{m}, \boldsymbol{\vartheta} \rangle \quad \forall \boldsymbol{\vartheta} \in \mathcal{Q}_h. \tag{5.40b}$$

While the discretization of the constraint equation (5.40b) with corresponding multiplier $\boldsymbol{\mu}_h(x,t) = \boldsymbol{N}_\mu(x)\boldsymbol{\lambda}(t)$ will be discussed in more detail below, we concentrate for the moment on carrying out the integrations with respect to the spatial variable x in (5.40a). The resulting equations of motion are displayed in Table 5.1.

Due to the splitting into rigid motion and elastic variables, the inertia term $\langle \rho \ddot{\boldsymbol{\varphi}}_h, \boldsymbol{\eta} \rangle$ has a rather complex structure and leads to a number of so-called *coupling integrals* $\boldsymbol{C}_1, \ldots, \boldsymbol{C}_5, \boldsymbol{s}_h, \boldsymbol{J}_h$, which express mutual coupling effects. A closer look at the corresponding expressions reveals that these integrals can be *evaluated a priori for a given discretization* $\mathcal{V}_h$. The main idea behind this is an adept extraction of the angular velocity.

For example, consider the last of the angular momentum terms from the right hand side of (4.17). It is transformed into

Table 5.1 Equations of motion based on the method of floating reference frames and Galerkin projection $u \mapsto u_h = N_u d$, $\mu \mapsto \mu_h = N_\mu \lambda$

Momentum equations

$$m\ddot{r} + A(\tilde{s}_r + \tilde{s}_h(d))^T \dot{\omega} + A C_1^T \ddot{d} =$$
$$-A\tilde{\omega}\tilde{\omega}(s_r + s_h(d)) - 2A\tilde{\omega}C_1^T \dot{d} + \mathbf{f}_T(\alpha, t) - \mathbf{G}_T^T \lambda$$

Angular momentum equations

$$(\tilde{s}_r + \tilde{s}_h(d)) A^T \ddot{r} + (J_r + J_h(d))\dot{\omega} + (C_2 + C_3(d))^T \ddot{d} =$$
$$-\tilde{\omega}(J_r + J_h(d))\omega - 2C_4^T(\dot{d})\omega - 2C_5^T(d, \dot{d})\omega + f_R(d, t) - G_R(\alpha, d)^T \lambda$$

Elastic deformation

$$C_1 A^T \ddot{r} + (C_2 + C_3(d))\dot{\omega} + M_h \ddot{d} =$$
$$\hat{C}_4(\omega)\omega + \hat{C}_5(\omega, d)\omega - 2C_3(\dot{d})\omega - A_h d + l_h(t) - G_h(\alpha)^T \lambda$$

Constraints $\quad G_T r + g_R(\alpha) + g_h(\alpha, d) = m_h(t)$

Coupling integrals $s_r := \int \rho x \, dx$, $s_h(d) := \int \rho N_u \, dx \, d$, $J_r := -\int \rho \tilde{x} \tilde{x} \, dx$,

$J_h(d) := -\sum_{i=1}^{n_d} d_i \int \rho(\tilde{x}\tilde{N}_{u,i} + \tilde{N}_{u,i}\tilde{x}) \, dx - \sum_{i,j=1}^{n_d} d_i d_j \int \rho \tilde{N}_{u,i} \tilde{N}_{u,j} \, dx$,

$C_1 := \int \rho N_u^T \, dx$, $C_2 := \int \rho N_u^T \tilde{x}^T \, dx$, $C_3(d) := \sum_{i=1}^{n_d} d_i \int \rho N_u^T \tilde{N}_{u,i} \, dx$,

$C_4(\dot{d}) := \sum_{i=1}^{n_d} \dot{d}_i \int \rho \tilde{N}_{u,i} \tilde{x}^T \, dx$, $\hat{C}_4(\omega) := \left(\omega^T \int \rho \tilde{N}_{u,i} \tilde{x}^T \, dx \right)_{i=1}^{n_d}$,

$C_5(d, \dot{d}) := \sum_{i,j=1}^{n_d} d_i \dot{d}_j \int \rho \tilde{N}_{u,j}^T \tilde{N}_{u,i} \, dx$,

$\hat{C}_5(\omega, d) := \left(\omega^T \sum_{j=1}^{n_d} d_j \int \rho \tilde{N}_{u,j}^T \tilde{N}_{u,i} \, dx \right)_{i=1}^{n_d}$;

Forces $\mathbf{f}_T(\alpha, t) := A \int \beta \, dx + A \int_{\Gamma_1} \tau \, ds$,

$f_R(d, t) := \int (\tilde{x} + \widetilde{N_u d}) \beta \, dx + \int_{\Gamma_1} (\tilde{x} + \widetilde{N_u d}) \tau \, ds$.

$$\int_\Omega 2\rho(\tilde{x} + \tilde{u})\tilde{\omega}\dot{u} \, dx \doteq \int_\Omega 2\rho(\tilde{x} + \tilde{u}_h)\tilde{\omega}\dot{u}_h \, dx$$
$$= \int_\Omega 2\rho(\tilde{x} + (\widetilde{N_u d}))(\widetilde{N_u \dot{d}})^T \, dx \, \omega$$
$$= 2 \sum_{i=1}^{n_d} \dot{d}_i \int \rho \tilde{x} \tilde{N}_{u,i}^T \, dx \, \omega$$
$$+ 2 \sum_{i,j=1}^{n_d} d_i \dot{d}_j \int \rho \tilde{N}_{u,j} \tilde{N}_{u,i}^T \, dx \, \omega$$
$$=: 2 C_4^T(\dot{d})\omega + 2 C_5^T(d, \dot{d})\omega.$$

5.2 Spatial Discretization of Flexible Multibody System

Here, the matrix $N_u = (N_{u,1}, \ldots, N_{u,n_d})$ is written column-wise and its entries define the coupling integrals C_4 and C_5. Note that integrals like $\int \rho \tilde{x} \tilde{N}_{u,i}^T \, dx$ do not depend on any of the state variables. Even more, since most finite element codes are able to evaluate such expressions for a given discretization, the coupling integrals do not involve any additional computational effort during a transient simulation. This observation holds also for the unconstrained model; in fact, the coupling integrals are in both cases the same.

Discretization of Constraints

The spatial discretization of the constraint equations (5.40b) is treated in a way similar to the coupling integrals above. For the right hand side of (5.40b) we obtain the contribution $m_h(t)$, compare (5.16b). Inserting $\vartheta = N_\mu \bar{\vartheta}$ and performing the variation for all $\bar{\vartheta} \in \mathbb{R}^{n_\lambda}$ yields for the left hand side

$$\int_{\Gamma_0} N_\mu^T \Big(r + A(x + N_u d) \Big) ds =: G_T r + g_R(\alpha) + g_h(\alpha, d) \tag{5.41}$$

where

$$G_T := \int_{\Gamma_0} N_\mu^T \, ds, \qquad g_R(\alpha) := \int_{\Gamma_0} N_\mu^T A x \, ds = \sum_{j=1}^{3} \int_{\Gamma_0} N_\mu^T x_j \, ds \, A_j$$

and

$$g_h(\alpha, d) := \int_{\Gamma_0} N_\mu^T A N_u d \, ds = \sum_{i=1}^{n_d} \left(\sum_{j=1}^{3} \int_{\Gamma_0} N_\mu^T N_{u,ij} \, ds \, A_j \right) d_i.$$

As above, the arising integrals can be evaluated in advance. All terms in g_R involving the rotation are combined with a columnwise evaluation of the direction cosine matrix $A = (A_1, A_2, A_3)$, and the elastic motion in g_h is split into the components $N_u = (N_{u,ij})$. Noteworthy, in case of basis functions with small support, very few basis functions contribute to g_h.

The derivatives of g_R and g_h, finally, are required for the Jacobian matrices

$$G_R(\alpha, d) := \frac{\partial}{\partial \alpha} \Big(g_R(\alpha) + g_h(\alpha, d) \Big), \qquad G_h(\alpha) := \frac{\partial}{\partial d} g_h(\alpha, d).$$

It holds

$$G_R(\alpha, d) = \int_{\Gamma_0} N_\mu^T A(\alpha) (\tilde{x} + \widetilde{N_u d})^T \, ds, \qquad G_h(\alpha) = \int_{\Gamma_0} N_\mu^T A(\alpha) N_u \, ds.$$

This completes the Galerkin projection of the equations of constrained motion for a floating reference frame.

In summary, the spatial discretization of a single body results in a differential-algebraic system of the form

$$M(\alpha, d) \begin{pmatrix} \ddot{r} \\ \dot{\omega} \\ \ddot{d} \end{pmatrix} = f(r, \alpha, \omega, d, \dot{d}, t) - G(\alpha, d)^T \lambda, \qquad (5.42a)$$

$$0 = G_T r + g_R(\alpha) + g_h(\alpha, d) - m_h(t). \qquad (5.42b)$$

This DAE is of index 3 if the constraint matrix

$$G(\alpha, d) := \Big(G_T, \ G_R(\alpha, d) S(\alpha), \ G_h(\alpha) \Big) \in \mathbb{R}^{n_\lambda \times (6 + n_d)}$$

with transformation matrix $S(\alpha)$ from (2.26) is of full rank n_λ. We may check this condition by inspecting the matrix B_h of the linear system (5.16a), (5.16b). Assume for this purpose that G is rank-deficient. Then there exists a vector $0 \neq (z, \gamma, \bar{v}) \in \mathbb{R}^{6+n_d}$ with

$$\bar{\vartheta}^T \Big(G_T, G_R, G_h \Big) \begin{pmatrix} z \\ \gamma \\ \bar{v} \end{pmatrix} = \bar{\vartheta}^T \int_{\Gamma_0} N_\mu^T \Big(z - A(\tilde{x} + \tilde{u}_h) \gamma + A v \Big) \, \mathrm{d}x = 0$$

for all $\bar{\vartheta} \in \mathbb{R}^{n_\lambda}$. If $v \in \mathcal{V}_h$ is not a variation of translation or rotation, it follows

$$\eta = z - A(\tilde{x} + \tilde{u}_h) \gamma + A v \neq 0.$$

Consequently, there exists $\eta \neq 0$ such that $b(\eta, \vartheta) = 0$ for all $\vartheta = N_\mu \bar{\vartheta} \in \mathcal{Q}_h$. After transformation into the body-fixed reference frame, this is equivalent to $b(v_\eta, \vartheta) = 0$ for $v_\eta \in \mathcal{V}_h$ different from zero. Therefore, B_h from (5.17) is also rank-deficient. The reverse reasoning yields the full rank of G when supposing that B_h is of full rank.

Special Cases

In Chap. 4, we have already discussed certain situations where the equations of motion simplify. This discussion is now taken up again. We start with the point constraint (4.30), which does not require a discrete space $\mathcal{Q}_h$. Applying the discrete variation (5.38) to (4.30) yields

$$r + A(\alpha)(x_p + N_u(x_p) d) = \varphi_0.$$

The evaluation of the basis functions obviously takes only place in the point x_p. If this point happens to be a node with corresponding nodal variable d_p, it additionally holds $N_u(x_p) d = d_p$.

For vanishing elastic deformation, i.e., $d \equiv 0$, in conjunction with a point constraint, the equations of motion (5.42a), (5.42b) coincide with the Newton–Euler formulation (4.31).

5.2 Spatial Discretization of Flexible Multibody System

Furthermore, we note that the discretization of the equations of motion simplifies considerably in the planar case. As introduced in (4.23), the floating reference frame then yields the representation

$$\varphi_h(x,t) = r(t) + A(\alpha(t))(x + u_h(x,t)), \quad A(\alpha) = \begin{pmatrix} \cos\alpha & -\sin\alpha \\ \sin\alpha & \cos\alpha \end{pmatrix}, \quad (5.43)$$

with translation $r(t) \in \mathbb{R}^2$, scalar angle $\alpha(t)$, rotation matrix $A(\alpha) \in SO(2)$, and discrete displacement field $u_h(\cdot,t) \in \mathcal{V}_h \subset H^1(\Omega)^2$. An admissible test function has the form

$$\eta = z + A'(\alpha)(x + u_h)\gamma + A(\alpha)v \quad (5.44)$$

where $A' = \partial A/\partial \alpha$ and $v \in \mathcal{V}_h \subset H^1(\Omega)^2$. The equations of planar motion follow by inserting (5.43) and (5.44) into the weak form (5.40a), (5.40b), with an added trivial zero component. Integration with respect to the third coordinate x_3 results in a common constant factor that should not be cancelled since it enters mass and inertia parameters. Table 5.2 displays the full set of discretized equations in detail and contains also a summary of the arising coupling integrals. The modeling and discretization of the elastic pendulum in the introductory Chap. 1 can be subsumed thereunder by omitting the translation and the constraint.

Like the Newton–Euler equations (2.25) and (4.31), the DAE (5.42a), (5.42b) is not a second order system in time due to the angular velocity ω. Referring to the reasoning in Sect. 2.2.2, we mention Euler parameters or the usage of relative coordinates as alternative in order to end up with a true second order structure. For ease of presentation, we build on the latter formulation in the following.

5.2.2 Flexible Multibody System

We next extend the spatial discretization to flexible multibody systems and provide a methodology for generating the corresponding equations of motion.

The starting point is the weak formulation (4.47a), (4.47b) derived in Sect. 4.3. As above for single bodies with a floating reference frame, the obvious idea is now to simultaneously apply the Galerkin projection to each elastic member of the multibody system. Assuming first absolute coordinates, each elastic body i can be treated individually based on its deformation

$$\varphi_{i,h}(x,t) := r_i(t) + A(\alpha_i(t))(x_i + u_{i,h}(x,t)) \quad (5.45)$$

and variation

$$\eta_i = z_i - A(\alpha_i)(\tilde{x}_i + \tilde{u}_{i,h})\gamma_i + A(\alpha_i)v_i. \quad (5.46)$$

Table 5.2 Planar equations of motion with floating reference frame

Momentum equations
$m\ddot{r} + A'(s_r + s_h(d))\ddot{\alpha} + AC_1^T\ddot{d} = A(s_r + s_h(d))\dot{\alpha}^2 - 2A'C_1^T\dot{\alpha}\dot{d} + f_T(\alpha, t) - G_T^T\lambda$
Angular momentum equation
$(s_r + s_h(d))^T A'^T \ddot{r} + (J_r + J_h(d))\ddot{\alpha} + (C_2 + C_3 d)^T \ddot{d} = -2(C_4 + M_h d)^T \dot{\alpha}\dot{d}$
$\qquad\qquad\qquad\qquad\qquad\qquad\qquad\qquad\qquad\qquad\qquad + f_R(d, t) - G_R^T(\alpha, d)\lambda$
Elastic deformation
$C_1 A^T \ddot{r} + (C_2 + C_3 d)\ddot{\alpha} + M_h \ddot{d} = (C_4 + M_h d)\dot{\alpha}^2 - 2C_3 \dot{\alpha}\dot{d} - A_h d + l_h - G_h^T(\alpha)\lambda$
Constraint $\quad G_T r + g_R(\alpha) + g_h(\alpha, d) = m_h(t),$
where $\;G_T := \int_{\Gamma_0} N_\mu^T \, ds, \quad g_R(\alpha) := \int_{\Gamma_0} N_\mu^T \begin{pmatrix} x_1 & -x_2 \\ x_2 & x_1 \end{pmatrix} ds \begin{pmatrix} \cos\alpha \\ \sin\alpha \end{pmatrix}$
and $\;g_h(\alpha, d) := \int_{\Gamma_0} N_\mu^T \begin{pmatrix} N_{u,1} & -N_{u,2} \\ N_{u,2} & N_{u,1} \end{pmatrix} ds \begin{pmatrix} d\cos\alpha \\ d\sin\alpha \end{pmatrix}.$
Coupling integrals $\quad s_r := \int \rho x \, dx, \; s_h(d) := \int \rho N_u \, dx \, d, \; J_r := \int \rho x^T x \, dx,$
$C_1 := \int \rho N_u^T \, dx, \; C_2 := \int \rho(x_1 N_{u,2} - x_2 N_{u,1}) \, dx,$
$C_3 := \int \rho(N_{u,2}^T N_{u,1} - N_{u,1}^T N_{u,2}) \, dx, \; C_4 := \int \rho(x_1 N_{u,1} + x_2 N_{u,2}) \, dx,$
$J_h(d) := 2C_4 d + d^T M_h d,$ where $N_u^T = (N_{u,1}^T, N_{u,2}^T).$
Forces $\;f_T(\alpha, t) := A \int \beta \, dx + A \int_{\Gamma_1} \tau \, ds,$
$f_R(d, t) := \int (x + N_u d)^T \hat{I} \beta \, dx + \int_{\Gamma_1} (x + N_u d)^T \hat{I} \tau \, ds, \quad \hat{I} := \begin{pmatrix} 0 & 1 \\ -1 & 0 \end{pmatrix}.$

The projection of the equations of motion (4.50a), (4.50b) yields the semi-discrete equations

$$\langle \rho_i \ddot{\varphi}_{i,h}, \eta_i \rangle + a_i(\varphi_{i,h}, \eta_i) + \sum_{(i,l) \in \mathcal{J}} b'_{il}(\eta_i, \cdot, \mu_{il,h}) = \ell_{\text{mbs},i}, \qquad (5.47a)$$

$$\sum_{(i,l) \in \mathcal{J}} b_{il}(\varphi_{i,h}, \varphi_{l,h}, \vartheta_{il}) = 0 \qquad (5.47b)$$

where $\mathcal{J}$ comprises the indices of all bodies connected to body i via joints and $l_{\text{mbs},i}$ the force couplings. Choices for the discrete multipliers $\mu_{il,h}$ will be discussed in the examples below.

As seen for the case of absolute coordinates, the structure of the equations of motion does not change under the Galerkin projection, and thus it is straightforward to state the most general model (4.47a), (4.47b) in semi-discrete form by simply

5.2 Spatial Discretization of Flexible Multibody System

replacing the motions $\boldsymbol{\varphi}_i$ by their discrete counterparts $\boldsymbol{\varphi}_{i,h}$:

$$\sum_{i=1}^{n_b} \left(\langle \rho_i \ddot{\boldsymbol{\varphi}}_{i,h}, \boldsymbol{\eta}_i \rangle + a_i(\boldsymbol{\varphi}_{i,h}, \boldsymbol{\eta}_i) \right) + \sum_{(k,l) \in \mathcal{J}} b'_{kl}(\boldsymbol{\eta}_k, \boldsymbol{\eta}_l, \boldsymbol{\mu}_{kl,h}) = \ell_{\text{mbs}}, \quad (5.48\text{a})$$

$$\sum_{(k,l) \in \mathcal{J}} b_{kl}(\boldsymbol{\varphi}_{k,h}, \boldsymbol{\varphi}_{l,h}, \boldsymbol{\vartheta}_{kl}) = 0, \quad (5.48\text{b})$$

with right hand side

$$\ell_{\text{mbs}} = \sum_{i=1}^{n_b} \langle \boldsymbol{l}_i, \boldsymbol{\eta}_i \rangle - \sum_{(i,j) \in \mathcal{K}} f_{ij} \frac{\boldsymbol{c}_{ij}^T}{\zeta_{ij}} (\boldsymbol{\eta}_i - \boldsymbol{\eta}_j). \quad (5.49)$$

In contrast to (5.47a), (5.47b), the choice of coordinates is arbitrary in (5.48a), (5.48b) and hidden in the definition of $\boldsymbol{\varphi}_{i,h}$, which may depend on some kind of generalized coordinates $\boldsymbol{y}$. The corresponding variations satisfy

$$\boldsymbol{\eta}_i = \frac{d}{d\theta} \boldsymbol{\varphi}'_{i,h}(\boldsymbol{y} + \theta \boldsymbol{w}) \Big|_{\theta=0}. \quad (5.50)$$

For rigid bodies, we have $\boldsymbol{\varphi}_{i,h} = \boldsymbol{\varphi}_i$, and the equations for the elastic deformation cancel.

Instead of further dwelling into the details of (5.48a), (5.48b), we take now a more general viewpoint and point out that all integrations and loops necessary to evaluate (5.48a), (5.48b) can be realized in a computer code, cf. [ZV98]. Geometry, coordinate system, and the coupling integrals in case of an elastic body are the main data that are required. The examples at the end of this chapter will provide more insight into this process.

In order to summarize the equations of motion for future usage and also in order to keep the notation concise, we introduce the following variables:

$\boldsymbol{q}(t) \in \mathbb{R}^{n_q}$: vector of all rigid (or gross) motion variables,
$\boldsymbol{d}(t) \in \mathbb{R}^{n_d}$: vector of all elastic variables,
$\boldsymbol{\lambda}(t) \in \mathbb{R}^{n_\lambda}$: vector of all Lagrange multipliers.

While the variables $\boldsymbol{q}$ comprise translations and rotations of both rigid and elastic bodies, the vector $\boldsymbol{d}$ is an assembly of all unknowns resulting from the spatial discretization, and the discrete Lagrange multipliers due to the bonds in the system form the vector $\boldsymbol{\lambda}$.

After these preparations, the outcome of the weak form (5.48a), (5.48b) can be summarized as a *partitioned differential-algebraic system*

$$\boldsymbol{M}(\boldsymbol{q}, \boldsymbol{d}) \begin{pmatrix} \ddot{\boldsymbol{q}} \\ \ddot{\boldsymbol{d}} \end{pmatrix} = \boldsymbol{f}(\boldsymbol{q}, \boldsymbol{d}, \dot{\boldsymbol{q}}, \dot{\boldsymbol{d}}, t) - \boldsymbol{G}(\boldsymbol{q}, \boldsymbol{d})^T \boldsymbol{\lambda}, \quad (5.51\text{a})$$

$$0 = \boldsymbol{g}(\boldsymbol{q}, \boldsymbol{d}). \quad (5.51\text{b})$$

Here, the mass matrix $M(q,d) \in \mathbb{R}^{(n_q+n_d) \times (n_q+n_d)}$ and the force vector $f(q,d,\dot{q},\dot{d},t) \in \mathbb{R}^{n_q+n_d}$ are arranged in a blockwise fashion corresponding to the unknowns and consist of the terms and integrals listed in Table 5.1. The Jacobian matrix $G := \partial g / \partial (q,d)$ contains the derivatives of the constraints with respect to both rigid motion and elastic deformation. Clearly, the equations of motion (5.51a), (5.51b) possess the same structure as the multibody model (2.9a), (2.9b) for rigid bodies, and even more, the equations of structural dynamics (5.16a), (5.16b) and the single body (5.42a), (5.42b) are also included as special cases. The latter, however, requires an extra treatment of the angular velocity ω.

Full Rank Condition

The transition from a single body under constraints to the complete multibody system does usually not affect the index of the differential-algebraic system. Since the mass matrix may become singular for certain choices of rigid motion coordinates, we proceed as in the condition (2.70) and postulate that

$$\begin{pmatrix} M(q,d) & G(q,d)^T \\ G(q,d) & 0 \end{pmatrix} \text{ is invertible.} \qquad (5.52)$$

This condition, along with sufficient smoothness, guarantees that (5.51a), (5.51b) is of index 3. Existence and uniqueness of the solution (q,d,λ), finally, follow in the same way as for rigid multibody systems, cf. Sect. 2.4.

5.2.3 Model Reduction

Though the discussion so far has mainly been associated with a finite element discretization of the elastic components, the methodology is actually not restricted to this framework and admits arbitrary approaches for approximating the deformation. Eigenfunctions are particularly attractive in combination with beam models and in widespread use since the number of unknowns can be substantially reduced by such an adept discretization. As soon as bodies with complex geometries arise, however, finite elements become the method of choice. The drawback of finite elements in the multibody context lies in a strong increase in the number of unknowns, and hence appropriate *model reduction techniques* are of great importance.

In contrast to the field of structural mechanics, multibody dynamics analysis is usually not valued for a high accuracy of local displacements or stresses. What practitioners are more interested in is high performance of the overall simulation, and for this goal it often suffices to take only a few elastic deformation modes into account. There are various examples where such a model reduction yields enormous savings in computing time with only little or no loss in accuracy. Nevertheless, engineering judgment is essential in model reduction, and so far there is a lack of automatic procedures with guaranteed error bounds.

5.2 Spatial Discretization of Flexible Multibody System

As a side remark we point out that model reduction is also related to the concept of a *stiff mechanical system*, see the next chapter.

In the following, we give a short survey of some standard model reduction techniques as discussed in Craig and Kurdila [CK06] and Gasch and Knothe [GK89]. The basic model is an unconstrained and discretized elastic body

$$M_h \ddot{d}(t) + A_h d(t) = l_h(t). \tag{5.53}$$

Compared to the linear DAE (5.16a), (5.16b), both constraints due to joints as well as force couplings are not explicitly taken into account. From an abstract point of view, model reduction is nothing else than a second Galerkin projection of the displacement field

$$u \underset{\text{FEM}}{\longmapsto} u_h = N_u d \underset{\text{condensation}}{\longmapsto} u_H = N_u T d_c \tag{5.54}$$

that results in *condensated* discrete displacements d_c of dimension $n_c \ll n_d$. Choices for the transformation matrix $T \in \mathbb{R}^{n_d \times n_c}$ will be discussed below.

Replacing d by $T d_c$ and premultiplying the equations of motion (5.53) by T^T yields the reduced system

$$T^T M_h T \ddot{d}_c + T^T A_h T d_c = T^T l_h \tag{5.55}$$

where the presence of an additional damping matrix D_h is easily included by means of the transformed matrix $T^T D_h T$.

Static Condensation

This method, also called *Guyan reduction* [Guy65], is based on partitioning the nodal variables into *master degrees of freedom* d_M and *slave degrees of freedom* d_S. Master nodes are, e.g., those nodes where forces are transmitted or where joints are attached. A second criterion for the division into master and slave variables compares the size of the diagonal entries of the mass and the stiffness matrix. It reads

$$\frac{A_h(i,i)}{M_h(i,i)} \leq c_M \quad \text{for all master nodes } i \tag{5.56}$$

with a threshold $c_M > 0$. In this way, all master nodes possess small local stiffness and large local mass. The slave nodes are then characterized by small local mass and large local stiffness and consequently, these mass terms are considered as negligible.

The rearrangement of the unknowns and equations in (5.53) according to the partitioning into master and slave nodes results in

$$\begin{pmatrix} M_{MM} & M_{MS} \\ M_{SM} & M_{SS} \end{pmatrix} \begin{pmatrix} \ddot{d}_M \\ \ddot{d}_S \end{pmatrix} + \begin{pmatrix} A_{MM} & A_{MS} \\ A_{SM} & A_{SS} \end{pmatrix} \begin{pmatrix} d_M \\ d_S \end{pmatrix} = \begin{pmatrix} l_M \\ l_S \end{pmatrix}. \tag{5.57}$$

At this point, a crucial assumption needs to be made. It states that master and slave nodes are related to each other via the static equation

$$A_{SM} d_M + A_{SS} d_S = 0. \tag{5.58}$$

In other words, in the second line of (5.57) the acceleration term and the right hand side are neglected, which is motivated by the criterion (5.56). The assumption (5.58) represents an additional *inner constraint* and implies the transformation formula

$$\begin{pmatrix} d_M \\ d_S \end{pmatrix} = T d_M, \qquad T := \begin{pmatrix} I_M \\ -A_{SS}^{-1} A_{SM} \end{pmatrix}, \tag{5.59}$$

for reducing the system to the master nodes d_M that play the role of the condensated degrees of freedom d_c in (5.54).

In summary, static condensation changes local finite element basis functions to functions with larger support. The savings with respect to the dimension rarely amount to more than 50 %, and band structures in the matrices are usually destroyed. In practice, the block matrix $A_{SS}^{-1} A_{SM}$ for the transformation (5.59) is computed by solving linear systems of the form $A_{SS} d_{S,i} = -A_{SM} e_i$ with unit vectors e_i of dimension $n_M = n_c$. The latter represent prescribed shapes of the master degrees of freedom and can be viewed as constraints for the discrete structure. The resulting column vectors $(e_i, d_{S,i})$ of T are the deformed shapes of the full structure and determine the subspace for the second Galerkin projection in (5.54).

Modal Condensation

Setting $d(t) = v \sin \omega t$ with a vector $v \in \mathbb{R}^{n_d}$ and considering the homogeneous case, we obtain from (5.53) the *generalized eigenvalue problem*

$$\omega^2 M_h v = A_h v. \tag{5.60}$$

The unknown eigenfrequencies ω along with the corresponding eigenvectors or *eigenmodes* v can be computed by standard eigenvalue solvers [GL96]. Of particular interest in this process are the lowest frequencies $\omega_1, \omega_2, \ldots, \omega_{n_c}$ with $n_c \ll n_d$ and the corresponding slow eigenmodes. Arranging the eigenmodes columnwise, we define the transformation matrix

$$T := (v_1, \ldots, v_{n_c}). \tag{5.61}$$

Due to the orthogonality of the eigenmodes, the transformed system (5.55) can be expressed in such a way that it possesses a diagonal structure,

$$T^T M_h T = \mathrm{diag}\,(1, \ldots, 1), \qquad T^T A_h T = \mathrm{diag}\,(\omega_1^2, \ldots, \omega_{n_c}^2).$$

As above for static condensation, the adept selection of eigenmodes and eigenfrequencies is crucial for the properties of the reduced system. Since the resulting basis functions have global support, locally acting forces and joints are often resolved

5.2 Spatial Discretization of Flexible Multibody System

poorly. On the other hand, the savings with respect to the number of unknowns easily exceed 90 %.

Mixed Static and Modal Condensation

Combining static and modal condensation takes local effects into account and still offers high savings. In the literature, this approach is also referred to as the *Craig–Bampton method* [CB68]. The technique is based on the same partitioning as in (5.57), with the master variables comprising all those nodes where forces from interconnection elements act. Next, the master variables are set to zero, $d_M = 0$, which means that the structure is kept fixed at the corresponding nodes. This yields

$$M_{SS}\ddot{d}_S + A_{SS}d_S = l_S$$

for the slave variables. The latter system is then condensated with respect to its eigenmodes v_S, which constitute the intermediate transformation matrix T_S. Finally, the full transformation matrix is assembled from the blocks

$$T := \begin{pmatrix} I_M & 0 \\ -A_{SS}^{-1}A_{SM} & T_S \end{pmatrix}. \tag{5.62}$$

This choice means that the slave variables are approximated by a linear combination

$$d_S = -A_{SS}^{-1}A_{SM}d_M + T_S\hat{d}_S,$$

and the reduced set of variables $d_c = (d_M, \hat{d}_S)$ consists of master variables d_M and a few modes $\hat{d}_S$ of the slave subsystem.

CMS Methods and Further Remarks

All three reduction methods discussed here are based on a transformation $d = T d_c$ that replaces the nodal variables by condensated coordinates d_c. In general, this transformation is applied in a preprocessing step before the equations of motion of the multibody system are generated, and thus the gain from a lower system dimension is immediately exploited.

There are various further approaches for the selection of deformed shapes of the finite element structure for the second Galerkin projection (5.54), and the common term in the literature for this procedure is *component-mode synthesis* (*CMS*) methods. While the Craig–Bampton transformation (5.62) is a so-called fixed-interface method where the master degrees of freedom are set to zero, there are also CMS methods that rely on free-interface modes. Since rigid body modes might still be present, in particular in these latter methods, they need to be eliminated before combining the flexible structure with a floating reference frame. Typically, an eigenvalue analysis (5.60) with the already reduced matrices is performed for this purpose, and the eigenvectors that correspond to the zero frequencies are then removed.

We stress that the condensation techniques discussed so far require a skilled user who has a good knowledge of the dynamics of the elastic body and its interaction with the other bodies. The equations of motion summarized in Tables 5.1 and 5.2, respectively, remain valid if one of the above condensations is performed since they still fit into the framework of the Galerkin projection. Note that the constraints are also affected by the model reduction, but their dimension remains unchanged because of $\boldsymbol{B}_h \boldsymbol{d} = \boldsymbol{B}_h \boldsymbol{T} \boldsymbol{d}_c$. Obviously, it makes no sense to represent an elastic body by very few, say less than 10, elastic modes and at the same time to formulate joints and their local interaction with the elastic body in full detail. This calls for a problem-specific and careful approach.

In recent years, alternative model reduction techniques, e.g., based on Krylov subspace methods, have become a promising alternative to the class of CMS methods. We refer to [Ant05, FE11, RS08] for further work in this field.

As a last remark, we turn briefly our attention to geometric stiffening terms, which render the equations of motion (5.53) nonlinear. It is still common place to extend the above linear model reduction techniques to this situation. By linearizing the dynamic equations and performing a corresponding condensation, new coordinates can be determined for the Galerkin projection of the nonlinear system [Sac96]. A promising alternative, which requires also less knowledge of the underlying dynamics, makes use of nonlinear model reduction techniques. The most prominent method in this field is the proper orthogonal decomposition (POD), see [Her08] for an application in multibody dynamics.

5.3 Examples

The examples of Sect. 4.5 are now continued in order to illustrate the spatial discretization process.

5.3.1 Slider Crank with Elastic Connecting Rod

The slider crank example provides already much insight into typical effects by carefully selecting a few elastic degrees of freedom. At the same time, the structure of the equations of motion remains transparent [Sim98a].

We start with the discretization of the connecting rod, which is modelled as an Euler–Bernoulli beam with longitudinal displacement w_1 and lateral displacement w_2 of the neutral fiber, cf. Sect. 4.5.1. Due to the boundary conditions, the eigenfunctions of w_2 are sinusoidal, and therefore the ansatz

$$w_2(\xi, t) \doteq \sin(\pi \xi / l_2) d_1(t) + \sin(2\pi \xi / l_2) d_2(t)$$

5.3 Examples

with l_2 as nominal distance between both pivots of the connecting rod represents a reasonable approximation by the first two eigenfunctions. Note that $\xi = \xi_1$ denotes the first coordinate with respect to the body-fixed reference frame. The coefficient d_1 stands for the displacement in the middle of the connecting rod at $\xi = l_2/2$.

The longitudinal displacement is discretized by a quadratic polynomial,

$$w_1(\xi,t) \doteq \frac{\xi^2}{l_2^2}(-4d_3(t) + 2d_4(t)) + \frac{\xi}{l_2}(4d_3(t) - d_4(t))$$

where the nodal variables d_3 and d_4 are the displacements in the middle and at the right end.

As a side effect, this discretization of the beam specifies a floating reference frame based on the secant defined via the boundary conditions $w_1(0,t) = 0$, $w_2(0,t) = 0$ and $w_2(l_2,t) = 0$.

Inserting the approximations for w_1 and w_2, applying the corresponding test functions, and carrying out the integrations leads in a straightforward way to the equations of motion of the elastic body. The mass and stiffness matrices read

$$M_h = \rho A l_2 \begin{pmatrix} 1/2 & 0 & 0 & 0 \\ 0 & 1/2 & 0 & 0 \\ 0 & 0 & 8/15 & 1/15 \\ 0 & 0 & 1/15 & 2/15 \end{pmatrix},$$

$$A_h = E l_2 \begin{pmatrix} 1/2 \mathrm{I} \pi^4 / l_2^2 & 0 & 0 & 0 \\ 0 & 8\mathrm{I}\pi^4/l_2^2 & 0 & 0 \\ 0 & 0 & 16/3 \cdot A & -8/3 \cdot A \\ 0 & 0 & -8/3 \cdot A & 7/3 \cdot A \end{pmatrix}.$$

If the nonlinear stiffness term from (4.59) is also taken into account, an extra term arises from the elastic potential $W_h(d)$ besides the product $A_h d$. The total discrete elastic forces then read

$$\nabla W_h(d) = A_h d + 1/2\pi^2 \, \mathrm{EA}/l_2^2 \begin{pmatrix} d_1 d_4 - \beta d_2(-4d_3 + 2d_4) \\ 4d_2 d_4 - \beta d_1(-4d_3 + 2d_4) \\ 4\beta d_1 d_2 \\ 1/2 d_1^2 + 2 d_2^2 - 2\beta d_1 d_2 \end{pmatrix} \quad (5.63)$$

where $\beta := 80/(9\pi^2)$. If the nonlinear term is omitted, the blockdiagonal structure of M_h and A_h indicates that longitudinal and lateral displacements do not interact.

For the coupling of rigid and elastic motion, the integrals of Table 5.2 are required. They can be computed a priori for a given spatial discretization:

$$C_1 = \rho A l_2 \begin{pmatrix} 0 & 0 & 2/3 & 1/6 \\ 2/\pi & 0 & 0 & 0 \end{pmatrix}^T,$$

$$C_2 = \rho A l_2^2 (1/\pi, -1/(2\pi), 0, 0)^T, \quad C_4 = \rho A l_2^2 (0, 0, 1/3, 1/6)^T,$$

$$C_3 = \rho A l_2 \begin{pmatrix} 0 & 0 & -16/\pi^3 & 8/\pi^3 - 1/\pi \\ 0 & 0 & 0 & 1/(2\pi) \\ 16/\pi^3 & 0 & 0 & 0 \\ 1/\pi - 8/\pi^3 & -1/(2\pi) & 0 & 0 \end{pmatrix}^T.$$

For completeness, we next state the equations of motion of this flexible multi-body system as partitioned DAE (5.51a), (5.51b), compare the rigid body model in Sect. 2.2.1.

The rigid body motion is expressed in terms of the variables

$$q := \begin{pmatrix} \alpha_1 \\ \alpha_2 \\ r_3 \end{pmatrix} \quad \begin{array}{l} \text{crank angle,} \\ \text{connecting rod angle,} \\ \text{displacement of sliding block;} \end{array}$$

while the elastic deformation of the connecting rod is described by

$$d := \begin{pmatrix} d_1 \\ d_2 \\ d_3 \\ d_4 \end{pmatrix} \quad \begin{array}{l} \text{lateral displacement, mode } \sin(\pi x/l_2), \\ \text{lateral displacement, mode } \sin(2\pi x/l_2), \\ \text{longitudinal displacement midpoint,} \\ \text{longitudinal displacement right endpoint.} \end{array}$$

The mass matrix in (5.51a), (5.51b) has the block structure

$$M(q,d) = \begin{pmatrix} M_r(q) + M_e(q,d) & C(q,d)^T \\ C(q,d) & M_h \end{pmatrix}$$

with rigid motion mass matrix

$$M_r(q) = \begin{pmatrix} J_1 + m_2 l_1^2 & 1/2\, l_1 l_2 m_2 \cos(\alpha_1 - \alpha_2) & 0 \\ 1/2\, l_1 l_2 m_2 \cos(\alpha_1 - \alpha_2) & J_2 & 0 \\ 0 & 0 & m_3 \end{pmatrix}$$

and elastic motion mass matrix M_h from above. Additionally, the floating reference frame yields the following extra blocks:

$$M_e(q,d) = \begin{pmatrix} 0 & l_1 \begin{pmatrix} \cos(\alpha_1 - \alpha_2) \\ \sin(\alpha_1 - \alpha_2) \end{pmatrix}^T C_1^T d & 0 \\ l_1 \begin{pmatrix} \cos(\alpha_1 - \alpha_2) \\ \sin(\alpha_1 - \alpha_2) \\ 0 \end{pmatrix}^T C_1^T d & d^T M_h d + 2 C_4^T d & 0 \\ & 0 & 0 \end{pmatrix}$$

and

$$C(q,d) = \left(l_1 C_1 \begin{pmatrix} -\sin(\alpha_1 - \alpha_2) \\ \cos(\alpha_1 - \alpha_2) \end{pmatrix},\ C_2 + C_3 d,\ 0 \right).$$

5.3 Examples

The revolute joint between sliding block and connecting rod leads to the point constraint

$$\mathbf{0} = \begin{pmatrix} r_3 \\ 0 \end{pmatrix} - A(\alpha_1) \begin{pmatrix} l_1 \\ 0 \end{pmatrix} - A(\alpha_2) \begin{pmatrix} l_2 + d_4 \\ 0 \end{pmatrix}.$$

A main feature of this constraint is its dependence on the longitudinal displacement d_4 of the right tip of the connecting rod.

The forces can be expressed as

$$f(q, d, \dot{q}, \dot{d}) = \begin{pmatrix} f_r(q, d, \dot{q}, \dot{d}) \\ f_h(q, d, \dot{q}, \dot{d}) \end{pmatrix} + \begin{pmatrix} 0 \\ \nabla W_h(d) \end{pmatrix}$$

where

$$f_r(q, d, \dot{q}, \dot{d}) = \begin{pmatrix} -1/2 l_1 (\gamma (m_1 + 2m_2) \cos\alpha_1 + l_2 m_2 \dot{\alpha}_2^2 \sin(\alpha_1 - \alpha_2)) \\ -1/2 l_2 \gamma m_2 \cos\alpha_2 + 1/2 l_1 l_2 m_2 \dot{\alpha}_1^2 \sin(\alpha_1 - \alpha_2) \\ 0 \end{pmatrix}$$

$$+ \begin{pmatrix} l_1 \dot{\alpha}_2^2 \begin{pmatrix} -\sin(\alpha_1 - \alpha_2) \\ \cos(\alpha_1 - \alpha_2) \end{pmatrix}^T C_1^T d - 2 l_1 \dot{\alpha}_2 \begin{pmatrix} \cos(\alpha_1 - \alpha_2) \\ \sin(\alpha_1 - \alpha_2) \end{pmatrix} C_1^T \dot{d} \\ l_1 \dot{\alpha}_1^2 \begin{pmatrix} \sin(\alpha_1 - \alpha_2) \\ -\cos(\alpha_1 - \alpha_2) \end{pmatrix}^T C_1^T d - 2 \dot{\alpha}_2 C_4^T \dot{d} - 2 \dot{\alpha}_2 \dot{d}^T M_h d \dots \\ \dots - \dot{d}^T C_3 \dot{d} - \gamma \begin{pmatrix} \cos\alpha_2 \\ -\sin\alpha_2 \end{pmatrix}^T C_1^T d \\ 0 \end{pmatrix}$$

and $f_h(q, d, \dot{q}, \dot{d})$ is given by the expression

$$\dot{\alpha}_2^2 M_h d + \left(\dot{\alpha}_2^2 C_4 + l_1 \dot{\alpha}_1^2 C_1 \begin{pmatrix} \cos(\alpha_1 - \alpha_2) \\ \sin(\alpha_1 - \alpha_2) \end{pmatrix} + 2 \dot{\alpha}_2 C_3 \dot{d} \right) - \gamma C_1 \begin{pmatrix} \sin\alpha_2 \\ \cos\alpha_2 \end{pmatrix}.$$

Overall, the partitioned structure of the equations of motion is clearly visible here, and it becomes also obvious that the floating reference frame results in various coupling terms that stem from differentiating the kinetic energy in the variational principle.

5.3.2 Loading Area of Planar Truck Model

We next study the discretization by finite elements as well as the model reduction techniques at the truck example of Sect. 4.5.2. Figure 5.3 shows a triangulation of the loading area with 125 elements and the interfaces to joints and force elements.

The origin of the body-fixed reference frame is placed in the node on top of the rear tire where the corresponding force element is attached. An alternative choice is the pivot of the revolute joint at the right end, which leads to a simplified treatment

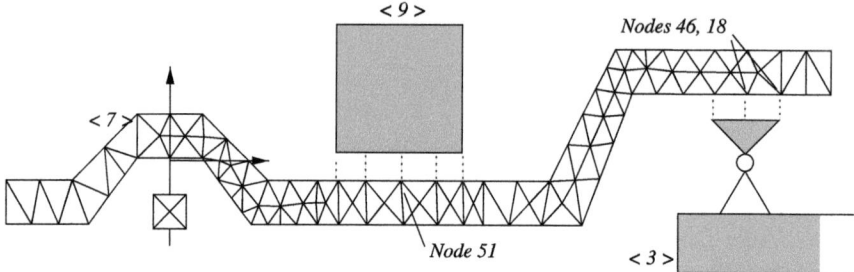

Fig. 5.3 Triangulation of loading area and interfaces

of the constraint arising from the boundary condition, cf. the discussion in Sect. 4.2. However, for demonstration purposes we pick the first approach and get thus two constraints for the coupling of the loading area with the load (body 9) and with the chassis (body 3).

The slim shape of the loading area suggests to use quadratic finite elements with six nodes per triangle in combination with the plane stress model. In this way, we obtain $n_d = 644$ unknown nodal variables d from the discretization of the displacement field. For the interface between loading area and load we take the trace of the basis functions on the affected boundary segment as natural discretization of the constraint, which means that the corresponding nodes and their 2 degrees of freedom each are directly coupled with the rigid load. Since there are five nodes from the original mesh on the interface and four extra nodes in the midpoints from the extended mesh of the quadratic elements, we obtain $9 \times 2 = 18$ discrete constraints from this coupling.

The attachment of the revolute joint to the chassis is accomplished by specifying a joint coordinate system based on nodes 46 and 18, which results in two additional constraints for the pivot. A finer resolution of this interface is of little influence on the overall behavior, as numerical experiments show.

Summarizing, the full finite element model of the loading area features $n_d = 644$ nodal variables and $n_\lambda = 20$ constraints. We next aim at determining a reduced model with less than 10 elastic degrees of freedom. Since the number of constraints is already beyond that threshold, the detailed interface between loading area and load must be weakened. Two rigid bodies in the plane can be coupled by three constraint equations, and this can be transferred to the elastic-rigid coupling. An example is the choice

$$N_\mu(x(s)) = \begin{pmatrix} 1 & 0 & 2s/l - 1 \\ 0 & 1 & 2s/l - 1 \end{pmatrix}, \quad s \in [0, l], \tag{5.64}$$

for the discretization of the Lagrange multipliers. Here, the straight line of length l between the two bodies is parametrized by the real variable s, and each translational and rotational degree of freedom is associated with one of the three multipliers. Note that this approach can be viewed as an averaging technique, compare (4.37a), (4.37b). For vanishing displacements, it reduces to the usual rigid body constraint.

5.3 Examples

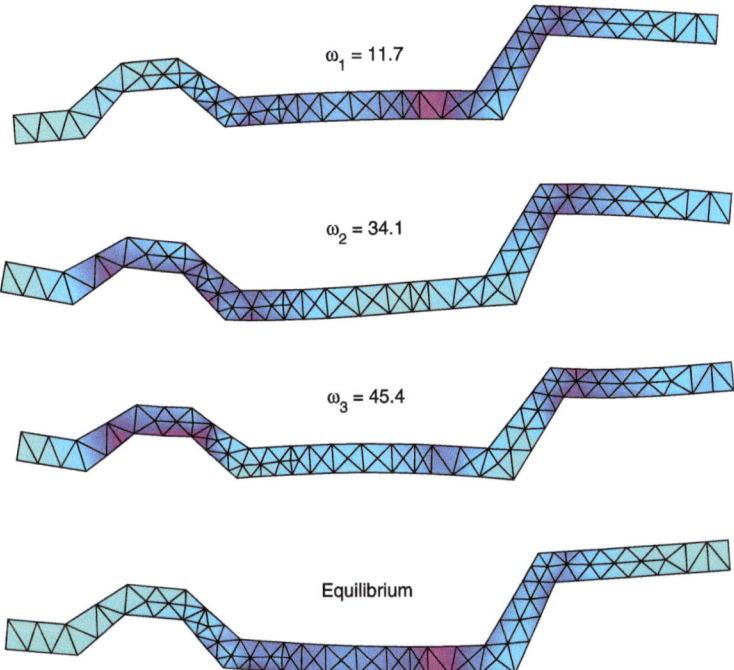

Fig. 5.4 Eigenmodes and static equilibrium of loading area. The displacements are magnified and the frequencies are given in Hz. Colors indicate the von Mises stress

The simplified constraint between loading area and load is now combined with the condensation methods from above. We start with the modal decomposition of the free structure according to (5.60). Figure 5.4 displays the first three eigenmodes and corresponding frequencies. Due to the specification of the body-fixed reference frame no rigid body modes occur. Adding two more frequencies and corresponding modes and arranging all eigenmodes columnwise in the transformation matrix T from (5.61) yields a reduced model for the elastic body which possesses solely $n_d = 5$ degrees of freedom. Its quality can be checked by computing the static equilibrium position of the complete multibody system and comparing it with the solution of the full model. In other words, we solve by Newton's method

$$0 = f(q,d,0,0,t_0) - G(q,d)^T \lambda, \quad (5.65a)$$
$$0 = g(q,d) \quad (5.65b)$$

for (q,d,λ) with d being the nodal displacements of the finite element model in the first run and thereafter the coefficient vector obtained from modal condensation. Note that the latter involves different force and constraint terms due to the coordinate transformation. Furthermore, this coefficient vector undergoes a back transformation to generate the displacement data presented in Table 5.3. It turns out that the

Table 5.3 Comparison of condensation methods at static equilibrium position of truck model. The notation $(d(46)_1, d(46)_2)$ stands for the horizontal and vertical displacements in node 46 (in m)

	full mesh	mixed condensation	modal condensation
$d(46)_1$	-2.4153e-03	-2.4153e-03	-2.4322e-03
$d(51)_1$	-9.1908e-04	-9.1908e-04	-9.0463e-04
$d(46)_2$	-2.2205e-03	-2.2205e-03	-2.2699e-03
$d(51)_2$	-6.7637e-03	-6.7637e-03	-6.7399e-03

Table 5.4 Comparison of condensation methods with respect to eigenfrequencies (in Hz)

mixed condensation	modal condensation
1.1706e+01	1.1706e+01
3.4063e+01	3.4059e+01
4.5415e+01	4.5380e+01
1.0369e+02	1.0241e+02
1.3633e+02	1.3119e+02
1.9339e+02	
2.2638e+02	
4.0731e+02	
9.1962e+02	

displacements with respect to the reduced model differ already in the second digit from the original reference solution.

An adept mixed static and modal condensation, however, leads to significantly better results, as can be seen in Table 5.3 in the middle column. For this approach, the horizontal and vertical displacements in nodes 46 and 51 serve as master variables, and five modes of the slave subsystem are added to form the transformation matrix with a total of $n_d = 9$ elastic degrees of freedom. This mixed condensation yields furthermore a satisfactory approximation of the low frequencies, see Table 5.4.

5.3.3 Pantograph and Catenary

The coupled system of pantograph and catenary contains two elastic bodies, which are the carrier wire and the contact wire as shown in Fig. 4.8. Both wires are subject to local effects resulting from the interaction with the droppers and the moving pantograph head, and for this reason a fine spatial mesh based on finite differences or finite elements is clearly superior to approaches based on global eigenfunctions [PEM+97].

As introduced in Sect. 4.5.3, we write $x = x_1$ for the single independent variable and discretize now the interval $[0, L]$ by the grid $0 = x_0 < \ldots < x_{N+1} = L$ with

5.3 Examples

stepsize $h_i = x_{i+1} - x_i$. A straightforward choice for the Galerkin projection are standard conformal finite elements, which means that linear elements are taken for the carrier wire and cubic elements for the contact wire. In each element $[x_i, x_{i+1}]$ we thus introduce the approximations

$$w_3(x,t) \doteq (1-\xi)d_{3,i}(t) + \xi d_{3,i+1}(t), \quad w_4(x,t) \doteq \sum_{j=0}^{3} \mathcal{N}_j(x) d_{4,2i+j}(t) \quad (5.66)$$

with $\xi := (x - x_i)/h_i$ and shape functions

$$\mathcal{N}_0(x) = (1-\xi)^2(1+2\xi), \quad \mathcal{N}_1(x) = \xi(1-\xi)^2 h_i,$$
$$\mathcal{N}_2(x) = \xi^2(3-2\xi), \quad \mathcal{N}_3(x) = -\xi^2(1-\xi)h_i.$$

The interface between pantograph head and contact wire requires particular attention. In (4.67), this constraint was formulated in a pointwise sense,

$$w_4(x_p(t), t) - r_1(t) = 0. \quad (5.67)$$

This is prone for numerical problems whenever the pantograph head passes over a grid point. To understand this phenomenon, we insert (5.66) into (5.67) and observe that the resulting discretized constraint is only of class C^1 with respect to time since it inherits the smoothness from the cubic finite element ansatz. The corresponding Lagrange multiplier μ_p, however, depends on the second time derivative of the constraint and thus exhibits a jump at each grid point. We remark that similar situations arise for sliding contact problems.

A remedy against this lack of smoothness can be constructed in the following way. The moving contact point is extended to a *contact zone* $[x_p - \epsilon, x_p + \epsilon]$ where we, instead of (5.67), postulate the weak constraint [AS00]

$$\int_{x_p - \epsilon}^{x_p + \epsilon} \delta_\epsilon(x - x_p(t)) \Big(w_4(x_p(t), t) - r_1(t)\Big) dx = 0. \quad (5.68)$$

Here, the function δ_ϵ stands for a given *load distribution* with the properties

$$\int_{x_p - \epsilon}^{x_p + \epsilon} \delta_\epsilon(\xi) d\xi = 1, \quad \delta'_\epsilon(-\epsilon) = \delta'_\epsilon(\epsilon) = 0.$$

An example is the piecewise cubic function of class C^1

$$\delta_\epsilon(\xi) := \frac{1}{\epsilon} - 3\frac{\xi^2}{\epsilon^3} + 2\frac{|\xi|^3}{\epsilon^4}, \quad |\xi| \leq \epsilon,$$

which is bell-shaped, Fig. 5.5.

It is easy to see that the contact zone leads to constraints of class C^3 and thus makes the discretized equations much smoother. Even more, from a practical view

Fig. 5.5 Bell-shaped function δ_ϵ for weak constraint (5.68)

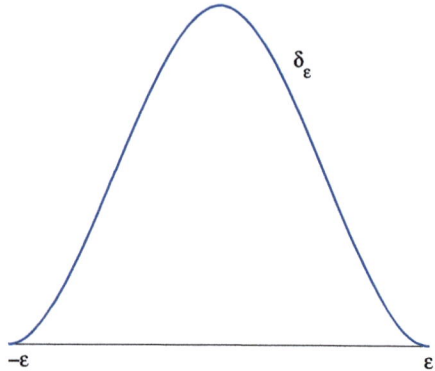

point, the pantograph head is modelled in a more precise way since the actual technical system features two contact stripes that are assembled with a certain horizontal distance. Finally, we remark that the weak constraint (5.68) can be refined by introducing a local subgrid with corresponding degrees of freedom and Lagrange multipliers.

We skip the display of the full set of equations for the system of pantograph and catenary. The partitioned equations (5.51a), (5.51b) hold also in this case, with rigid motion variables $\boldsymbol{q} = (r_1, r_2)$ for the pantograph head and elastic degrees of freedom $\boldsymbol{d}$ for the displacements $d_{3,i}$ and $d_{4,i}$ in the grid points of the catenary. A specific feature of this system is its linearity with constant matrices except for the constraint matrix $\boldsymbol{G}(t)$, whose dependence on time stems from the moving contact point. The condensation methods for reducing the model are not applicable here, as explained above. For more details see [AS00, SA00, TSS05].

Chapter 6
Stiff Mechanical System

In the previous chapter we have seen how the spatial discretization of a flexible multibody system leads to a differential-algebraic equation in time. The partitioning into two types of state variables, namely, those for the gross motion, on the one hand, and those for the elastic deformations, on the other, quite often involves widely different time scales. This chapter is devoted to such *stiff mechanical systems*.

In numerical analysis, the adjective "stiff" typically characterizes an ordinary differential equation whose eigenvalues have strongly negative real parts. However, numerical stiffness may also arise in case of second order differential equations with large eigenvalues on or close to the imaginary axis. If such high frequencies are viewed as a parasitic effect which perturbs a slowly varying smooth solution, implicit time integration methods with adequate numerical dissipation are an option and usually superior to explicit methods. A well-known example is the "stiff beam" of Hairer and Wanner [HW96, §IV.1]. For a mechanical system, this form of numerical stiffness is directly associated with *large stiffness forces*, and thus the notion of a stiff mechanical system has a twofold meaning.

If the high frequencies carry physical significance and need to be resolved by the time integration, even implicit methods are compelled to take tiny stepsizes. Here, however, we restrict ourselves to tracking a smooth motion where the high frequency part represents a singular perturbation. Based on the work of Lubich [Lub93], we show that asymptotic methods can be developed to provide both a firm analysis and a computational method for obtaining an average solution. The latter is related to standard engineering approaches as, for instance, the so-called linear theory of elasto-dynamics [Sha98]. Moreover, as introduced in [RS99], this asymptotic expansion is extended to the general differential-algebraic case by using suitable local ODE representations. In this way, consistent approximations can be computed that satisfy all constraints and are close to the smooth motion.

6.1 Elastic Pendulum Revisited

In the following, the elastic pendulum, which has been discussed in the introductory Sect. 1.1, illustrates some basic properties of a stiff mechanical system. For simplicity, we set for the constants $\rho A = 1$ and $l = 1$ in the equations of motion (1.8a), (1.8b), which yields

$$(2/3 + d^2)\ddot{\alpha} + \bar{\pi}\ddot{d} = -\gamma \cos\alpha + 2\bar{\pi}\gamma d \sin\alpha - 2d\dot{\alpha}\dot{d},$$
$$\bar{\pi}\ddot{\alpha} + \ddot{d} = -2\bar{\pi}\gamma \cos\alpha + d\dot{\alpha}^2 - \frac{1}{\epsilon^2}d, \qquad \bar{\pi} = 2/\pi, \tag{6.1}$$

with unknown angle α and elastic displacement d, see Fig. 1.1. In (6.1), the frequency ω has been replaced by the perturbation parameter $\epsilon := 1/\omega$. If ω is large— or ϵ very small, respectively, the equations of motion are singularly perturbed.

An *outer expansion* consists of the series [O'M88]

$$\alpha^\epsilon(t) = \alpha^0(t) + \epsilon^2\alpha^1(t) + \epsilon^4\alpha^2(t) + \cdots, \qquad d^\epsilon(t) = d^0(t) + \epsilon^2 d^1(t) + \epsilon^4 d^2(t) + \cdots$$

with coefficient functions α^i and d^i that do not depend on the parameter ϵ. Inserting $(\alpha^\epsilon, d^\epsilon)$ into (6.1), applying Taylor expansion, and sorting out powers of ϵ leads to a chain of equations that define the coefficient functions for this particular case:

$$\epsilon^{-2}: \qquad\qquad\qquad 0 = d^0,$$
$$\epsilon^0: \quad (2/3 + (d^0)^2)\ddot{\alpha}^0 + \bar{\pi}\ddot{d}^0 = -\gamma\cos\alpha^0 + 2\bar{\pi}\gamma d^0 \sin\alpha^0 - 2d^0\dot{\alpha}^0\dot{d}^0, \tag{6.2}$$
$$\bar{\pi}\ddot{\alpha}^0 + \ddot{d}^0 = -2\bar{\pi}\gamma\cos\alpha^0 + d^0(\dot{\alpha}^0)^2 - d^1.$$

Analogously, higher powers of ϵ result in further conditions on the subsequent coefficient functions.

What are the implications of the expansion (6.2)? The first two coefficient functions α^0 and d^0 are part of a differential-algebraic equation where d^1 plays the role of a Lagrange multiplier. Due to the semi-explicit structure, it is easy to see that its index equals 3. If we insert the trivial constraint $d^0 = 0$, we get

$$2/3\ddot{\alpha}^0 = -\gamma\cos\alpha^0, \qquad d^1 = -2\bar{\pi}\gamma\cos\alpha^0 - \bar{\pi}\ddot{\alpha}^0.$$

Thus, the coefficient function α^0 satisfies the equation of a rigid pendulum and represents the limit case $\epsilon \to 0$. On the other hand, the elastic deformation vanishes in the limit. For small $\epsilon > 0$, the elastic approximation

$$d^a := d^0 + \epsilon^2 d^1 = \frac{-1}{\omega^2}(2\bar{\pi}\gamma\cos\alpha^0 + \bar{\pi}\ddot{\alpha}^0)$$

contains the dominant term of the expansion up to $\mathcal{O}(\epsilon^4)$.

Since the coefficient functions in the outer expansion do not depend on ϵ, the series for α^ϵ and for d^ϵ are slowly varying and represent the smooth motion. In the limit $\epsilon \to 0$, the smooth motion tends to the solution (α^0, d^0) of the DAE (6.2).

Whether the smooth motion also solves the singularly perturbed system (6.1) is mainly determined by the initial values, see below.

We remark that the application of an *inner expansion* with respect to the fast time scale t/ϵ leads to additional high frequency modes, which overlay the smooth motion d^ϵ and its approximation d^a. More precisely, for small angular velocity $\dot{\alpha}$ and vanishing initial velocities $\dot{d}(t_0) = \dot{d}^a(t_0) = 0$ it can be shown that

$$d(t) = d^a(t) + (d(t_0) - d^a(t_0)) \cos \frac{\zeta(t-t_0)}{\epsilon} + \mathcal{O}(\epsilon^4) \qquad (6.3)$$

for the elastic component. Notice that in (6.3), the frequency $\omega = 1/\epsilon$ is shifted to ζ/ϵ with $\zeta = \sqrt{\pi^2/(\pi^2-6)}$. Figure 1.2 on page 8 displays both the smooth and the oscillatory solution.

6.2 General Framework

The expansion steps for the elastic pendulum raise several questions. Due to the partitioned structure of the equations of motion, the outer expansion turned out to be very simple. Is this procedure also feasible at more general systems, in particular in combination with constraints and a differential-algebraic framework? Another aspect are the model reduction methods of Sect. 5.2.3 such as the modal condensation that eliminates high frequency components a priori. With regard to numerical analysis, the crucial issue is how this specific type of singular perturbation affects the time integration methods.

The answer to the last item is presented in depth in Chap. 7. For the remainder of this chapter, we concentrate on an asymptotic analysis of the partitioned equations of motion (5.51a), (5.51b). It is very useful to introduce now a refined notation that displays the structure of the equations in more detail. This notation reads

$$M(q,d) \begin{pmatrix} \ddot{q} \\ \ddot{d} \end{pmatrix} = \begin{pmatrix} f_r(q,d,\dot{q},\dot{d}) \\ f_h(q,d,\dot{q},\dot{d}) \end{pmatrix} - \frac{1}{\epsilon^2} \begin{pmatrix} 0 \\ \nabla W_s(d) \end{pmatrix} - G(q,d)^T \lambda, \quad (6.4a)$$

$$0 = g(q,d) \qquad (6.4b)$$

with parameter $\epsilon \ll 1$. Compared to (5.51a), (5.51b), the force vector is split into rigid motion terms f_r and elastic terms f_h, and the mass matrix is also partitioned into blocks that match the dimensions of q and d, respectively,

$$M(q,d) = \begin{pmatrix} M_r(q,d) & C(q,d)^T \\ C(q,d) & M_h \end{pmatrix}.$$

Here, $M_r(q,d) \in \mathbb{R}^{n_q \times n_q}$ stands for the block that corresponds to the gross motion, $M_h \in \mathbb{R}^{n_d \times n_d}$ for the block of the elastic deformation, and $C(q,d) \in \mathbb{R}^{n_d \times n_q}$ contains the coupling integrals, cf. the slider crank example of Sect. 5.3.1.

The forces on the right hand side are assumed to be autonomous, and in addition to f_r and f_h, the stiff forces stemming from the spatial discretization are written as extra term $\epsilon^{-2}\nabla W_s$ with potential $W_s(d)$. In case of linear elasticity, we have

$$\epsilon^{-2}\nabla W_s(d) = \nabla W_h(d) = A_h d$$

with stiffness matrix A_h. Apparently, the size of ϵ depends on model parameters such as the modulus of elasticity and geometry data.

In order to include nonlinear stiffness terms in the analysis, as described in (5.37), we employ here a general model with the discretized potential W_s whose Hessian $\nabla^2 W_s(d)$ is assumed to be positive definite. This requires that the elastic structure has no rigid body mode, which is not a severe restriction as the modeling with a floating reference frame guarantees this by construction. Additionally, we suppose that the elastic potential takes its minimum in the undeformed reference position,

$$\nabla W_s(d) = 0 \quad \Rightarrow \quad d = 0.$$

In practice, not all deformation modes of an elastic body necessarily induce a singular perturbation. For instance, let $\omega_1 \leq \omega_2 \leq \ldots \leq \omega_{n_d}$ be the eigenvalues of the elastic members in the linear case; that is, the solutions of the eigenvalue problem $\omega^2 M_h v = A_h v$. In the modal condensation technique, an index $n_c \ll n_d$ is selected such that the frequencies $\omega_1, \ldots, \omega_{n_c}$ are in the time scale of the gross or rigid body motion q, and the higher frequencies $\omega_{n_c+1}, \ldots, \omega_{n_d}$ represent stiff modes. A standard example is here a beam with both bending (slow) and lengthening (fast) modes. Rearranging the unknowns such that q contains exactly the slow deformation modes and d only the fast or stiff ones, we can view the equations of motion again as a perturbed system. In the ODE case, this has been an approach used in [Sac96]. In such a partitioned system where d contains only the fast modes, the process of going to the limit $\epsilon \to 0$ corresponds to the frequently used alternative of reducing the dimension and complexity of the model by chopping off simply all higher frequencies [KH90].

As additional prerequisite, the mappings M, f_r, f_h, W_s and g are assumed to be sufficiently smooth with bounded derivatives. Finally, since dissipation may appear not only in such interconnection elements as dampers but also in elastic body models, we shall assume that the damping terms are small in comparison to the force term $\epsilon^{-2}\nabla W_s(d)$.

Under these assumptions, the partitioned differential-algebraic system (6.4a), (6.4b) is amenable to a thorough asymptotic analysis using outer expansions

$$q^\epsilon(t) = q^0(t) + \epsilon^2 q^1(t) + \ldots, \quad d^\epsilon(t) = d^0(t) + \epsilon^2 d^1(t) + \ldots \qquad (6.5)$$

for both the low frequency variables q and the high frequency or stiff variables d. We start with the case of unconstrained equations of motion, i.e., the state space form.

6.3 State Space Form

Assume that the equations of motion (6.4a), (6.4b) are given as an unconstrained system

$$\begin{pmatrix} M_r(q,d) & C^T(q,d) \\ C(q,d) & M_h \end{pmatrix} \begin{pmatrix} \ddot{q} \\ \ddot{d} \end{pmatrix} = \begin{pmatrix} f_r(q,d,\dot{q},\dot{d}) \\ f_h(q,d,\dot{q},\dot{d}) \end{pmatrix} - \frac{1}{\epsilon^2} \begin{pmatrix} 0 \\ \nabla W_s(d) \end{pmatrix} \quad (6.6)$$

with a symmetric positive definite mass matrix on the left and a stiff elastic potential $\epsilon^{-2}W_s(d)$ on the right hand side. Introducing new variables $p(t) = (q(t), d(t)) \in \mathbb{R}^{n_p}$ where $n_p = n_q + n_d$, we can subsume the partitioned ODE (6.6) under the class of singularly perturbed systems

$$M(p)\ddot{p} = f_n(p, \dot{p}) - \frac{1}{\epsilon^2} \nabla U(p) \quad (6.7)$$

considered in [Lub93]. Here, the term f_n comprises the non-stiff forces, and the gradient $\epsilon^{-2} \nabla U$ stands for a separable stiff force.

6.3.1 Asymptotic Expansion

The following assumptions are required for an asymptotic expansion of (6.6):

$M(p) \in \mathbb{R}^{n_p \times n_p}$ is symmetric positive definite for all $p \in \mathbb{R}^{n_p}$. (6.8a)

There exists a nonempty open set $E \subset \mathbb{R}^{n_p}$ such that

$$\mathcal{U} = \{p \in E : \nabla U(p) = 0\} = \{p \in E : U(p) = \min_{s \in E} U(s)\} \quad (6.8b)$$

is a submanifold of $\mathbb{R}^{n_p}$.

There exists $\alpha > 0$ such that $s^T \nabla^2 U(p) s \geq \alpha s^T M(p) s$ (6.8c)

for all s with $s^T M(p) v = 0 \; \forall v \in T_p \mathcal{U}$.

The third assumption states that the stiff potential U is strongly convex with respect to the $M(p)$-orthogonal complement of the tangent space

$$T_p \mathcal{U} = \ker \nabla^2 U(p).$$

As shown in [Lub93], under these three conditions the smooth motion p^ϵ is a sufficiently differentiable solution of (6.7) which, together with its derivatives, is bounded independently of ϵ and can be represented in the form of an outer expansion

$$p^\epsilon(t) = p^0(t) + \epsilon^2 p^1(t) + \epsilon^4 p^2(t) + \ldots + \epsilon^{2N} p^N(t) + \mathcal{O}(\epsilon^{2N+2}). \quad (6.9)$$

Here, the coefficient functions p^i are independent of ϵ and are generally specified as the solutions of a chain of DAE's, each of index 3.

The conditions (6.8a)–(6.8c) are easily verified for the partitioned system (6.6) since the Hessian $\nabla^2 W_s(d)$ is positive definite and since

$$\mathcal{U} = \{(q,d) \in \mathbb{R}^{n_p} : d = 0\}.$$

Now, when the partitioned expansion (6.5) is applied to (6.6), we obtain a chain of equations that define the coefficient functions for this particular case.

By Taylor expansion, the coefficient of ϵ^{-2} in (6.6) vanishes if

$$0 = \nabla W_s(d^0) \quad \Rightarrow \quad d^0 = 0,$$

and the coefficient ϵ^0 provides the equation

$$M_r(q^0, 0) \ddot{q}^0 = f_r(q^0, 0, \dot{q}^0, 0). \tag{6.10}$$

In other words, if q is identified with the gross motion and d with the elastic motion, then, as expected, the first coefficient functions (q^0, d^0) define the rigid body motion with zero deformation.

In addition, when comparing coefficients of ϵ^0 in the second line of (6.6), we obtain a linear system for d^1,

$$\nabla^2 W_s(0) d^1 = f_h(q^0, 0, \dot{q}^0, 0) - C(q^0, 0) \ddot{q}^0. \tag{6.11}$$

The next steps follow recursively and are omitted here since they involve higher derivatives that are typically not available in simulation programs.

Decoupled Quasi-Static Analysis

The expansion steps derived thus far form the basis of a standard approach in the engineering literature, the so-called *decoupled quasi-static analysis* [JPD93]. Let

$$q^a(t) := q^0(t), \qquad d^a(t) := d^0(t) + \epsilon^2 d^1(t) = \epsilon^2 d^1(t) \tag{6.12}$$

denote the approximations given by the ODE of rigid motion (6.10) and the linear system (6.11), respectively. The second order system (6.10) can be solved for q^a by a standard ODE method and then, in a post-processing step, the corresponding elastic response d^a can be computed from the linear system

$$\frac{1}{\epsilon^2} \nabla^2 W_s(0) d^a = f_h(q^a, 0, \dot{q}^a, 0) - C(q^a, 0) \ddot{q}^a \tag{6.13}$$

involving the Hessian matrix, which satisfies $\epsilon^{-2} \nabla^2 W_s(0) = A_h$ in linear elasticity. Obviously, $d^a = \epsilon^2 d^1$, and the decoupled analysis is equivalent to the outer expansion from above. In practice, this approach can be implemented as a post-processing step and thus requires little computational effort.

6.3 State Space Form

It leaves, however, the question of how well q^a and d^a approximate a solution. Both q^a and d^a are smooth and, by construction, we know that

$$q^a(t) - q^{\epsilon}(t) = \mathcal{O}(\epsilon^2), \qquad d^a(t) - d^{\epsilon}(t) = \mathcal{O}(\epsilon^4).$$

More interesting is a comparison of (q^a, d^a) with a possibly oscillatory solution (q, d) of (6.6). The following result, derived essentially from Theorem 2.2 of [Lub93], provides an estimate.

Theorem 6.1 *Let (q^a, d^a) be the approximation of the decoupled quasi-static analysis and (q, d) a solution of the partitioned ODE (6.6) with initial values $q(t_0) - q^a(t_0) = \mathcal{O}(\epsilon^3)$, $\dot{q}(t_0) - \dot{q}^a(t_0) = \mathcal{O}(\epsilon^2)$, $d(t_0) - d^a(t_0) = \mathcal{O}(\epsilon^3)$ and $\dot{d}(t_0) - \dot{d}^a(t_0) = \mathcal{O}(\epsilon^2)$. Then we have on bounded time intervals*

$$q(t) - q^a(t) = \mathcal{O}(\epsilon^2), \quad \dot{q}(t) - \dot{q}^a(t) = \mathcal{O}(\epsilon^2),$$
$$d(t) - d^a(t) = \mathcal{O}(\epsilon^2), \quad \dot{d}(t) - \dot{d}^a(t) = \mathcal{O}(\epsilon^2).$$

Proof We use the main ideas of [Lub93] but specialize, where necessary, to the system (6.6). First, a coordinate change is applied to simplify the structure of the potential W_s. Locally we have

$$W_s(d) - W_s(0) = \frac{1}{2} d^T \nabla^2 W_s(0) d + r(d), \qquad r(d) = \mathcal{O}(\|d\|^3).$$

Let $LL^T = \nabla^2 W_s(0)$ be the Cholesky decomposition of the Hessian, then with the new coordinates

$$w := \omega(d) L^T d, \qquad \omega(d) := \left(1 + \frac{r(d)}{d^T \nabla^2 W_s(0) d}\right)^{1/2},$$

it follows that

$$W_s(d) = \frac{1}{2} w^T w + \text{const}.$$

Without loss of generality we assume from now on that $\nabla W_s(d) = d$ but otherwise retain the notation. The approximation (q^a, d^a) inserted into (6.6) has a defect of order $\mathcal{O}(\epsilon^2)$, and hence the differences $\Delta q := q - q^a$ and $\Delta d := d - d^a$ satisfy

$$M(t) \begin{pmatrix} \ddot{\Delta q} \\ \ddot{\Delta d} \end{pmatrix} = \mathcal{O}(\Delta q) + \mathcal{O}(\Delta d) + \mathcal{O}(\dot{\Delta q}) + \mathcal{O}(\dot{\Delta d}) - \frac{1}{\epsilon^2} \begin{pmatrix} 0 \\ \Delta d \end{pmatrix} + \mathcal{O}(\epsilon^2) \quad (6.14)$$

where

$$M(t) = \begin{pmatrix} M_r(q^a, d^a) & C(q^a, d^a)^T \\ C(q^a, d^a) & M_h \end{pmatrix}.$$

Let $Q(t)$ be an orthogonal matrix such that

$$Q(t)^T M(t)^{-1/2} \begin{pmatrix} 0 & 0 \\ 0 & I_{n_d} \end{pmatrix} M(t)^{-1/2} Q(t) = \begin{pmatrix} 0 & 0 \\ 0 & B(t) \end{pmatrix},$$

then, clearly, the $n_d \times n_d$ matrix $\boldsymbol{B}(t)$ is positive definite and both $\boldsymbol{B}$ and $\boldsymbol{Q}$ are smooth functions of t. With

$$\begin{pmatrix} \boldsymbol{\delta} \\ \boldsymbol{\eta} \end{pmatrix} := \boldsymbol{Q}(t)^T \boldsymbol{M}(t)^{1/2} \begin{pmatrix} \Delta \boldsymbol{q} \\ \Delta \boldsymbol{d} \end{pmatrix}$$

it follows from (6.14)

$$\ddot{\boldsymbol{\delta}} = \mathcal{O}(\|\boldsymbol{\delta}\| + \|\dot{\boldsymbol{\delta}}\| + \|\boldsymbol{\eta}\| + \|\dot{\boldsymbol{\eta}}\|) + \mathcal{O}(\epsilon^2) \tag{6.15}$$

$$\ddot{\boldsymbol{\eta}} = -\epsilon^{-2} \boldsymbol{B}(t) \boldsymbol{\eta} + \mathcal{O}(\|\boldsymbol{\delta}\| + \|\dot{\boldsymbol{\delta}}\| + \|\boldsymbol{\eta}\| + \|\dot{\boldsymbol{\eta}}\|) + \mathcal{O}(\epsilon^2). \tag{6.16}$$

Let $\boldsymbol{B}(t) = \boldsymbol{R}(t) \boldsymbol{R}(t)^T$ be a smooth Cholesky decomposition of $\boldsymbol{B}(t)$. Then in terms of the new variable

$$\boldsymbol{\xi} := \begin{pmatrix} \boldsymbol{\eta} \\ \epsilon \boldsymbol{R}(t)^{-1} \dot{\boldsymbol{\eta}} \end{pmatrix}$$

we are finally led to the first order system

$$\dot{\boldsymbol{\xi}} = \frac{1}{\epsilon} \begin{pmatrix} 0 & \boldsymbol{R}(t) \\ -\boldsymbol{R}(t)^T & 0 \end{pmatrix} \boldsymbol{\xi} + \mathcal{O}(\|\boldsymbol{\xi}\| + \epsilon(\|\boldsymbol{\delta}\| + \|\dot{\boldsymbol{\delta}}\|)) + \mathcal{O}(\epsilon^3).$$

The skew-symmetry of the matrix on the right allows here for an energy estimate (Gronwall inequality). More specifically, for initial values $\boldsymbol{\xi}(t_0) = \mathcal{O}(\epsilon^3)$, we obtain

$$\|\boldsymbol{\xi}(t)\| \leq C\epsilon \int_{t_0}^{t} (\|\boldsymbol{\delta}(\tau)\| + \|\dot{\boldsymbol{\delta}}(\tau)\|) d\tau + \mathcal{O}(\epsilon^3)$$

and

$$\|\boldsymbol{\eta}(t)\| + \|\dot{\boldsymbol{\eta}}(t)\| \leq C \int_{t_0}^{t} (\|\boldsymbol{\delta}(\tau)\| + \|\dot{\boldsymbol{\delta}}(\tau)\|) d\tau + \mathcal{O}(\epsilon^2).$$

With the latter bound inserted into (6.15) and with $\dot{\boldsymbol{\delta}} = \boldsymbol{\sigma}$ and the Dini derivative D_+ this leads to

$$D_+ \|\boldsymbol{\delta}(t)\| \leq \|\boldsymbol{\sigma}(t)\|,$$
$$D_+ \|\boldsymbol{\sigma}(t)\| \leq C_1(\|\boldsymbol{\delta}(t)\| + \|\boldsymbol{\sigma}(t)\|) + C_2 \int_{t_0}^{t} (\|\boldsymbol{\delta}(\tau)\| + \|\boldsymbol{\sigma}(\tau)\|) d\tau + \mathcal{O}(\epsilon^2).$$

When the inequality is replaced by an equality, then the resulting integro-differential system satisfies the standard monotonicity requirements of [Wal70, p. 122]. Thus from the results of [Wal70] we conclude that, on bounded time intervals, it follows that

$$\boldsymbol{\delta}(t) = \mathcal{O}(\epsilon^2), \qquad \boldsymbol{\sigma}(t) = \mathcal{O}(\epsilon^2),$$

for any initial values such that $\boldsymbol{\delta}(t_0) = \mathcal{O}(\epsilon^2)$, $\boldsymbol{\sigma}(t_0) = \mathcal{O}(\epsilon^2)$. □

The initial values play a crucial role in Theorem 6.1 since they are required to be close enough to the smooth motion. In general, the approximation $\boldsymbol{d}^a$ is worse

6.3 State Space Form

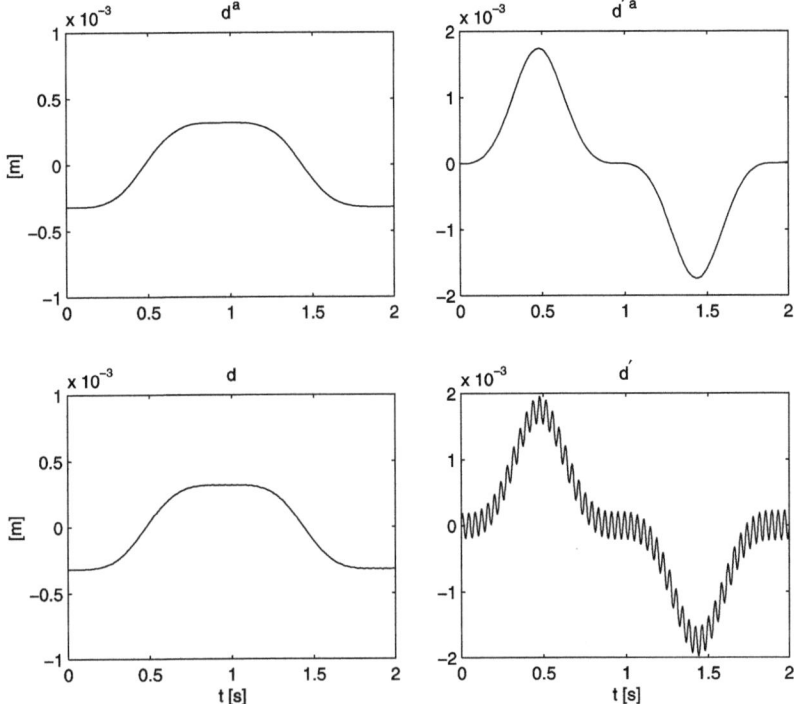

Fig. 6.1 Solutions of the elastic pendulum. The approximation d^a of the elastic component is compared to the reference d and the corresponding velocity

than q^a due to $d^a = \epsilon^2 d^1 = \mathcal{O}(\epsilon^2)$. This means that the solution d oscillates with a frequency $\mathcal{O}(\epsilon^{-1})$ and an amplitude $\mathcal{O}(\epsilon^2)$ around d^a.

6.3.2 Elastic Pendulum

The asymptotic behavior of the elastic pendulum illustrates the statements of Theorem 6.1. Figure 6.1 shows a comparison of the approximation d^a and its velocity $\dot{d}^a$ for $\epsilon = 0.01$ and $\gamma = 10$, compare Fig. 1.2(b). For the underlying rigid motion, the equation $2/3\,\ddot{\alpha}^a = -\gamma\cos\alpha^a$ was integrated with zero initial values $\alpha^a(t_0) = \dot{\alpha}^a(t_0) = 0$. The initial value $d^a(t_0)$ describes the stationary bending of the pendulum due to the gravitational force.

When comparing the approximation (α^a, d^a) with the solution (α, d), the initial values are, as explained above, of decisive influence. The choice

$$\alpha(t_0) = \dot{\alpha}(t_0) = 0, \quad d(t_0) = d^a(t_0) - \epsilon^3, \quad \dot{d}(t_0) = \dot{d}^a(t_0) + \epsilon^2$$

leads to the solution displayed in Fig. 6.1 at the bottom. While the position variable d is almost smooth with only tiny oscillations, the velocity $\dot{d}$ exhibits larger amplitudes and fluctuates around $\dot{d}^a$. Initial values that exactly coincide with $d^a(t_0)$ and $\dot{d}^a(t_0)$ result in almost negligible oscillations. Notice that Fig. 1.2(c) on page 8 contains the solution (α, d) for initial values

$$\alpha(t_0) = \dot{\alpha}(t_0) = 0, \quad d(t_0) = \dot{d}(t_0) = 0.$$

Since the initial deformation of the elastic rod is not taken into account, the oscillations are there much stronger and noticeable already in the position variable. Nevertheless, the asymptotic approximation (6.3) still covers this case quite well and underlines the influence of the initial values. More precisely, we have

$$\dot{d}(t) = \dot{d}^a(t) - \frac{\zeta}{\epsilon}(d(t_0) - d^a(t_0)) \sin \frac{\zeta(t - t_0)}{\epsilon} + \mathcal{O}(\epsilon^3),$$

and thus the difference of the initial values with respect to the position must be one power of ϵ smaller in order to yield an estimate with an error of size $\mathcal{O}(\epsilon^2)$ with respect to the velocity.

In conclusion, the decoupled quasi-static analysis demands for a detailed knowledge of the multibody system under consideration. Employing it as a black box solver instead of a fully coupled simulation is prone to errors. In situations where the smooth motion of the stiff components is of interest in its own, however, this approach yields a satisfactory approximation. In any case, it represents a means to determine physically meaningful initial values that are close to the smooth motion.

6.3.3 Remarks

Singularly perturbed systems of the form (6.7) arise also in the field of molecular dynamics. Among others, we mention here the work of Bornemann [Bor98] and Reich [Rei00] on *homogenization*, i.e., on the limit process $\epsilon \to 0$. The term *averaging* is also common. In short, the stiff potentials in molecular dynamics may possess a complex structure where a naively applied outer expansion results in a wrong asymptotic behavior. So-called correcting potentials are then introduced as a remedy. For the problem class discussed here such correcting potentials are not required since even in case of geometric stiffening terms, the Hessian $\nabla^2 W_s$ possesses a constant spectrum on the manifold $\mathcal{U}$ from (6.8b). The homogenized system is, in other words, exactly the rigid body motion.

In the engineering literature, an outer expansion is often referred to as "perturbation analysis" [Sac96]. It is particularly popular for reducing models of single elastic bodies with geometric stiffening effects. Nonlinear beam structures are a typical example where an adept choice of state variables enables the derivation of explicit expressions for the stiff components.

A variant of the decoupled quasi-static analysis is the *linear theory of elastodynamics* [Sha98]. In addition to the smooth motion terms, this approach also considers the dynamics of the elastic components. Based on the rigid motion trajectory

q^a, one solves

$$M_h \ddot{d}^b = f_h(q^a, 0, \dot{q}^a, 0) - A_h d^b - C(q^a, 0) \ddot{q}^a \qquad (6.17)$$

by a time integration method. In this way, the approximation d^b also contains oscillations in the fast time scale. However, we point out that the coupling blocks C in the full mass matrix are not correctly taken into account in (6.17), which may lead to erroneous frequencies. For the elastic pendulum, solving (6.17) results in the frequency $\omega = 1/\epsilon$, but, as given in (6.3), the coupled system has actually the frequency $\zeta\omega$.

A final remark concerns the dissipation, which may be also present in a multibody system. As noted above, we assumed here small dissipative terms when compared to the stiff forces. The reverse situation with small stiffness and large dissipation is also amenable to a singular perturbation analysis. The limit case is then a differential-algebraic equation of index 2 [Stu08].

6.4 The Differential-Algebraic Case

The asymptotic analysis of the unconstrained stiff mechanical system (6.7) revealed hidden constraints that follow from carrying out the expansion steps and letting $\epsilon \to 0$. Any singularly perturbed system of the form (6.7) is thus related to a chain of differential-algebraic equations of index 3 each. The situation becomes now even more complex when we turn to the constrained equations of motion (6.4a), (6.4b) and extend the asymptotic approach, as introduced in [RS99].

Using again the notation $p(t) = (q(t), d(t))$, the partitioned system (6.4a), (6.4b) can be written as

$$M(p)\ddot{p} = f_n(p, \dot{p}) - \frac{1}{\epsilon^2} \nabla U(p) - G(p)^T \lambda, \qquad (6.18)$$
$$0 = g(p),$$

which is a stiff mechanical system on the manifold

$$\mathcal{M} = \{p \in \mathbb{R}^{n_p} : 0 = g(p)\}.$$

Concentrating on the conservative case and applications in molecular dynamics, the behavior of this dynamical system has been studied in [Bor98]. If the submanifold

$$\mathcal{N} = \{p \in \mathbb{R}^{n_p} : 0 = g(p), \ 0 = \nabla U(p)\} \subset \mathcal{M}$$

is non-empty, the solutions will, in general, oscillate on a time scale of order $\mathcal{O}(\epsilon)$ around $\mathcal{N}$. For a large class of stiff potentials, there exists a homogenized system with solutions in $\mathcal{N}$.

In the case of the partitioned equations (6.4a), (6.4b) with $U(p) = W_s(d)$, the set $\mathcal{N}$ turns out to be the rigid motion space,

$$\mathcal{N} = \{(q, d) : 0 = g(q, d), \ 0 = d\}$$

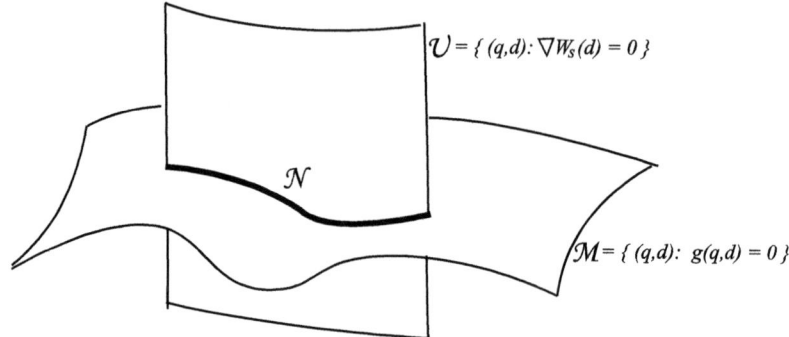

Fig. 6.2 Transversality condition. The intersection $\mathcal{N}$ is required to be non-empty

since we assumed that $\nabla W_s(d) = 0$ implies that $d = 0$. In other words, when (6.4a), (6.4b) is viewed as a stiff mechanical system, its limit is—as naturally expected—the rigid body system.

These observations suggest that we should not consider the case $\epsilon \to 0$ but, instead, seek to approximate solutions for nonzero, but small ϵ. In this way, we gain a better understanding of the stiff components in the DAE (6.4a), (6.4b) and furthermore obtain an approach to generate reduced models and to compute consistent initial values close to the smooth motion.

6.4.1 Structure-Preserving Local Parametrization

Clearly, in order to utilize the expected asymptotic behavior, the submanifold $\mathcal{N} \subset \mathcal{M}$ will have to be non-empty. A sufficient condition for this is that the $(n_d + n_\lambda) \times (n_d + n_\lambda)$ matrix

$$\begin{pmatrix} 0 & \nabla^2 W_s(d) \\ \partial g(q,d)/\partial q & \partial g(q,d)/\partial d \end{pmatrix} \tag{6.19}$$

is invertible. This means in particular that $\partial g(q,d)/\partial q$ must have full rank n_λ. The *transversality condition* (6.19) for the configuration space and the constraining stiff potential is illustrated in Fig. 6.2.

Under the full-rank condition for the matrix (6.19), it is possible to derive a local parametrization that preserves the partitioned structure and reduces the differential-algebraic system to an ordinary differential equation. The expansion steps are then applied to this local state space form, and by transforming back to the original variables, an approximation of the smooth motion for the DAE case is obtained. To this end, let $E_q \subset \mathbb{R}^{n_q}$ and $E_d \subset \mathbb{R}^{n_d}$ be non-empty open sets such that E_d contains the origin and $\mathcal{M} = \{(q,d) \in E_q \times E_d : 0 = g(q,d)\}$ the rigid motion space of interest. As in the preceding chapter, we require that the matrix (5.52) is invertible for any $(q,d) \in \mathcal{M}$ whence $\mathcal{M}$ is a submanifold of $\mathbb{R}^{n_q} \times \mathbb{R}^{n_d}$ of dimension $n_q + n_d - n_\lambda$.

6.4 The Differential-Algebraic Case

In addition, as indicated above, the matrix (6.19) is assumed to have full rank for $(q,d) \in \mathcal{M}$ which implies, in particular, that rank $\partial g(q,d)/\partial q = n_\lambda$.

Let $(q_0, 0) \in \mathcal{M}$ and $((q_0, 0), (v_0, 0)) \in \mathcal{TM}$ be some given rigid motion points on the manifold $\mathcal{M}$ and its tangent bundle $\mathcal{TM}$, respectively. Due to the above full rank condition for the Jacobian $\partial g/\partial q$, the implicit function theorem guarantees that there exists a local parametrization $(\mathcal{W}, \psi)$ of $\mathcal{M}$ near $(q_0, 0)$,

$$\psi: \mathcal{W} \subset \mathbb{R}^{n_q-n_\lambda} \times E_d \to \psi(\mathcal{W}) \subset \mathcal{M}, \quad \psi(s,d) = (q,d) \tag{6.20}$$

where $\mathcal{W}$ is an open neighborhood of the origin. Then

$$(\mathcal{W} \times \mathbb{R}^{n_q-n_\lambda} \times \mathbb{R}^{n_d}, (\psi, \Psi))$$

with the Jacobian $\Psi := (\partial \psi/\partial s, \partial \psi/\partial d)$ is a local parametrization of the tangent bundle near $((q_0, 0), (v_0, 0))$ [Rhe96].

The Jacobian Ψ has the block structure

$$\Psi(s,d) = \begin{pmatrix} H(s,d) & K(s,d) \\ 0 & I_{n_d} \end{pmatrix}$$

with $H(s,d) \in \mathbb{R}^{n_q \times (n_q-n_\lambda)}$, $K(s,d) \in \mathbb{R}^{n_q \times n_d}$. By construction, it spans the null space of the constraint Jacobian G, i.e.,

$$G(\psi(s,d)) \cdot \Psi(s,d) = 0 \quad \text{for all } (s,d) \in \mathcal{W}.$$

The transformation of the partitioned DAE (6.4a), (6.4b) to local coordinates (s,d) proceeds now like in the rigid body case, compare the state space form (2.12) in Sect. 2.1.2 and the discussion on differential equations on a manifold in Sect. 2.3.3.

First, the expressions for the velocity $(\dot{q}, \dot{d})$ and the acceleration $(\ddot{q}, \ddot{d})$ depending on $(\dot{s}, \dot{d})$ and $(\ddot{s}, \ddot{d})$ are substituted, and then the differential equations of (6.4a), (6.4b) are multiplied by $\Psi(s,d)^T$ from the left. We skip these steps and, omitting the arguments for simplicity, state the resulting unconstrained system in the form

$$\begin{pmatrix} H^T M_r H & H^T(M_r K + C^T) \\ (K^T M_r + C)H & K^T C^T + CK + M_h \end{pmatrix} \begin{pmatrix} \ddot{s} \\ \ddot{d} \end{pmatrix} = b - \frac{1}{\epsilon^2} \begin{pmatrix} 0 \\ \nabla W_s \end{pmatrix}. \tag{6.21}$$

Here, the right hand side vector b is given by

$$b = \Psi^T \begin{pmatrix} f_r \\ f_h \end{pmatrix} - \Psi^T M \frac{\partial \Psi}{\partial (s,d)}((\dot{s}, \dot{d}), (\dot{s}, \dot{d})).$$

Due the particular construction of the parametrization, the local ODE (6.21) preserves the partitioned structure of (6.6) with a symmetric positive definite mass matrix.

In summary, this proves the following result:

Theorem 6.2 *Suppose that the matrices (5.52) and (6.19) have full rank. Then, near any $(q_0, 0) \in \mathcal{M}$ and $((q_0, 0), (v_0, 0)) \in \mathcal{TM}$ there exists a local parametrization such that the DAE (6.4a), (6.4b) can be transformed to a partitioned ODE (6.6).*

It should be noted that there is a slight difference between the local ODE (6.21) and the partitioned system (6.6). In fact, in (6.21) the lower right block of the mass matrix depends on the states (s, d) while in (6.6) the corresponding block M_h is a constant matrix. This difference, however, plays no role here.

6.4.2 Computational Method

Theorem 6.2 guarantees that the decoupled quasi-static analysis (6.13) can be generalized to the differential-algebraic framework. In a computational method, this means that first the partition-preserving local parametrization (6.20) needs to be constructed and then the linear solve (6.13) is performed in local coordinates. The first task is easily accomplished by using the MANPAK algorithms of Rheinboldt [Rhe96].

Given a rigid motion trajectory and a point $(q, 0) \in \mathcal{M}$ on this trajectory with velocity $(v_q, 0)$, the method consists of the following three principal steps:

$$\begin{pmatrix} (q,0) \\ (v_q,0) \end{pmatrix} \xrightarrow{\psi^{-1}} \begin{pmatrix} (0,0) \\ (v_s,0) \end{pmatrix} \xrightarrow{(6.13)} \begin{pmatrix} (0,d^a) \\ (v_s,v_d^a) \end{pmatrix} \xrightarrow{\psi} \begin{pmatrix} (q^a,d^a) \\ (v_q^a,v_d^a) \end{pmatrix}. \quad (6.22)$$

After the transition to local coordinates by means of ψ^{-1}, the approximation d^a of the stiff variables is computed from (6.13), with the right hand side evaluated in local coordinates. Additionally, the corresponding velocity $v_d^a = \dot{d}^a$ must be computed since it is part of the back transformation to global coordinates. Numerical experience shows that simple finite differences work well for this purpose. After the final transformation, in general we have $q \neq q^a$ and $v_q \neq v_q^a$, which means that the rigid motion coordinates are slightly changed. The computational results presented in [RS99] demonstrate that this extension of the quasi-static analysis to the differential-algebraic case represents an efficient approach whenever the assumption of strongly differing time scales holds.

We close this chapter with several remarks. The transversality condition (6.19) excludes constraints that depend only on the stiff variables. In a multibody context, constraints are typically expressed in terms of the gross motion variables, i.e., translation and rotation, plus an elastic offset, and they reduce to rigid body constraints if the elastic displacement vanishes. Thus, this condition is not a severe restriction in practice.

The second remark concerns the applicability of the above results and algorithms to stiff mechanical systems in general. Though we have concentrated here on flexible multibody systems with a given spatial discretization and the resulting partitioned DAE (6.4a), (6.4b), other stiff mechanical systems, e.g., with stiff springs, can also be analysed in the same fashion.

This applies in particular to multibody applications where certain subsystems feature small masses when compared to the force terms. A quasi-static approximation becomes again attractive in such cases, and we refer to [WAV12] for recent work in this field.

Chapter 7
Time Integration Methods

In flexible multibody dynamics, the time integration of the semi-discretized equations of motion represents a challenging problem due to the simultaneous presence of constraints and different time scales. This combination leads, as analyzed in the foregoing chapter, to a stiff differential-algebraic system. We investigate here the behavior of numerical methods for such problems, with particular focus on the well-established implicit integrators that are either based on the BDF (Backward Differentiation Formulas) methods or on implicit Runge–Kutta methods of collocation type.

The chapter starts, however, with an overview on time integration methods for constrained mechanical systems. Using a model equation with smooth and highly oscillatory solution parts, we then show that stiff methods suffer from order reductions which are directly related to the limiting DAE of index 3 to which the stiff mechanical system converges.

At the end of this chapter, we also introduce extensions of the generalized-α method and of the implicit midpoint rule to the differential-algebraic case and investigate their potential for mechanical multibody systems.

7.1 Overview on Time Integration Methods

We have already seen that the semi-discretized equations (5.51a), (5.51b) possess the same structure as the equations of motion (2.9a), (2.9b) in the rigid body case. This means that we can build upon the framework of Sect. 2.4 throughout this chapter. Since most time integration methods are designed for first order systems, we introduce velocity variables $v(t) := \dot{p}(t)$ in addition to the partitioned vector $p(t) = (q(t), d(t))$ of rigid body motion and elastic deformation. Mostly we will not exploit the partitioned structure of p, and the discussion of Sect. 2.4.4 on alternative formulations can be carried over directly.

7.1.1 Alternative Formulations of the Equations of Motion

Using the velocities variables v, the equations of motion (5.51a), (5.51b) are rewritten as first order system

$$\dot{p} = v, \tag{7.1a}$$

$$M(p)\dot{v} = f(p, v, t) - G(p)^T \lambda, \tag{7.1b}$$

$$0 = g(p). \tag{7.1c}$$

Assuming the full rank condition (5.52) to hold, we know from Sect. 5.2.2 that this differential-algebraic system is of index 3, and thus index reduction is a standard remedy to alleviate the implications for the integration schemes.

Semi-explicit systems of index 2 can be solved by various implicit and also half-explicit methods, as will be discussed below, and for this reason we present next appropriate re-formulations of the equations of motion.

Replacing the position constraint (7.1c) by the velocity constraint

$$0 = G(p)v \tag{7.2}$$

as in (2.82) reduces the index to 2 and is thus one of the available options. However, one should be aware that the original position constraint is an *invariant* of the system that will be, in general, not conserved by a time integration method. Though a linear *drift off* might thus occur, in practice this drawback is rarely noticeable, in particular if higher order methods are used. On the other hand, replacing the position constraint by the acceleration constraint

$$0 = G(p)\dot{v} + \kappa(p, v), \quad \kappa(p, v) := \frac{\partial G(p)}{\partial p}(v, v), \tag{7.3}$$

is prone for substantial drift off with quadratic growth whenever the constraint is nonlinear. A first explanation of this phenomenon was already given in (2.79), and in the next section we will study the drift off in more detail.

The Gear–Gupta–Leimkuhler formulation (2.83) does not suffer from drift off. For convenience, we state it here again in terms of the variables p:

$$\dot{p} = v - G(p)^T \mu,$$
$$M(p)\dot{v} = f(p, v, t) - G(p)^T \lambda, \tag{7.4}$$
$$0 = G(p)v,$$
$$0 = g(p).$$

As shown in Sect. 2.4.4, the index of this augmented system with additional Lagrange multiplier μ is also 2.

For some implementations, a state-dependent mass matrix $M(p)$ on the left hand side of the dynamics equation is not admitted. One way out is to formally invert the

7.1 Overview on Time Integration Methods

mass matrix and move it to the right hand side. An alternative procedure is the introduction of additional *acceleration variables* $w(t) := \ddot{p}(t)$, which in case of the formulation of index 2 results in an enlarged semi-explicit system

$$\begin{aligned}\dot{p} &= v, \\ \dot{v} &= w, \\ 0 &= M(p)w - f(p, v, t) + G(p)^T \lambda, \\ 0 &= G(p) v.\end{aligned} \qquad (7.5)$$

At first sight, the extra variables w and also v involve a significant increase in the number of unknowns, but the corresponding additional computational effort can be minimized by exploiting the block structure in the linear algebra, see Sect. 7.2.5 below. Additional acceleration variables may also be employed in case of the GGL formulation (7.4).

7.1.2 Basic Discretization Schemes

We discuss next a selection of popular time integration methods that are typically used for solving the constrained mechanical system (7.1a)–(7.1c). For this purpose, we assume consistent initial values p_0, v_0 as defined in (2.73), i.e.,

$$0 = g(p_0), \qquad 0 = G(p_0) v_0,$$

and denote by $t_0 < t_1 < \ldots < t_n$ the time grid, with $\tau_i = t_{i+1} - t_i$ being the stepsize. Though the methods and results below are valid also in the variable stepsize case, we mostly assume $\tau_i = \tau$ as constant and comment whenever necessary on the variable stepsize case. The approximation of $p(t_n)$ is written as p_n, and analogously v_n and λ_n stand for the discrete velocity vector and the Lagrange multiplier.

The Drift Off Phenomenon

In Chap. 2 it was mentioned that by solving the linear system (2.69), the formulation (2.77) of index 1 can be reduced to an ODE for the position and velocity variables. Here, we take up this idea and write the linear system in the form

$$\begin{pmatrix} M(p) & G(p)^T \\ G(p) & 0 \end{pmatrix} \begin{pmatrix} \Psi \\ \Upsilon \end{pmatrix} = \begin{pmatrix} f(p, v, t) \\ -\kappa(p, v) \end{pmatrix} \qquad (7.6)$$

with vectors $\Psi \in \mathbb{R}^{n_p}$ and $\Upsilon \in \mathbb{R}^{n_\lambda}$. Given p and v, the solution of (7.6) defines the right hand side of

$$\dot{p} = v, \qquad \dot{v} = \Psi(p, v, t), \qquad (7.7)$$

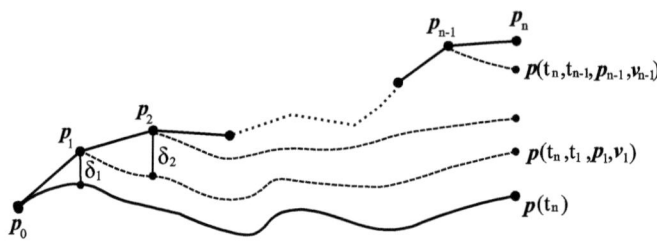

Fig. 7.1 Sketch of drift off analysis

which means that any standard ODE integrator can be applied in this way to solve the equations of constrained mechanical motion. At the same time, the Lagrange multiplier follows directly from evaluating $\lambda = \Upsilon(p, v, t)$. If the time integration method is convergent of order k, we have

$$p(t_n) - p_n = \mathcal{O}(\tau^k), \quad v(t_n) - v_n = \mathcal{O}(\tau^k), \quad \lambda(t_n) - \lambda_n = \mathcal{O}(\tau^k).$$

This error estimate, however, is misleading since (p_n, v_n, λ_n) is not the numerical solution of the original problem (7.1a)–(7.1c). The position and velocity constraints are in general not preserved under discretization of (7.7) and consequently, these constraints are violated,

$$g(p_n) \neq 0, \qquad G(p_n) v_n \neq 0.$$

In detail, this behavior can be analyzed as follows.

Let $\tilde{p}_0$ and $\tilde{v}_0$ denote arbitrary (possibly inconsistent) initial values of the ODE (7.7). By integration of the constraints at acceleration level (7.3) we obtain

$$g(p(t)) = g(\tilde{p}_0) + (t - t_0) G(\tilde{p}_0) \tilde{v}_0, \qquad G(p(t)) v(t) = G(\tilde{p}_0) \tilde{v}_0. \tag{7.8}$$

For consistent initial data $\tilde{p}_0 = p_0$, $\tilde{v}_0 = v_0$, the exact solution of (7.7) thus satisfies the constraints at all levels. Otherwise, there is a linear growth in the position constraints and a constant deviation in the velocity constraints.

If we consider now a numerical time integration method, the analysis becomes more involved and resembles the typical approach taken for a convergence proof. Figure 7.1 illustrates the situation. The exact solution $p(t)$ of (7.7) proceeds from the consistent initial value p_0 to the final value $p(t_n)$ and satisfies $g(p(t)) = 0$, and analogously we have for the exact velocity $G(p(t)) v(t) = 0$ for consistent initial velocity v_0.

We introduce next a specific notation that allows to keep track of the different trajectories required for the analysis. Let $p(t; t_i, p_i, v_i)$ denote the exact solution of (7.7) for initial data (p_i, v_i) at time t_i. As example, $p(t; t_1, p_1, v_1)$ stands for the solution trajectory that starts in the numerical approximation (p_1, v_1) of the first step. For the local error after one step, we have

$$\delta_1 = p(t_1) - p_1 = \mathcal{O}(\tau^{k+1})$$

7.1 Overview on Time Integration Methods

for a method of order k, see Fig. 7.1, and analogously

$$\delta_i := p(t_i; t_{i-1}, p_{i-1}, v_{i-1}) - p_i = \mathcal{O}(\tau^{k+1}) \tag{7.9}$$

for $i = 1, \ldots, n$. Based on this framework, the following result can be shown [HW96, §VII.2]:

Theorem 7.1 *If the ODE (7.7) stemming from the index-1 formulation is integrated by a method of order k, the numerical solution p_n and v_n satisfies after n time steps the bound*

$$\|g(p_n)\| \leq \tau^k (A(t_n - t_0) + B(t_n - t_0)^2), \quad \|G(p_n)v_n\| \leq \tau^k C(t_n - t_0) \tag{7.10}$$

with constants A, B, C.

Proof From Fig. 7.1, we conclude that the error in the position constraint is bounded by

$$\|g(p_n)\| \leq \sum_{j=0}^{n-1} \|g(p(t_n; t_{j+1}, p_{j+1}, v_{j+1})) - g(p(t_n; t_j, p_j, v_j))\| \tag{7.11}$$

where $g(p_n) = g(p(t_n; t_n, p_n, v_n))$ and $g(p(t_n)) = g(p(t_n; t_0, p_0, v_0)) = 0$. From (7.8) it follows that

$$g(p(t_n; t_{j+1}, p_{j+1}, v_{j+1})) = g(p_{j+1}) + (t_n - t_{j+1})G(p_{j+1})v_{j+1}$$

and

$$g(p(t_n; t_j, p_j, v_j)) = g(p_j) + (t_n - t_j)G(p_j)v_j.$$

Inserting the local error as $p_{j+1} = p(t_{j+1}; t_j, p_j, v_j) - \delta_{j+1}$, applying Taylor expansion and one more time (7.8), we arrive at

$$g(p_{j+1}) = g(p_j) + (t_{j+1} - t_j)G(p_j)v_j + \mathcal{O}(\tau^{k+1}).$$

The same steps imply that

$$G(p_{j+1})v_{j+1} = G(p_j)v_j + \mathcal{O}(\tau^{k+1}),$$

and overall

$$\|g(p(t_n; t_{j+1}, p_{j+1}, v_{j+1})) - g(p(t_n; t_j, p_j, v_j))\| \leq \tau^{k+1}(A + (t_n - t_{j+1})D)$$

with constants A and D. Summing up the right hand side of (7.11), the desired estimate is obtained via

$$\|g(p_n)\| \leq \tau^{k+1} \sum_{j=0}^{n-1} (A + (t_n - t_{j+1})D) \leq \tau^k A n \tau + \frac{1}{2}\tau^k D n^2 \tau^2.$$

Since $n\tau = t_n - t_0$, the result for the drift off in the position constraint follows with $B := D/2$. For the velocity constraint, the bound is derived in the same fashion. □

The drift off from the position constraints grows thus quadratically with the length of the integration interval, compare (2.79). It depends, however, also on the order of the method. If the constraints are linear, there is no drift off since the corresponding invariants are preserved by linear integration methods. Note that for the system (2.82) of index 2, a similar estimate with linear drift off can be shown. In case of variable stepsizes, a corresponding result holds where the stepsize τ in the bound (7.10) is replaced by the maximum stepsize $\tau_{\max}$.

Projection Methods

A very common cure for the drift off is a two-stage projection method where after each integration step, the numerical solution is projected onto the manifold of position and velocity constraints. Let p_{n+1} and v_{n+1} denote the numerical solution of the system (7.7), obtained by integration from consistent values p_n and v_n. Then, the projection consists of the following steps:

$$\text{solve} \quad \begin{cases} 0 = M(\tilde{p}_{n+1})(\tilde{p}_{n+1} - p_{n+1}) + G(\tilde{p}_{n+1})^T \mu, \\ 0 = g(\tilde{p}_{n+1}) \end{cases} \text{for } \tilde{p}_{n+1}, \mu; \quad (7.12a)$$

$$\text{solve} \quad \begin{cases} 0 = M(\tilde{p}_{n+1})(\tilde{v}_{n+1} - v_{n+1}) + G(\tilde{p}_{n+1})^T \eta, \\ 0 = G(\tilde{p}_{n+1}) \tilde{v}_{n+1} \end{cases} \text{for } \tilde{v}_{n+1}, \eta. \quad (7.12b)$$

A simplified Newton method can be used to solve the nonlinear system (7.12a), and since the corresponding iteration matrix is just a saddle point matrix as in (7.6), evaluated at p_{n+1} and already available in decomposed form due to the previous integration step, this projection is inexpensive to compute. Furthermore, (7.12b) represents a linear system for $\tilde{v}_{n+1}$ and η with similar structure where the corresponding matrix decomposition can be re-used for solving (7.6) in the next integration step [Sch99]. The global convergence of the underlying time integration method is not destroyed by the projection step since the corresponding correction is of size $\mathcal{O}(\tau^{k+1})$ and can thus be subsumed under the leading local error term.

As the projection (7.12a), (7.12b) reflects a metric that is induced by the mass matrix M [Lub89], the projected value $\tilde{p}_{n+1}$ is the point on the constraint manifold that has minimum distance to p_{n+1} in this metric. An analysis of the required number of Newton iterations and of the relation to alternative stabilization techniques including the classical Baumgarte method (2.80) is provided by [ACPR95].

Projection methods are particularly attractive in combination with explicit ODE integrators. The combination with implicit methods is nonetheless possible but not as efficient as the direct discretization by DAE integrators discussed below.

7.1 Overview on Time Integration Methods

Half-Explicit Methods

Half-explicit methods for DAEs discretize the differential equations explicitly while the constraint equations are enforced in an implicit fashion. Due to the linearity of the velocity constraint (7.2), the formulation of index 2 is a good candidate for this method class. Several one-step and extrapolation methods have been tailored to the needs and peculiarities of mechanical systems. The half-explicit Euler method as generic algorithm for the method class reads

$$\begin{aligned} p_{n+1} &= p_n + \tau v_n, \\ M(p_n) v_{n+1} &= M(p_n) v_n + \tau f(p_n, v_n, t_n) - \tau G(p_n)^T \lambda_n, \\ 0 &= G(p_{n+1}) v_{n+1}. \end{aligned} \qquad (7.13)$$

Similar to the index-1 case above, only a linear system of the form

$$\begin{pmatrix} M(p_n) & G(p_n)^T \\ G(p_{n+1}) & 0 \end{pmatrix} \begin{pmatrix} v_{n+1} \\ \tau \lambda_n \end{pmatrix} = \begin{pmatrix} M(p_n) v_n + \tau f(p_n, v_n, t_n) \\ 0 \end{pmatrix}$$

arises in each step. The scheme (7.13) forms the basis for a class of half-explicit Runge–Kutta methods [HLR89, AM98] and extrapolation methods [Lub89]. These methods have in common that only information of the velocity constraints is required. As remedy for the drift off, which grows only linearly here but might still be noticeable, the projection (7.12a), (7.12b) can be applied.

Implicit Methods

The approaches discussed so far rely mainly on explicit discretization schemes, with appropriate extensions to treat constraints and to preserve invariants. Yet, if stiffness is an issue, implicit methods become mandatory. For this reason, we focus in the following on the two most successful classes of implicit integration methods that are used today in multibody dynamics and also in general DAE problems.

It is convenient to write the different formulations of the equations of motion from above as linear implicit system (2.33), i.e., as

$$E \dot{x} = \phi(x, t)$$

with singular matrix E, right hand side ϕ, and with the vector $x(t)$ collecting the position and velocity coordinates as well as the Lagrange multipliers. A state-dependent mass matrix $M(p)$ can be (formally) inverted and moved to the right hand side or, alternatively, treated by introducing additional acceleration variables as in (7.5).

The prototype of an implicit method is the implicit Euler method

$$E \frac{x_{n+1} - x_n}{\tau} = \phi(x_{n+1}, t_{n+1}), \qquad (7.14)$$

which involves the solution of a system of nonlinear equations in each time step. The corresponding Jacobian matrix in Newton's method reads

$$\frac{1}{\tau}\mathbf{E} - \frac{\partial \boldsymbol{\phi}}{\partial \mathbf{x}}(\mathbf{x}_n, t_n). \tag{7.15}$$

Despite the singularity of $\mathbf{E}$, this matrix is invertible for a well-defined problem, which means that the linearization of (7.14) has a regular matrix pencil $\mu \mathbf{E} - \partial \boldsymbol{\phi}/\partial \mathbf{x}$, compare the discussion on the index in Sect. 2.3.2. In practice, however, higher index problems may result in ill-conditioned iteration matrices and require additional scaling techniques.

We will discuss this issue in the next section where BDF and implicit Runge–Kutta methods are treated in more detail. For the moment, we just remark that the computation of the Jacobian (7.15) by finite differences and the subsequent matrix decomposition are the bottleneck of implicit DAE solvers in multibody dynamics, in particular for large-scale applications.

7.2 BDF and Implicit Runge–Kutta Methods

In this section, we study the two most important method classes for constrained mechanical systems. We start with the BDF methods, analyze their convergence behavior, and continue then with the corresponding steps for implicit Runge–Kutta methods. The main references are [BCP96] and [HW96].

7.2.1 Backward Differentiation Formulas

The Backward Differentiation Formulas (BDFs) give an approximation to the derivative of a function $\mathbf{x}(t)$ at a time t_{n+k} in terms of the function values $\mathbf{x}_n = \mathbf{x}(t_n), \ldots, \mathbf{x}_{n+k} = \mathbf{x}(t_{n+k})$. They are derived by constructing a polynomial $\mathbf{u} \in \mathbb{P}_k$ that interpolates the data

$$(t_n, \mathbf{x}_n), \ldots, (t_{n+k}, \mathbf{x}_{n+k}).$$

Differentiating $\mathbf{u}$ and evaluating it at t_{n+k} yields a difference operator that approximates $\dot{\mathbf{x}}(t_{n+k})$. In case of constant stepsize $\tau_i = \tau$, this difference operator has the representation

$$\dot{\mathbf{u}}(t_{n+k}) = \frac{1}{\tau} \sum_{i=0}^{k} \alpha_i \mathbf{x}_{n+i} \doteq \dot{\mathbf{x}}(t_{n+k}) \tag{7.16}$$

with the BDF coefficients α_i, $i = 0, \ldots, k$. In particular, the first two BDF formulas of order 1 and order 2 read

$$\frac{1}{\tau}(\mathbf{x}_{n+1} - \mathbf{x}_n) = \dot{\mathbf{x}}(t_{n+1}) + \mathcal{O}(\tau) \qquad \text{implicit Euler}, \tag{7.17}$$

7.2 BDF and Implicit Runge–Kutta Methods

$$\frac{1}{\tau}\left(\frac{3}{2}x_{n+2} - 2x_{n+1} + \frac{1}{2}x_n\right) = \dot{x}(t_{n+2}) + \mathcal{O}(\tau^2) \qquad \text{BDF-2.} \qquad (7.18)$$

For the BDF method of order k, we have analogously

$$\frac{1}{\tau}\sum_{i=0}^{k}\alpha_i x_{n+i} = \dot{x}(t_{n+k}) + \mathcal{O}(\tau^k). \qquad (7.19)$$

The BDFs are mainly used as a multistep discretization of stiff and differential-algebraic equations [Gea71]. For the linear-implicit system (2.33), the BDF discretization with fixed stepsize simply replaces $\dot{x}(t_{n+k})$ by the difference operator,

$$E\frac{1}{\tau}\sum_{i=0}^{k}\alpha_i x_{n+i} = \phi(x_{n+k}, t_{n+k}). \qquad (7.20)$$

Since the difference operator on the left is nothing else than $\dot{u}(t_{n+k})$ from (7.16), this discretization can be interpreted as a *collocation method*, and the definition immediately extends to variable stepsizes.

The new solution x_{n+k} is thus given by the nonlinear system

$$\frac{\alpha_k}{\tau}E x_{n+k} - \phi(x_{n+k}, t_{n+k}) + \frac{1}{\tau}E\sum_{i=0}^{k-1}\alpha_i x_{n+i} = 0, \qquad (7.21)$$

where α_k is the *leading coefficient* of the method.

Zero-Stability and A-Stability

A linear multistep method is called *zero-stable* if its application to the trivial scalar test equation $\dot{y} = 0$ always leads to a bounded numerical solution. The BDFs are zero-stable for $k \leq 6$. In combination with consistency, zero-stability is a necessary and sufficient condition to show the convergence of multistep methods for ODEs.

Stiff problems, however, require *A-stable* methods. If we apply a BDF method to Dahlquist's test equation $\dot{y} = \lambda y$ where $\lambda \in \mathbb{C}^-$ stands for a system eigenvalue in the left half of the complex plane, we obtain the recursion

$$\sum_{i=0}^{k}\alpha_i y_{n+i} - \mu y_{n+k} = 0, \quad \mu = \tau\lambda \in \mathbb{C}^-. \qquad (7.22)$$

The stability of this recursion depends on the k roots $\zeta_1, \ldots, \zeta_k$ of

$$\sum_{i=0}^{k}\alpha_i \zeta^i - \mu\zeta^k = 0. \qquad (7.23)$$

The *stability region* of the BDF method is the set

$$\mathcal{S} = \{\mu \in \mathbb{C} : |\zeta_i(\mu)| \le 1,\ i = 1, \ldots, k\}$$

with those roots ζ_i lying on the unit circle being simple. The method is then A-stable if $\mathbb{C}^- \subset \mathcal{S}$. An A-stable method may take, for λ fixed, arbitrarily large stepsizes τ while still giving a bounded solution.

Due to Dahlquist's second barrier [Dah63], linear multistep methods are A-stable only up to order 2. This implies that the BDFs are not A-stable for $k \ge 3$. While the left half of the real line is still included in their stability domain even for $k \ge 3$, two segments of the imaginary axis above and below the origin are no more part of it. This property is of importance when we discuss the numerical solution of stiff mechanical systems below.

7.2.2 Error Analysis

Our aim in this section is to understand the basics of an error analysis for the BDFs, in particular the order reduction phenomenon with respect to the local error in the case of an index-2 system. For this purpose, it is convenient to consider the semi-explicit DAE (2.36a), (2.36b) where we can distinguish between differential variables y and algebraic variables z. The BDF method (7.20) applied to (2.36a), (2.36b) leads to

$$\sum_{i=0}^{k} \alpha_i y_{n+i} = \tau a(y_{n+k}, z_{n+k}), \tag{7.24a}$$

$$0 = b(y_{n+k}, z_{n+k}). \tag{7.24b}$$

In the index-1 case, the implicit function theorem can be applied to formally solve the constraint (7.24b) for $z_{n+k} = z(y_{n+k})$. Thus, the solution of (7.24a), (7.24b) is equivalent to solving the unconstrained system

$$\sum_{i=0}^{k} \alpha_i y_{n+i} = \tau a(y_{n+k}, z(y_{n+k})). \tag{7.25}$$

Both local and global error behavior are accordingly the same as in the ODE case.

The situation changes if we consider the index-2 case where the constraint (7.24b) reads

$$0 = b(y_{n+k}) \quad \text{with } \partial b/\partial y \cdot \partial a/\partial z \text{ invertible.} \tag{7.26}$$

We set $n = 0$ and assume that the previously computed values of the differential variable are exact, i.e., $y_i = y(t_i)$ for $i = 0, \ldots, k-1$. The algebraic variables from

7.2 BDF and Implicit Runge–Kutta Methods

previous steps do not enter the current step and thus do not matter—these variables are said *to possess no memory*. For the BDF method of order k, we have from (7.19)

$$\sum_{i=0}^{k} \alpha_i \, y(t_i) = \tau \, \dot{y}(t_k) + \tau \Delta = \tau a(y(t_k), z(t_k)) + \tau \Delta \tag{7.27}$$

with $\Delta = \mathcal{O}(\tau^k)$. The exact solution thus satisfies the difference equation with a defect Δ.

Subtracting (7.24a) from (7.27) and using $y_i = y(t_i)$ for $i = 0, \ldots, k-1$, we arrive at

$$\alpha_k (y(t_k) - y_k) = \tau \big(a(y(t_k), z(t_k)) - a(y_k, z_k) \big) + \tau \Delta.$$

A Lipschitz condition on a with respect to both y and z yields

$$\|y(t_k) - y_k\| \leq \tau \frac{L}{\alpha_k} \big(\|y(t_k) - y_k\| + \|z(t_k) - z_k\| \big) + \frac{\tau}{\alpha_k} \|\Delta\|. \tag{7.28}$$

In order to obtain an estimate for the difference $\|z(t_k) - z_k\|$ on the right hand side, a discrete analogue for differentiating the constraint (7.26) is required. A vector-valued extension of the mean value theorem provides the right means for this purpose. More specifically, setting $\eta := -\sum_{i=0}^{k-1} \frac{\alpha_i}{\alpha_k} y_i$, we have

$$0 = b(y_k) - b(\eta) + b(\eta)$$

$$= \int_0^1 \frac{\partial b}{\partial y}(\eta + \theta(y_k - \eta)) \, d\theta \cdot (y_k - \eta) + b(\eta)$$

$$= \int_0^1 \frac{\partial b}{\partial y}(\eta + \theta(y_k - \eta)) \, d\theta \cdot \frac{\tau}{\alpha_k} a(y_k, z_k) + b(\eta). \tag{7.29}$$

For the exact solution, the "differentiated" constraint follows in the same way, including the extra defect term,

$$0 = \int_0^1 \frac{\partial b}{\partial y}(\eta + \theta(y(t_k) - \eta)) \, d\theta \cdot \left(\frac{\tau}{\alpha_k} a(y(t_k), z(t_k)) + \frac{\tau}{\alpha_k} \Delta \right) + b(\eta). \tag{7.30}$$

Defining

$$r(y, z) := \int_0^1 \frac{\partial b}{\partial y}(\eta + \theta(y - \eta)) \, d\theta \cdot a(y, z),$$

we can express the difference of (7.30) and (7.29) as

$$0 = r(y(t_k), z(t_k)) - r(y_k, z_k) + D(y(t_k)) \Delta \tag{7.31}$$

where the matrix $D(y)$ is given by

$$D(y) := \int_0^1 \frac{\partial b}{\partial y}(\eta + \theta(y(t_k) - \eta)) \, d\theta.$$

Since $\partial \boldsymbol{b}/\partial \boldsymbol{y} \cdot \partial \boldsymbol{a}/\partial \boldsymbol{z}$ is invertible, the Jacobian $\partial \boldsymbol{r}/\partial \boldsymbol{z}$ is also invertible in a neighborhood of the solution, and thus (7.31) leads to the estimate

$$\|\boldsymbol{z}(t_k) - \boldsymbol{z}_k\| \le c_1 \left(\|\boldsymbol{y}(t_k) - \boldsymbol{y}_k\| + \|\boldsymbol{\Delta}\| \right). \tag{7.32}$$

Inserting this estimate into (7.28), we get

$$(1 - \tau \frac{L}{\alpha_k} - c_1 \tau \frac{L}{\alpha_k}) \|\boldsymbol{y}(t_k) - \boldsymbol{y}_k\| \le c_2 \tau \|\boldsymbol{\Delta}\|.$$

Due to $\boldsymbol{\Delta} = \mathcal{O}(\tau^k)$, this implies the desired estimate $\boldsymbol{y}(t_k) - \boldsymbol{y}_k = \mathcal{O}(\tau^{k+1})$ for the differential component. From the bound (7.32), however, it follows that one power of τ is lost with respect to the algebraic component. We summarize these findings in the following theorem.

Theorem 7.2 *For the local error of the BDF method (7.20) applied to the semi-explicit system (2.36a), (2.36b), we have in the index-2 case*

$$\boldsymbol{y}(t_k) - \boldsymbol{y}_k = \mathcal{O}(\tau^{k+1}), \quad \boldsymbol{z}(t_k) - \boldsymbol{z}_k = \mathcal{O}(\tau^k). \tag{7.33}$$

The *order reduction* that affects the algebraic variable is a typical phenomenon when solving DAEs of higher index. Intuitively speaking, the differential operator d/dt that is applied when analyzing the algebraic component, compare Sect. 2.3.2, is replaced by the difference operator, and this involves a factor of $1/\tau$ in the corresponding numerical approximation.

Despite the order reduction in the local error, BDF methods still converge for semi-explicit systems of index 2 as shown in [GGL85, LP86]. An important property in this context is that the algebraic variables have no memory, which makes it possible to study the global error in $\boldsymbol{y}$ and $\boldsymbol{z}$ separately. Assuming starting values $\boldsymbol{y}_0, \ldots, \boldsymbol{y}_{k-1}$ that satisfy $\boldsymbol{y}(t_i) - \boldsymbol{y}_i = \mathcal{O}(\tau^{k+1})$ for $i = 0, \ldots, k-1$, the convergence result reads

$$\boldsymbol{y}(t_n) - \boldsymbol{y}_n = \mathcal{O}(\tau^k), \quad \boldsymbol{z}(t_n) - \boldsymbol{z}_n = \mathcal{O}(\tau^k) \tag{7.34}$$

for fixed final time t_n and $n \to \infty$. The result applies also to variable stepsizes. In conclusion, this means that the global convergence order k from the ODE case carries over to the DAE case although the local approximation of the algebraic variables suffers from the order reduction.

Practical Aspects

The nonlinear system (7.20) or (7.24a), (7.24b), respectively, represents the main computational burden of a BDF method, and its solution demands for extra care in case of a system of index 2. To better understand this, we rewrite (7.24a), (7.24b) in the form

$$\boldsymbol{F}(\boldsymbol{y}, \boldsymbol{z}) := \begin{pmatrix} \alpha_k \boldsymbol{y} - \boldsymbol{\eta} - \tau \boldsymbol{a}(\boldsymbol{y}, \boldsymbol{z}) \\ \boldsymbol{b}(\boldsymbol{y}) \end{pmatrix} = \boldsymbol{0}. \tag{7.35}$$

7.2 BDF and Implicit Runge–Kutta Methods

Here, $y = y_{n+k}$ and $z = z_{n+k}$ stand for the numerical solution to be computed while $\eta = -\sum_{i=0}^{k-1} \alpha_i y_{n+i}$ contains the previously computed data. The standard method for solving (7.35) is a simplified Newton scheme with starting values $(y^{(0)}, z^{(0)})$ that are obtained from extrapolating a predictor polynomial. This scheme can be written as fixed-point iteration

$$\begin{pmatrix} y^{(\ell+1)} \\ z^{(\ell+1)} \end{pmatrix} = \Phi(y^{(\ell)}, z^{(\ell)}) \tag{7.36}$$

where

$$\Phi(y, z) = \begin{pmatrix} y \\ z \end{pmatrix} - DF(y^{(0)}, z^{(0)})^{-1} F(y, z)$$

and where DF is the Jacobian matrix of F,

$$DF(y, z) = \begin{pmatrix} \alpha_k I - \tau \partial a(y,z)/\partial y & -\tau \partial a(y,z)/\partial z \\ \partial b(y)/\partial y & 0 \end{pmatrix}.$$

For convergence, we require contractivity, i.e., $\|D\Phi\| < 1$ for the Jacobian of Φ. If the starting values $(y^{(0)}, z^{(0)})$ are close enough to the fixed point, which is guaranteed by the predictor step, it can be shown that the Jacobian possesses blocks of different orders. More precisely, we have [HW96, §VII.3]

$$D\Phi = \begin{pmatrix} \mathcal{O}(\tau) & \mathcal{O}(\tau^2) \\ \mathcal{O}(1) & \mathcal{O}(\tau) \end{pmatrix}.$$

To achieve contractivity, a *scaling* of (7.36) is hence necessary, and the choice

$$T := \begin{pmatrix} I & 0 \\ 0 & \tau I \end{pmatrix}$$

turns out to yield the desired property $\|T D\Phi(y, z) T^{-1}\| = \mathcal{O}(\tau)$. For stepsize τ small enough, the simplified Newton iteration (7.36) then converges. The scaling corresponds to a multiplication of the algebraic variables by the stepsize τ. We measure the convergence of (7.36) accordingly with respect to the norm

$$\|y\| + \tau \|z\|, \tag{7.37}$$

and in this norm, each iteration yields one power of τ. Note that this scaling applies also to other implicit time integration schemes, in particular to the implicit Runge–Kutta methods of collocation type to be discussed in the next section.

The scaling (7.37) comes also into play when a variable stepsize algorithm is used. BDF codes such as DASSL [BCP96] and IDA from SUNDIALS [HBG+05], which are designed for fully implicit systems (2.32) and represent the state-of-the-art in the field, estimate the error in all state variables x via the difference of predictor and corrector,

$$e \doteq c_{n+1} \left(x_{n+1}^{(0)} - x_{n+1} \right) = \mathcal{O}(\tau_n^{k+1}), \tag{7.38}$$

where c_{n+1} is a constant that depends on stepsize ratios and on the leading coefficient of the method. Observe that the subscripting in (7.38) for annotating the current step $n \to n+1$ is different from the multistep notation in (7.21).

In case of a semi-explicit index-2 system, we have $x = (y, z)$ and the error in z behaves as $\mathcal{O}(\tau_n^k)$ due to the order reduction. E.g., put $k = 2$ and assume the code has an error test failure with new suggested stepsize $\tau_n^{\text{new}} = \tau_n/2$. When the code repeats the step with halved stepsize, the error in y will be reduced by a factor of 8 while the error in z decreases by a factor of 4 only. Consequently, the algorithm tends to *overestimate* the error in the algebraic variable and repeatedly decreases the stepsize, leading eventually to a failure of the integration process.

A simple cure for this problem is to employ the scaled norm (7.37) simultaneously in the error estimator and in the convergence criterion of the simplified Newton method. In this way, asymptotically correct estimates are provided. In case of a linear-implicit system (2.33) or a fully implicit system (2.32), however, the partitioning into differential and algebraic variables is in general unknown and hence the scaling (7.37) cannot be applied. This is one of the reasons why it is typically recommended to use the above codes solely at systems of index 1.

7.2.3 Implicit Runge–Kutta Methods of Collocation Type

We come now to the second prominent method class that plays an important role in multibody dynamics, in particular for stiff systems. Most of the relevant higher order implicit Runge–Kutta methods are collocation methods and derived as follows [GS69, Wri70]. Consider the ODE system $\dot{x} = \phi(x, t)$ with initial value $x(t_0) = x_0$ in the interval $[t_0, t_0 + \tau]$. The polynomial $u(t) \in \mathbb{P}_s$ collocates the differential equation if

$$u(t_0) = x_0,$$
$$\dot{u}(t_0 + c_i \tau) = \phi(u(t_0 + c_i \tau), t_0 + c_i \tau), \quad i = 1, \ldots, s. \tag{7.39}$$

In other words, in the s distinct points or nodes $0 \le c_1 < \ldots < c_s \le 1$ the polynomial u satisfies the differential equation. The points $t_0 + c_i h$ are called *collocation points*. Depending on the properties of the collocation polynomial, the evaluation of $u(t_0 + \tau)$ yields a numerical approximation of $x(t_0 + \tau)$.

It is easy to show that, given the collocation points, such a collocation method defines an implicit Runge–Kutta scheme with s stages. For this purpose, we define the *stage derivatives* k_i and the *internal stages* X_i as

$$k_i := \dot{u}(t_0 + c_i \tau), \qquad X_i := u(t_0 + c_i \tau), \tag{7.40}$$

along with the coefficients

$$a_{ij} := \int_0^{c_i} l_j(\theta) \, d\theta, \quad b_j := \int_0^1 l_j(\theta) \, d\theta, \quad i, j = 1, \ldots, s.$$

7.2 BDF and Implicit Runge–Kutta Methods

Here, $l_j(\theta) = \prod_{l \neq j}(\theta - c_l)/(c_j - c_l)$ for $j = 1, \ldots, s$ are the Lagrange polynomials [SB02]. It then holds

$$\dot{u}(t_0 + \theta\tau) = \sum_{j=1}^{s} k_j l_j(\theta),$$

and by integration we obtain

$$X_i - x_0 = \tau \int_0^{c_i} \dot{u}(t_0 + \theta\tau)\, d\theta = \tau \sum_{j=1}^{s} a_{ij} k_j$$

as well as

$$x_1 - x_0 = \tau \int_0^1 \dot{u}(t_0 + \theta\tau)\, d\theta = \tau \sum_{j=1}^{s} b_j k_j.$$

Finally, we insert $k_j = \phi(u(t_0 + c_j\tau), t_0 + c_j\tau) = \phi(X_j, t_0 + c_j\tau)$ and arrive at the implicit Runge–Kutta scheme

$$X_i = x_0 + \tau \sum_{j=1}^{s} a_{ij}\phi(X_j, t_0 + c_j\tau), \quad i = 1, \ldots, s; \tag{7.41a}$$

$$x_1 = x_0 + \tau \sum_{j=1}^{s} b_j \phi(X_j, t_0 + c_j\tau). \tag{7.41b}$$

The coefficients of a collocation method possess the property

$$\sum_{j=1}^{s} a_{ij} c_j^{q-1} = \frac{c_i^q}{q}, \quad 1 \leq q \leq s, \tag{7.42}$$

and moreover, by construction, a collocation method has *stage order s*, which means that the coefficients also satisfy the condition

$$\sum_{i=1}^{s} b_i c_i^{q-1} = \frac{1}{q}, \quad 1 \leq q \leq s. \tag{7.43}$$

If the last node satisfies $c_s = 1$, the method is called *stiffly accurate*. For such methods, we have $b_j = a_{sj}$ for $j = 1, \ldots, s$, and the last internal stage yields directly the numerical solution $x_1 = X_s$ after one step.

Specific method classes are, among others, the Gauss methods where the nodes c_i are the zeros of the Legendre polynomial

$$\frac{d^s}{dt^s}\left(t^s(t-1)^s\right)$$

and the Radau IIa methods [But64] where the c_i are the zeros of

$$\frac{d^{s-1}}{dt^{s-1}}\left(t^{s-1}(t-1)^s\right).$$

As an example, the Gauss method with $s=2$ stages possesses the nodes

$$c_1 = \frac{1}{2} - \frac{\sqrt{3}}{6}, \qquad c_2 = \frac{1}{2} + \frac{\sqrt{3}}{6},$$

while the Radau IIa method with $s=3$ stages is given by

$$c_1 = \frac{2}{5} - \frac{\sqrt{6}}{10}, \qquad c_2 = \frac{2}{5} + \frac{\sqrt{6}}{10}, \qquad c_3 = 1.$$

The optimality property of the Gauss–Legendre quadrature rule carries over to the time integration scheme [HNW93], which implies that the Gauss methods are of optimal order $2s$ but not stiffly accurate since $c_s < 1$. The Radau IIa methods, on the other hand, are of order $2s - 1$ and stiffly accurate. For stiff and differential-algebraic equations, the latter method class is particularly attractive.

A-Stability and L-Stability

If we apply an implicit Runge–Kutta method (7.41a), (7.41b) to Dahlquist's test equation $\dot{y} = \lambda y$ where $\lambda \in \mathbb{C}^-$, we obtain a recursion of the form $y_{n+1} = R(\mu) y_n$ with $\mu = \tau\lambda \in \mathbb{C}^-$. The stability function R has the form [But08]

$$R(\mu) = 1 + \mu \boldsymbol{b}^T (\boldsymbol{I}_s - \mu \boldsymbol{A})^{-1} \mathbf{1}. \tag{7.44}$$

Here, the method coefficients are denoted by $\boldsymbol{A} = (a_{ij})_{i,j=1}^s$, $\boldsymbol{b} = (b_1, \ldots, b_s)^T$, and $\boldsymbol{c} = (c_1, \ldots, c_s)^T$ while $\boldsymbol{I}_s$ stands for the $s \times s$ identity matrix and $\mathbf{1} = (1, \ldots, 1)^T$.

A-stable methods satisfy $|R(\mu)| \leq 1$ for all $\mu \in \mathbb{C}^-$, and both the Gauss and the Radau IIa methods meet this requirement. The Radau IIa methods are also *L-stable*, which means that

$$\lim_{\mu \to \infty} R(\mu) = 0. \tag{7.45}$$

This property will play an important role when we study implicit Runge–Kutta methods at stiff mechanical systems in Sect. 7.3.

7.2.4 Application to Differential-Algebraic Equations

Implicit Runge–Kutta methods (7.41a), (7.41b) are extended to linear-implicit systems $\boldsymbol{E}\dot{\boldsymbol{x}} = \boldsymbol{\phi}(\boldsymbol{x}, t)$ by assuming for the moment that the matrix $\boldsymbol{E}$ is invertible and

7.2 BDF and Implicit Runge–Kutta Methods

discretizing $\dot{x} = E^{-1}\phi(x,t)$. Multiplying the resulting scheme by E, we get the method definition

$$EX_i = Ex_0 + \tau \sum_{j=1}^{s} a_{ij}\phi(X_j, t_0 + c_j\tau), \quad i = 1, \ldots, s; \tag{7.46a}$$

$$x_1 = \left(1 - \sum_{i,j=1}^{s} b_i \gamma_{ij}\right) x_0 + \tau \sum_{i,j=1}^{s} b_i \gamma_{ij} X_j \tag{7.46b}$$

where $(\gamma_{ij}) = (a_{ij})^{-1}$ is the inverse of the coefficient matrix. Obviously, (7.46a), (7.46b) makes sense also in the case where E is singular.

The advantage of using stiffly accurate methods for differential-algebraic equations is more evident if we consider the equivalent discretization of the semi-explicit system (2.36a), (2.36b). The method (7.46a), (7.46b) then becomes

$$Y_i = y_0 + \tau \sum_{j=1}^{s} a_{ij} a(Y_j, Z_j), \quad i = 1, \ldots, s, \tag{7.47a}$$

$$0 = b(Y_i, Z_i), \tag{7.47b}$$

for the internal stages and

$$y_1 = y_0 + \tau \sum_{j=1}^{s} b_j a(Y_j, Z_j), \tag{7.48a}$$

$$z_1 = \left(1 - \sum_{i,j=1}^{s} b_i \gamma_{ij}\right) z_0 + \tau \sum_{i,j=1}^{s} b_i \gamma_{ij} Z_j \tag{7.48b}$$

as update for the numerical solution after one step. For stiffly accurate methods, we have $\sum_{i,j=1}^{s} b_i \gamma_{ij} = 1$ and $y_1 = Y_s, z_1 = Z_s$. The update (7.48a), (7.48b) is hence superfluous and furthermore, the constraint $0 = b(y_1, z_1)$ is satisfied by construction.

Convergence Results

Starting with [Pet86, HLR89], implicit Runge–Kutta methods for DAEs have been the subject of extensive research, and we restrict our exposition here to a summary of the established results for the Radau IIa methods as given in [HLR89, HW96].

For systems of index 1, these methods retain their order $2s - 1$ in all variables, which follows by the implicit function theorem in the same fashion as for the BDF methods in (7.25). Systems of index 2 require a deeper analysis since the order reduction comes again into play. Taking a similar approach as in the multistep case in Sect. 7.2.2, the local error of a Radau IIa method with s stages turns out to be

$$y(t_1) - y_0 = \mathcal{O}(\tau^{2s}), \quad z(t_1) - z_1 = \mathcal{O}(\tau^s). \tag{7.49}$$

The substantial order drop in the algebraic variable is related to the stage order s of the method. Despite this order reduction, the methods still converge with error bounds

$$y(t_n) - y_n = \mathcal{O}(\tau^{2s-1}), \quad z(t_n) - z_n = \mathcal{O}(\tau^s). \tag{7.50}$$

With respect to the differential variable, the classical order is thus preserved.

Practical Aspects and Remarks

Efficient methods for solving the nonlinear system (7.46a), (7.46b) by means of simplified Newton iterations and a transformation of the coefficient matrix are discussed in [HW96]. In practice, the code RADAU5, based on the 5th order Radau IIa method with $s = 3$ stages, has stood the test as versatile and robust integration scheme for stiff and differential-algebraic equations and represents an alternative to the BDF codes. An extension to a variable order method called RADAU with $s = 3, 5, 7$ stages and corresponding order $5, 9, 13$ is presented in [HW99].

In case of an index-2 system, both codes RADAU5 and RADAU employ the scaled norm (7.37) in order to absorb the effect of order reduction on the variable stepsize algorithm and in order to guarantee convergence of the simplified Newton iteration.

Recently, a new class of collocation methods called *SAFERK* (Strongly A-stable First stage Explicit Runge–Kutta methods) has been found that is also well-suited for the integration of stiff and differential-algebraic systems [GPHAM10]. Computationally, a SAFERK method with s stages is equivalent to the $(s - 1)$-stage Radau IIA method and has the same order $2s - 3$, whereas the stage order is one unit higher. Although there are no L-stable schemes in this method family, there exists a free parameter that can be selected in order to minimize the error coefficients or to maximize the numerical dissipation.

7.2.5 Solving Constrained Mechanical Systems in Practice

We have seen above that for both the BDF and the Radau IIa methods, the solution of DAEs of index 2 comes at the price of an order reduction in the local error of the algebraic variable. The equations of motion (7.1a)–(7.1c), however, are of index 3 and involve even stronger effects. E.g., for a BDF method of order k applied to (7.1a)–(7.1c), the local error is of the form

$$p(t_k) - p_k = \mathcal{O}(\tau^{k+1}), \quad v(t_k) - v_k = \mathcal{O}(\tau^k), \quad \lambda(t_k) - \lambda_k = \mathcal{O}(\tau^{k-1}). \tag{7.51}$$

7.2 BDF and Implicit Runge–Kutta Methods

In other words, two powers of τ are lost in the algebraic variable, which is the Lagrange multiplier λ, and also the velocity variable, which is a differential component, is affected. Moreover, global convergence can only be shown for constant stepsizes [BCP96]. For the Radau IIa method with s stages, the local error bound reads [HLR89]

$$p(t_1) - p_1 = \mathcal{O}(\tau^{2s}), \quad v(t_1) - v_1 = \mathcal{O}(\tau^s), \quad \lambda(t_1) - \lambda_1 = \mathcal{O}(\tau^{s-1}). \quad (7.52)$$

Similar to the BDF methods, the velocity components are now affected by the order reduction, and overall, the direct solution of (7.1a)–(7.1c) is not advisable for both method classes.

We discuss instead the relevant aspects when applying either BDF or Radau IIa methods to a formulation with reduced index, i.e., the GGL-formulation (7.4) or the formulation (7.5) of index 2.

Applying BDF and Radau IIa Methods

The codes DASSL and IDA solve DAEs in the implicit form (2.32), i.e., $F(\dot{x}, x, t) = 0$. In case of the GGL-formulation (7.4), we thus need to move all terms to the left hand side and provide a subroutine that evaluates the equations in this *residual form* for given states $x = (p, v, \lambda, \mu)$ and derivatives $\dot{x}$. The differential variables are (p, v) in this setup, the algebraic ones (λ, μ). Similarly, Eqs. (7.5) lead to $x = (p, v, w, \lambda)$ where the algebraic variables consist of (w, λ).

Alternatively, the codes RADAU and RADAU5 are based on the linear-implicit DAE (2.33), i.e., $E\dot{x} = \phi(x, t)$, and the user provides the constant matrix E and the right hand side vector ϕ. If the mass matrix M depends on p, this involves the introduction of extra acceleration variables w as in (7.5). Additionally, for DAEs of index 2 the partitioning of the states into differential and algebraic variables is specified such that the scaled norm (7.37) can be applied.

Block Gaussian Elimination

Both the BDFs and the implicit Runge–Kutta methods require a formulation of the equations of motion as first order system, which obviously increases the size of the linear systems within Newton's method. For an efficient and robust implementation, it is crucial to apply block Gaussian elimination to reduce the dimension in such a way that only a system similar to (7.6) remains to be solved in each iteration. We illustrate this procedure with implicit Euler, which belongs to both the BDF methods ($k = 1$) and the Radau IIa methods ($s = 1$).

If we apply implicit Euler to the system (7.5) of index 2 with additional acceleration variables, we get the nonlinear system

$$\begin{aligned}
p_1 - p_0 - \tau v_1 &= 0, \\
v_1 - v_0 - \tau w_1 &= 0, \\
M(p_1)w_1 - f(p_1, v_1, t_1) + G(p_1)^T \lambda_1 &= 0, \\
G(p_1) v_1 &= 0
\end{aligned} \qquad (7.53)$$

with $(3n_p + n_\lambda) \times (3n_p + n_\lambda)$ Jacobian matrix

$$\begin{pmatrix} I & & -\tau I & 0 & 0 \\ 0 & & I & -\tau I & 0 \\ (Mw - f + G^T\lambda)_p & -f_v & M & G^T \\ (Gv)_p & & G & 0 & 0 \end{pmatrix}. \qquad (7.54)$$

Here, partial derivatives are written in short as $f_v = \partial f / \partial v$ and so forth. By block elimination of the first two lines, it is possible to just solve a linear system of size $n_p + n_\lambda$ with the matrix

$$\begin{pmatrix} M - \tau f_v + \tau^2 (Mv - f + G^T\lambda)_p & G^T \\ G + \tau(Gv)_p & 0 \end{pmatrix} \qquad (7.55)$$

for the new iterates of w_1 and λ_1. The updates for position and velocity are then given explicitly by the first two block lines of (7.54).

The codes RADAU5 and RADAU offer this block elimination for second order systems as extra options. In case of the GGL-formulation (7.4), the first line in (7.53) is replaced by

$$p_1 - p_0 - \tau v_1 + \tau G(p_1)^T \mu_1 = 0,$$

and the position constraint $g(p_1) = 0$ is added. At first sight, the term with the extra multiplier μ_1 perturbs the block structure and renders the block elimination more expensive. However, from the exact solution we know that $\mu(t_1) = 0$, and this justifies to neglect the partial derivative $(G(p)^T \mu)_p$ in the Jacobian matrix. In this way, the block structure of (7.54) carries over to the stabilized formulation, and only the extra position constraint needs to be taken into account.

Ill-Conditioned Linear Systems

Besides the computational savings, the above block elimination technique has also benefits with respect to the condition number. Ill-conditioned matrices are an important issue in the simulation of constrained mechanical systems. We distinguish between two sources of ill-conditioning: One is due to stiffness, i.e., widely different time scales in the equations of motion, and this situation is often present in flexible multibody dynamics. Recall from the partitioned equations of motion (6.4a), (6.4b) with rigid (slow) motion q and elastic (fast) motion d that the force vector f in a stiff mechanical system reads

$$f(p,\dot{p},t) = \begin{pmatrix} f_r(q,d,\dot{q},\dot{d},t) \\ f_h(q,d,\dot{q},\dot{d},t) \end{pmatrix} - \frac{1}{\epsilon^2} \begin{pmatrix} 0 \\ \nabla W_s(d) \end{pmatrix}.$$

For increasing stiffness, i.e., for $\epsilon \to 0$, the Hessian of the stiff potential W_s will dominate the Jacobian (7.54) and hence lead to an ill-conditioned system [Lub93].

The other source stems from the discretization of a DAE of higher index. When discussing the simplified Newton iteration (7.36), we saw already that the iteration matrix contains blocks of different orders, and this structure may also lead to ill-conditioning. While the situation is not that severe for systems of index 2, the direct discretization of the index-3-formulation (7.1a)–(7.1c) demands for an appropriate scaling. In this context, we mention [Bau10, BBC07] for approaches that further develop the ideas of [LP86] and [HLR89] and lead to stepsize-independent condition numbers. This does, however, not cure the issue of order reduction and impaired global convergence.

As a general rule of thumb, the Lagrange multipliers are the variables that require most attention when dealing with such ill-conditioned linear systems. E.g., the matrix (7.55) that has been obtained from block Gaussian elimination features entries that refer to either small displacement variables or large constraint forces, i.e., Lagrange multipliers. If we scale the constraint $0 = G(p)v$ by a large factor $c_s \gg 1$ and replace it by $0 = c_s G(p)v$, this helps to equilibrate the system and has a positive effect on the condition number. Besides the above references, we also mention [Arn95] for additional results in this context.

7.3 Order Reduction for Stiff Mechanical Systems

While the previous section has provided a summary of established numerical methods for differential-algebraic equations in general and constrained mechanical systems in particular, this section continues the discussion of Chap. 6 and deals with the issues that occur when integrating stiff mechanical systems by implicit methods. A general theory on the behavior of Runge–Kutta methods in such situations has been elaborated by Lubich [Lub93]. Here, however, we concentrate mainly on a simplified test problem that generalizes the Prothero–Robinson model [PR74] to second order form [Sim98b].

7.3.1 Beam under Point Load

We start with a numerical example, a planar beam that is subjected to a time-dependent point load, Fig. 7.2. The beam has fixed ends, and its vertical displacement $w = w(x_1, t)$ satisfies

$$\rho A \ddot{w} + EI w'''' = F,$$

Fig. 7.2 Beam under point load, discretized by four cubic finite elements

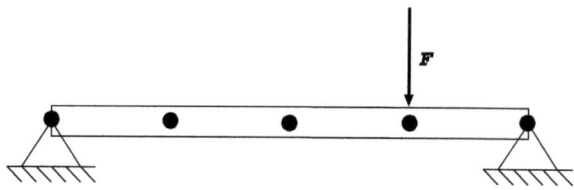

which is the dispersive wave equation (1.5) with modified right hand side, cf. also the beam model in Sect. 4.4.2.

Spatial discretization by four cubic finite elements and incorporation of the boundary conditions leads to the second order system

$$M_h \ddot{d} = -A_h d + l_h(t)$$

in $n_d = 8$ nodal variables d, cf. the equations of structural dynamics (5.4). Performing an eigenvalue analysis and transforming the equations via (5.61) to modal coordinates η_i, we finally arrive at the decoupled system

$$\ddot{\eta}_i = -\omega_i^2 \eta_i + \beta_i(t), \quad i = 1, \ldots, 8 \tag{7.56}$$

for the vertical displacement of the beam's neutral fiber.

The material constants and lengths chosen are typical for practical applications in multibody dynamics and result in frequencies ranging from $\omega_1 = 1.277 \cdot 10^3$ to $\omega_8 = 1.019 \cdot 10^5$ (in dimensionless form). As excitation, we define $\beta_i(t) := \zeta_i \sin(\Omega t)$ where $\Omega = 75$, which allows to determine the exact solution of (7.56). It reads

$$\eta_i(t) = \frac{\zeta_i}{\omega_i^2 - \Omega^2} \sin(\Omega(t-t_0)) + c_1 \cos(\omega_i(t-t_0)) + c_2 \sin(\omega_i(t-t_0)). \tag{7.57}$$

Here, the first term defines the smooth motion, and the following two terms represent an overlay of additional oscillations. Depending on the constants c_1 and c_2, which in turn follow from the initial values, the oscillations show up or not. Note also that $\Omega = \omega_i$ for some i would result in a resonance instability.

We now solve (7.56) by an implicit Runge–Kutta method of type Radau IIa with $s = 3$ stages and constant stepsize. The initial values are chosen on the smooth motion trajectory, and we evaluate the global error in the different modes when integrating over $[t_0, t_1] = [0, 0.16]$.

Figure 7.3 displays, in double logarithmic scale, the observed global error depending on the stepsize τ for the components η_i, $i = 1, 2, 3, 5$. Since the fourth mode η_4 vanishes identically due to $\zeta_4 = 0$, it is not taken into account in the diagrams.

This numerical example reveals an order reduction in both position and velocity variables. The reduction mainly shows up for *large stepsizes* and *higher frequencies* ω_i. With respect to the position coordinates η_i, the order drops from $2s - 1 = 5$ to $s - 1 = 2$, with respect to the velocities $\dot{\eta}_i$ to $s = 3$. On the other hand, for fixed

7.3 Order Reduction for Stiff Mechanical Systems

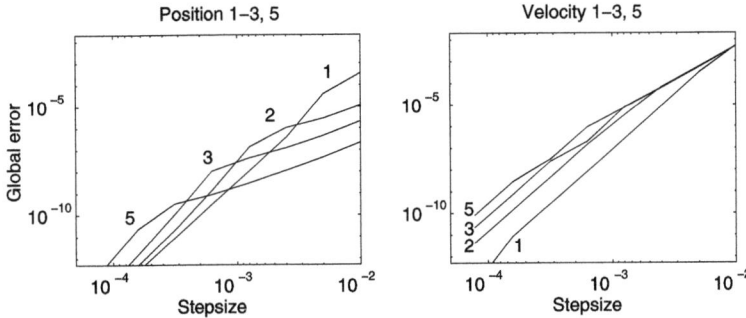

Fig. 7.3 Order reduction Radau IIa, $s = 3$, when solving (7.56)

and large stepsize τ, the error in the position components decreases with increasing frequency, despite the order reduction.

Clearly, the observed order reduction will also arise if this particular beam is part of a flexible multibody system and the time integration proceeds with relatively large stepsizes. In order to better understand this phenomenon, we next study the behavior of implicit time integrators in more detail. Note that we do not solve a DAE problem here but a singular singularly perturbed ODE system as in (6.7). In the limit, the singularly perturbed system, however, turns into a constrained mechanical system and this process is reflected by the behavior of numerical methods.

7.3.2 A Model Problem

Consider the second order initial value problem

$$\ddot{d} = -\omega^2(d - \phi) + \ddot{\phi}, \qquad d(t_0) = \phi(t_0) + \zeta, \quad \dot{d}(t_0) = \dot{\phi}(t_0) + \gamma, \qquad (7.58)$$

with given smooth function ϕ and frequency ω. This analogue of the famous Prothero–Robinson problem [PR74]

$$\dot{y} = \lambda(y - \phi) + \dot{\phi}$$

has been introduced by van der Houwen and Sommeijer [vdHS87] as a test example for Runge–Kutta–Nyström methods. The exact solution of (7.58) reads

$$d(t) = \phi(t) + \frac{\gamma}{\omega}\sin(\omega(t - t_0)) + \zeta \cos(\omega(t - t_0)),$$

compare the beam modes (7.57) in the example above.

In contrast to [vdHS87], we are mainly interested in smooth solutions of (7.58) where the initial values lie on the slowly varying function ϕ, i.e., where $\zeta = \gamma = 0$, but the following analysis covers also to some extent the case of oscillatory perturbations. If additionally the frequency ω is large, the model equation (7.58) provides much insight into the behavior of numerical methods at stiff mechanical systems [Sim98b].

Written as first order system, (7.58) reads

$$\dot{y} = J(y - \phi) + \dot{\phi}, \qquad y(t_0) = \phi(t_0) + \begin{pmatrix} \zeta \\ \gamma \end{pmatrix} \qquad (7.59)$$

where

$$y = \begin{pmatrix} d \\ \dot{d} \end{pmatrix}, \quad \phi = \begin{pmatrix} \phi \\ \dot{\phi} \end{pmatrix}, \quad J = \begin{pmatrix} 0 & 1 \\ -\omega^2 & 0 \end{pmatrix}.$$

If we apply a numerical method to (7.59), the recursion for the error propagation follows in a straightforward way due to the linearity of the problem [Sch89, SW95]. However, when studying the influence of the local error, we will in the sequel not diagonalize but rather derive explicit formulas which distinguish between the errors with respect to position d and velocity $\dot{d}$.

7.3.3 Runge–Kutta Methods

One step of an implicit Runge–Kutta method (7.41a), (7.41b) with s stages and stepsize τ for the model problem (7.59) is given by

$$y_1 = y_0 + \tau \sum_{j=1}^{s} b_j \left[J\left(Y_j - \phi(t_0 + c_j \tau)\right) + \dot{\phi}(t_0 + c_j \tau)\right], \qquad (7.60)$$

$$Y_i = y_0 + \tau \sum_{j=1}^{s} a_{ij} \left[J\left(Y_j - \phi(t_0 + c_j \tau)\right) + \dot{\phi}(t_0 + c_j \tau)\right], \quad i = 1, \ldots, s.$$

Here, the method coefficients are again denoted by $A = (a_{ij})_{i,j=1}^{s}$, $b = (b_1, \ldots, b_s)^T$, and $c = (c_1, \ldots, c_s)^T$. The smooth solution ϕ satisfies (7.60) with a defect or truncation error Δ_0,

$$\phi(t_0 + h) = \phi(t_0) + \tau \sum_{j=1}^{s} b_j \dot{\phi}(t_0 + c_j \tau) + \Delta_0, \qquad (7.61)$$

$$\phi(t_0 + c_i \tau) = \phi(t_0) + \tau \sum_{j=1}^{s} a_{ij} \dot{\phi}(t_0 + c_j \tau) + \Delta_{0,i}, \quad i = 1, \ldots, s.$$

In order to get an explicit expression for the difference $y_1 - \phi(t_0 + \tau)$, the corresponding right hand sides are now subtracted and the internal stages Y_i and $\phi(t_0 + c_i \tau)$ eliminated. After some calculations, we find the relation

$$y_1 - \phi(t_0 + \tau) = R(\tau J)(y_0 - \phi(t_0)) + \delta_0 \qquad (7.62)$$

7.3 Order Reduction for Stiff Mechanical Systems

with the *stability matrix*

$$R(hJ) = I_2 + (b^T \otimes \tau J)(I_{2s} - A \otimes \tau J)^{-1}(1 \otimes I_2),$$

which is the stability function (7.44) of the method for the matrix argument τJ. Moreover, I_k denotes the $k \times k$ identity matrix, $1 = (1, \ldots, 1)^T$ the vector of ones, and the Kronecker product $\otimes$ is used to obtain a compact notation. In (7.62), the local error δ_0 consists of defect terms,

$$\delta_0 = -(b^T \otimes \tau J)(I_{2s} - A \otimes \tau J)^{-1}\Delta_{0,Y} - \Delta_0 \qquad (7.63)$$

with $\Delta_{0,Y} = (\Delta_{0,1}, \ldots, \Delta_{0,s})^T$.

Finally, we apply (7.62) to compute recursively the global error after $n+1$ steps and obtain

$$y_{n+1} - \phi(t_{n+1}) = R(\tau J)^{n+1}(y_0 - \phi(t_0)) + \sum_{j=0}^{n} R(\tau J)^{n-j}\delta_j. \qquad (7.64)$$

The formula (7.64) is completely analogous to the one characterizing Runge–Kutta methods at the scalar Prothero–Robinson model. However, as the matrix J is not normal, one has to be careful when deriving bounds on the error propagation in terms of eigenvalues of the stability matrix $R(\tau J)$. We therefore rescale (7.64) by $S = \mathrm{diag}\,(1, 1/\omega)$ and get in this way

$$S(y_{n+1} - \phi(t_{n+1})) = R(\tau \tilde{J})^{n+1}S(y_0 - \phi(t_0)) + \sum_{j=0}^{n} R(\tau \tilde{J})^{n-j}S\delta_j \qquad (7.65)$$

where $\tilde{J}$ stands for the skew symmetric and thus normal matrix

$$\tilde{J} = SJS^{-1} = \begin{pmatrix} 0 & \omega \\ -\omega & 0 \end{pmatrix}.$$

A closer look at the scaled stability matrix in (7.65) shows that $R(\tau \tilde{J})$ is, like $\tilde{J}$, a normal matrix and given by

$$R(\tau \tilde{J}) = \begin{pmatrix} 1 - \tau^2\omega^2 b^T W^{-1} A1 & \tau\omega b^T W^{-1}1 \\ -\tau\omega b^T W^{-1}1 & 1 - \tau^2\omega^2 b^T W^{-1} A1 \end{pmatrix}$$

where $W := I_s + \tau^2\omega^2 A^2$. This can be easily seen by a permutation of rows and columns of the matrix $I_{2s} - A \otimes \tau \tilde{J}$ such that the position and velocity entries are grouped together. Due to the structure of $\tilde{J}$, the corresponding permutation of $(I_{2s} - A \otimes \tau \tilde{J})^{-1}$ has then the form

$$\begin{pmatrix} W^{-1} & \tau\omega W^{-1} A \\ -\tau\omega W^{-1} A & W^{-1} \end{pmatrix},$$

and the above formula for $R(\tau \tilde{J})$ follows by taking the permutation in $b^T \otimes \tau \tilde{J}$ and $\mathbf{1} \otimes I_2$ into account.

The stability matrix $R(\tau \tilde{J})$ has eigenvalues $R(\pm i\tau\omega)$ which are given by evaluating the scalar stability function $R(z) = 1 + zb^T(I - zA)^{-1}\mathbf{1}$ for imaginary arguments $z = \pm i\tau\omega$. Furthermore, its eigenvectors are $(1, i)^T$ and $(1, -i)^T$, and consequently we have

$$R(h\tilde{J})^k = T \begin{pmatrix} R(i\tau\omega)^k & 0 \\ 0 & R(-i\tau\omega)^k \end{pmatrix} T^{-1}, \quad T = \begin{pmatrix} 1 & 1 \\ i & -i \end{pmatrix}. \tag{7.66}$$

Thus, the scaled global error (7.65) is determined by the behavior of the stability function on the imaginary axis. In particular for large stepsizes $\tau \gg 1/\omega$, L-stable methods *strongly damp errors of previous steps* and the current local error δ_n dominates.

We analyze the situation in more detail for the class of Radau IIa methods, which are of classical order $2s - 1$ in the ODE case. Since the coefficients satisfy then $a_{si} = b_i$ for all i, the last internal stage yields already the numerical solution, and the local error (7.63) simplifies to

$$\delta_n = -(e_s^T \otimes I_2)(I_{2s} - A \otimes hJ)^{-1}\Delta_{n,Y}$$

with unit vector $e_s = (0, \ldots, 0, 1)^T$. Like for the stability matrix above, the rows and columns of $I_{2s} - A \otimes \tau J$ are rearranged such that the position and velocity entries are grouped together. With corresponding permutations in $e_s^T \otimes I_2$ and $\Delta_{n,Y}$, the local error becomes

$$\delta_n = \begin{pmatrix} e_s^T(I_s + \tau^2\omega^2 A^2)^{-1}(-\Delta_{n,d} - \tau A \Delta_{n,\dot{d}}) \\ e_s^T(I_s + \tau^2\omega^2 A^2)^{-1}(\tau\omega^2 A \Delta_{n,d} - \Delta_{n,\dot{d}}) \end{pmatrix}. \tag{7.67}$$

Note that $\Delta_{n,d}$ and $\Delta_{n,\dot{d}}$ stand for the stage defects with respect to position d and velocity $\dot{d}$ and just contain the permuted entries of $\Delta_{n,Y}$.

As mentioned above, we are in particular interested in the situation where the stepsize τ is appropriate for tracking the smooth motion ϕ but rather large compared to the frequency ω. In other words, we want to study global and local error when $\tau \to 0$ and $\tau\omega \gg 1$. In this case, the stability function of the Radau IIa methods tends to zero, and from the global error formula (7.65) it follows that the local error δ_n dominates the global error.

Since Radau IIa methods have stage order s, the defects $\Delta_{n,d}$ and $\Delta_{n,\dot{d}}$ in tracking the smooth motion are of order $\mathcal{O}(\tau^{s+1})$. Furthermore, the coefficient matrix A is invertible, and for $\tau\omega \gg 1$ we have

$$e_s^T(I_s + \tau^2\omega^2 A^2)^{-1} = \mathcal{O}(\tau^{-2}\omega^{-2}).$$

In terms of the stiffness parameter $\epsilon = 1/\omega$, the behavior of this class of Runge–Kutta methods for $\tau/\epsilon \gg 1$ can now be summarized as follows:

7.3 Order Reduction for Stiff Mechanical Systems

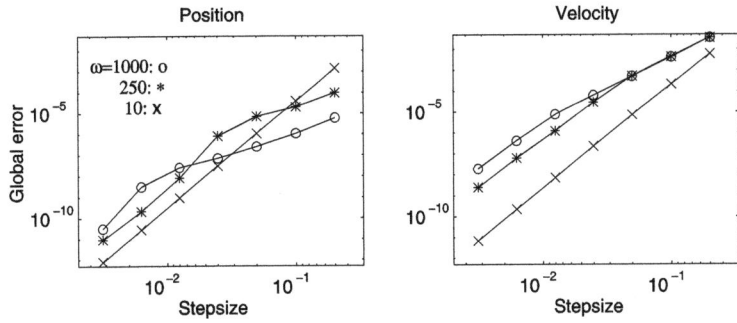

Fig. 7.4 Order reduction of Radau IIa, $s = 3$, when solving (7.59)

Theorem 7.3 *Assume $\tau \to 0$ and $\tau \gg \epsilon$ when applying the Radau IIa method to the model equation (7.59). Then, the local error $\delta_{n,d}$ with respect to the position variable d satisfies*

$$\delta_{n,d} = \mathcal{O}(\epsilon^2 \tau^{s-1}) \qquad (7.68)$$

and for the local error $\delta_{n,\dot{d}}$ with respect to the velocity variable $\dot{d}$ we have

$$\delta_{n,\dot{d}} = \mathcal{O}(\epsilon^2 \tau^{s-1}) + \mathcal{O}(\tau^s). \qquad (7.69)$$

Moreover, the local error $\delta_n = (\delta_{n,d}, \delta_{n,\dot{d}})$ dominates in the global error recursion (7.65).

This result proves that the order of both the local and the global error drops considerably in case of stepsizes $\tau \gg \epsilon$. Nevertheless, we still obtain an asymptotically correct result. Note that this order reduction is much larger than the one occurring at the standard Prothero–Robinson test equation. In particular, the $\mathcal{O}(\tau^s)$ term in (7.69) characterizes also the global velocity error in case of a DAE of index 3 [HLR89, Chap. 6], cf. (7.52), and indicates that the stiff problem (7.58) is very close to a constrained mechanical system. The $\mathcal{O}(\epsilon^2 \tau^{s-1})$ term in (7.68), on the other hand, leads due to $\epsilon \ll \tau$ to a decreasing error and is thus not relevant.

Figure 7.4 shows order diagrams of the Radau IIa method with $s = 3$ stages, which is the basis of the RADAU5 code [HW96, HW99]. As smooth solution, $\phi(t) = \cos(6t)$ is taken, the initial values $d_0 = 1$ and $\dot{d}_0 = 0$ lie on ϕ, and the global error at $t_1 = 2.2$ is plotted against the stepsize for different frequencies ω. Looking first at the position d, we notice classical order 5 for the low frequency $\omega = 10$. With increasing ω, the order drops for large stepsizes to $s - 1 = 2$, which is in perfect agreement with the local error behavior (7.68). On the other hand, because of the factor ϵ^2, the global error decreases with increasing ω for large stepsizes and the results are then very accurate. The situation is different for the velocity $\dot{d}$. Here, the $\mathcal{O}(h^s)$ term dominates for large stepsizes due to $\tau \gg \epsilon$. We observe order $s = 3$ and also an error growth with increasing ω. Summarizing, this numerical test confirms the results observed for the beam example (7.56).

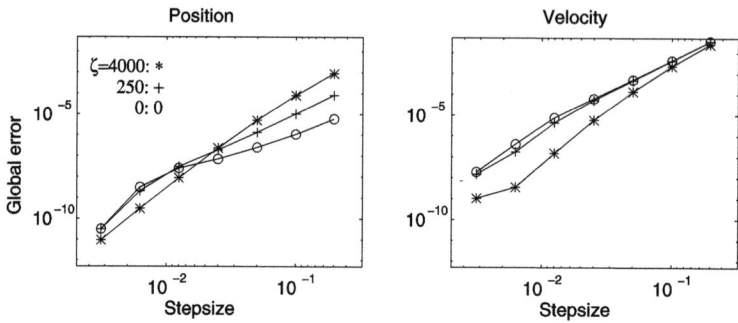

Fig. 7.5 Radau IIa, $s = 3$, for the model (7.70) with $\omega = 1000$ and different damping parameters

Dissipation

The above analysis concentrated on a purely oscillatory test case, omitting any influence of an additional physical damping term. It is easy to analyze this situation by modifying the model (7.58) to

$$\ddot{d} = -\omega^2(d - \phi) - 2\zeta(\dot{d} - \dot{\phi}) + \ddot{\phi}, \tag{7.70}$$

which includes a damping term with parameter $\zeta \geq 0$ and leads to eigenvalues

$$\kappa_{1,2} = -\zeta \pm \sqrt{\zeta^2 - \omega^2}.$$

Accordingly, the ratio ζ/ω determines the solution behavior and either leads to a damped oscillation or an exponential decay to the smooth motion. In case of Rayleigh damping (3.20), we have $\zeta = \zeta_1 \omega^2/2 + \zeta_2/2$, and beyond a certain threshold where $\zeta = \omega$, the *critical damping*, the oscillation is suppressed.

To study the influence of the damping term, we fix the frequency ω and vary the parameter ζ. The limit case $1/\zeta \to 0$ leads to a constraint equation $0 = \dot{d} - \dot{\phi}$ for the velocity, which corresponds to a system of index 2 and thus diminishes the order reduction. Figure 7.5 displays this effect for $\omega = 1000$ and different values of ζ. Increasing the dissipation results in a better approximation of the velocity variable, in particular in the medium stepsize range.

The bottom line is thus that physical dissipation not only influences the stability of the differential equation at hand but also has a beneficial effect on the quality of the discretization. This leaves the question unanswered whether or not it makes, from an engineering viewpoint, sense to include dissipation in the flexible members of a multibody system. If dissipation is omitted, e.g., in order to trace a resonance phenomenon, and high frequencies are present, one has to be aware of the implications of a stiff mechanical system.

A detailed numerical analysis of strongly damped mechanical systems is given in [Stu08].

7.3 Order Reduction for Stiff Mechanical Systems

The General Case

The above analysis can be easily extended to other Runge–Kutta methods. The more general approach by Lubich [Lub93], on the other hand, studies the discretization of the singularly perturbed system (6.7) under mild assumptions on the method, i.e., the coefficient matrix A is required to be invertible and $|R(\infty)| < 1$. Using asymptotic expansions and an invariant manifold theorem, it is shown there that in case of the Radau IIa methods for $\tau \gg \epsilon$, the global error satisfies

$$p_{n+1}^\epsilon - p^\epsilon(t_{n+1}) = p_{n+1}^0 - p^0(t_{n+1}) + \mathcal{O}(\epsilon^2 \tau^{s-2}), \tag{7.71a}$$

$$\dot{p}_{n+1}^\epsilon - \dot{p}^\epsilon(t_{n+1}) = \dot{p}_{n+1}^0 - \dot{p}^0(t_{n+1}) + \mathcal{O}(\epsilon^2 \tau^{s-2}). \tag{7.71b}$$

Here, p_{n+1}^ϵ stands for the numerical approximation to the smooth motion $p^\epsilon(t_{n+1})$, which is given by the outer expansion (6.9). On the right hand side, we observe the first term $p^0(t_{n+1})$ of the expansion and its numerical counterpart p_{n+1}^0 plus an additional term of order $\mathcal{O}(\epsilon^2 \tau^{s-2})$, which is, like above, of minor relevance.

Most important, however, is the behavior of the method for the limiting system of index 3, which defines the function $p^0(t)$. Since this system is linear with respect to the Lagrange multiplier, we have [HLR89]

$$p_{n+1}^0 - p^0(t_{n+1}) = \mathcal{O}(\tau^{2s-1})$$

with respect to the position variables and

$$\dot{p}_{n+1}^0 - \dot{p}^0(t_{n+1}) = \mathcal{O}(\tau^s)$$

with respect to the velocities. Like in Theorem 7.3, this implies a severe order reduction in the latter variables.

The results of [Lub93] are in some sense surprising since the stability condition $|R(\infty)| < 1$, which stems from a linear test equation, turns out to be powerful enough to derive advanced convergence estimates in the fashion of (7.71a), (7.71b).

Additional Notes

The scaled error recursion (7.65) shows that the global velocity error contains, besides the current local error, a series with terms of the form $c \cdot \omega R(i\tau\omega)^{n-j} \delta_{j,d}$. This suggests that local errors from previous steps still play a role here. However, due to the ϵ^2-factors in $\delta_{j,d}$ and the strong decay of the stability function, these terms are of minor influence in the stiff case.

If the initial values are chosen off the smooth motion trajectory, the *numerical dissipation* of the method results in a decay of the corresponding high frequency oscillations where the convergence rate to the smooth motion depends on the ratio τ/ϵ. We will come back to this point at the end of the next section.

7.3.4 Rosenbrock–Wanner Methods

Rosenbrock–Wanner (ROW) methods are very popular for solving stiff ODEs [SW95]. It is thus interesting to study this method class for the model problem (7.59). In non-autonomous form, one step of a ROW method reads

$$y_1 = y_0 + \tau \sum_{j=1}^{s} b_j K_j, \qquad (7.72)$$

$$K_i = J\left(y_0 + \sum_{j=1}^{i-1} \alpha_{ij} K_j - \boldsymbol{\phi}(t_0 + \alpha_i \tau)\right) + \dot{\boldsymbol{\phi}}(t_0 + \alpha_i \tau)$$

$$+ \gamma_i \tau(-J\dot{\boldsymbol{\phi}}(t_0) + \ddot{\boldsymbol{\phi}}(t_0)) + J \sum_{j=1}^{i} \gamma_{ij} K_j, \quad i = 1, \ldots, s.$$

Here, b_i, α_{ij}, and γ_{ij} for $i = 1, \ldots, s$ and $j = 1, \ldots, i$ denote the method coefficients, with additionally $\alpha_i := \sum_{j=1}^{i-1} \alpha_{ij}$ and $\gamma_i := \sum_{j=1}^{i} \gamma_{ij}$.

Straightforward calculations yield, like in the Runge–Kutta case, a local error relation (7.62) and a global error recursion of the form (7.64). But the stability matrix $R(\tau J)$ and the local error δ_n are now given by different expressions. We skip the details and summarize in the following the results.

Let $b = (b_1, \ldots, b_s)^T$ and $B = (\beta_{ij})_{i,j=1}^{s}$ where $\beta_{ij} = \alpha_{ij} + \gamma_{ij}$, then the stability matrix takes a form analogous to the Runge–Kutta case:

$$R(\tau J) = \begin{pmatrix} 1 - \tau^2 \omega^2 b^T W^{-1} B\mathbf{1} & \tau b^T W^{-1} \mathbf{1} \\ -\tau \omega^2 b^T W^{-1} \mathbf{1} & 1 - \tau^2 \omega^2 b^T W^{-1} B\mathbf{1} \end{pmatrix}$$

where $W := I_s + \tau^2 \omega^2 B^2$. The calculation of the local error leads to

$$\delta_n = \boldsymbol{\phi}(t_n) - \boldsymbol{\phi}(t_n + \tau) + (b^T \otimes I_2)(I_{2s} - B \otimes \tau J)^{-1}(\mathbf{1} \otimes \tau J \boldsymbol{\phi}(t_n) + r_n),$$

with the components of the $2s$-dimensional vector r_n given by

$$\begin{pmatrix} r_{n,2i-1} \\ r_{n,2i} \end{pmatrix} = -\tau J\left(\boldsymbol{\phi}(t_n + \alpha_i \tau) + \tau \gamma_i \dot{\boldsymbol{\phi}}(t_n)\right) + \tau \dot{\boldsymbol{\phi}}(t_n + \alpha_i \tau) + \tau^2 \gamma_i \ddot{\boldsymbol{\phi}}(t_n)$$

for $i = 1, \ldots, s$.

Again, the local error can be further investigated if we distinguish between position error $\delta_{n,d}$ and velocity error $\delta_{n,\dot{d}}$. Exploiting the structure of J, we arrive at the following expressions:

$$\delta_{n,d} = \phi(t_n) - \phi(t_n + \tau) \qquad (7.73)$$

$$+ b^T W^{-1}\left(-\tau^2 \omega^2 B \mathbf{1} \phi(t_n) + \mathbf{1}\tau \dot{\phi}(t_n) + \tau B r_{n,\dot{d}}\right),$$

7.3 Order Reduction for Stiff Mechanical Systems

Table 7.1 Local error coefficients of Rosenbrock–Wanner method. The notation $(\alpha_i)^k$ stands for an s-dimensional vector with entries α_i^k, $i = 1, \ldots, s$

$\delta_{n,d}$	$\xi_2 = b^T W^{-1} B (\tau^2 \omega^2 (\alpha_i)^2 + 2\mathbf{1}) - 1$
	$\xi_3 = b^T W^{-1} B (\tau^2 \omega^2 (\alpha_i)^3 + 6(\alpha_i + \gamma_i)) - 1$
	$\xi_4 = b^T W^{-1} B (\tau^2 \omega^2 (\alpha_i)^4 + 12(\alpha_i)^2) - 1$
$\delta_{n,\dot{d}}$	$\zeta_2 = b^T W^{-1} (\tau^2 \omega^2 (\alpha_i)^2/2 + \mathbf{1}) - 1$
	$\zeta_3 = b^T W^{-1} (\tau^2 \omega^2 (\alpha_i)^3/3 + 2(\alpha_i + \gamma_i)) - 1$
	$\zeta_4 = b^T W^{-1} (\tau^2 \omega^2 (\alpha_i)^4/4 + 3(\alpha_i)^2) - 1$
	$\zeta_5 = b^T W^{-1} (\tau^2 \omega^2 (\alpha_i)^5/5 + 4(\alpha_i)^3) - 1$

$$\delta_{n,\dot{d}} = \dot{\phi}(t_n) - \dot{\phi}(t_n + \tau) \tag{7.74}$$
$$+ b^T W^{-1} \Big(-\mathbf{1} \tau \omega^2 \phi(t_n) - \tau^2 \omega^2 B \mathbf{1} \dot{\phi}(t_n) + r_{n,\dot{d}} \Big).$$

The s-dimensional vector $r_{n,\dot{d}}$ comprises all velocity entries of r_n and reads

$$r_{n,\dot{d}} = \Big(\tau \omega^2 \big(\phi(t_n + \alpha_i \tau) + \tau \gamma_i \dot{\phi}(t_n) \big) + \tau \ddot{\phi}(t_n + \alpha_i \tau) + \tau^2 \gamma_i \phi^{(3)}(t_n) \Big)_{i=1}^s.$$

Note that the position entries of r_n vanish identically. Taylor expansion of all smooth motion terms with arguments $t_n + \tau$ and $t_n + \alpha_i \tau$ in (7.73) and (7.74) results finally for a consistent method with $\sum b_j = 1$ in the error formulas

$$\delta_{n,d} = \sum_{i \geq 2} \frac{\tau^i}{i!} \xi_i \phi^{(i)}(t_n), \qquad \delta_{n,\dot{d}} = \sum_{i \geq 2} \frac{\tau^{i-1}}{(i-1)!} \zeta_i \phi^{(i)}(t_n) \tag{7.75}$$

with the first coefficients ξ_i and ζ_i given in Table 7.1. All these error coefficients depend on the method coefficients, in particular on $W = I_s + \tau^2 \omega^2 B^2$. The representation (7.75) holds also in the stiff case $\tau \omega \gg 1$ since the stepsize τ is assumed to be small enough to track the smooth motion ϕ.

The Scholz Conditions

Scholz derived in [Sch89] additional conditions on Rosenbrock–Wanner methods such that the order reduction observed at the scalar Prothero–Robinson equation can be avoided. These conditions play a role in the sequel when we study the RODAS method [HW96] and the extension RODASP [RS97] at the model problem (7.59).

The RODAS method consists of $s = 6$ stages and is L-stable. Moreover, the coefficients satisfy the additional order condition

$$b^T B^{-1} (\alpha_i^2)_{i=1}^s = 1$$

for singularly perturbed systems and systems of index 1. If $\tau \omega$ is small, RODAS has classical order 4, and the coefficients ξ_i and ζ_i in the error formulas (7.75) are of

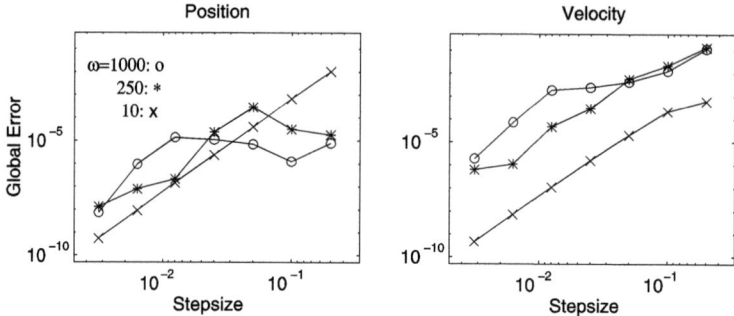

Fig. 7.6 Order reduction for RODAS

higher order in τ. If $\tau\omega \gg 1$, errors of previous steps are strongly damped, and the local error becomes again dominant in the global error (7.64). Also, the order drops now considerably. E.g., the coefficient ξ_2 tends to $\mathcal{O}(\epsilon^2\tau^{-2})$ since

$$\boldsymbol{b}^T \boldsymbol{W}^{-1} \boldsymbol{B} \left(\tau^2\omega^2(\alpha_i^2)_{i=1}^s + 2\mathbf{1}\right) - \mathbf{1} \xrightarrow{\tau\omega \to \infty} \underbrace{\boldsymbol{b}^T \boldsymbol{B}^{-1}(\alpha_i^2)_{i=1}^s - \mathbf{1}}_{=0} + \underbrace{\boldsymbol{b}^T \boldsymbol{B}^{-1}\mathbf{1}}_{=1} \frac{2\epsilon^2}{\tau^2}.$$

The other coefficients show a similar behavior, and in total this means

$$\delta_{n,d} = \mathcal{O}(\epsilon^2) + \mathcal{O}(\epsilon^2\tau), \qquad \delta_{n,\dot{d}} = \mathcal{O}(\epsilon^2) + \mathcal{O}(\tau) + \mathcal{O}(\epsilon^2\tau).$$

Figure 7.6 illustrates the global error behavior of RODAS with the function ϕ and other data taken from the Radau IIa order diagram of Fig. 7.4. For large values of $\tau\omega$, the method returns an asymptotically correct result despite the order reduction. Also, the position error decreases with increasing ω but the velocity error grows. Since the numerical damping of RODAS is less strong than that of Radau IIa for $s = 3$, cf. Fig. 7.8 below, the errors δ_{n-1} and δ_{n-2} of previous steps still have some influence here on the global error.

Based on the Scholz conditions, an alternative coefficient set is constructed in [RS97] that aims at the treatment of parabolic systems by the method of lines. This variant called RODASP has also $s = 6$ stages and classical order 4, and we have

$$\xi_2 \equiv \xi_3 \equiv 0 \quad \text{and} \quad \zeta_2 \equiv \zeta_3 \equiv 0 \quad \text{for all } \omega > 0.$$

An analysis of the coefficients ξ_4 and ζ_4 shows that the local error of RODASP satisfies for $\tau/\epsilon \gg 1$

$$\delta_{n,d} = \mathcal{O}(\epsilon^2\tau^2), \qquad \delta_{n,\dot{d}} = \mathcal{O}(\epsilon^2\tau) + \mathcal{O}(\tau^3).$$

Thus, we still have order 2 in the position variable, and additionally the local error in both position and velocity is small due to the ϵ^2 factor. The numerical experiment depicted in Fig. 7.7 confirms this result.

7.3 Order Reduction for Stiff Mechanical Systems

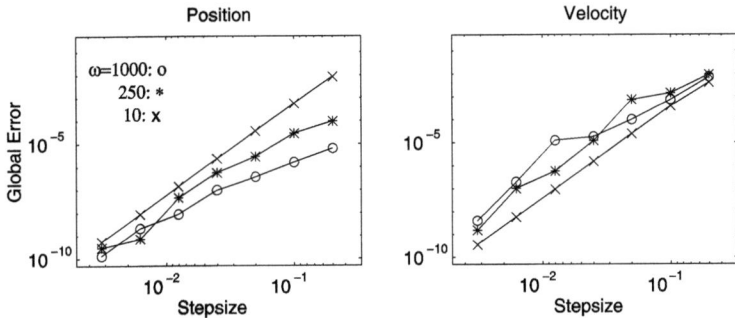

Fig. 7.7 Order reduction for RODASP

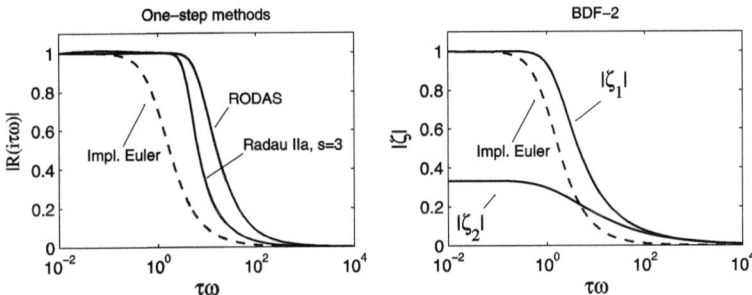

Fig. 7.8 Spectral radii, i.e., the decay of the stability functions on the imaginary axis. *On the right*, the roots of the stability polynomial of the BDF-2 method are plotted for comparison

Spectral Radius

We have seen that numerical dissipation is an important property when tracking the smooth motion of a stiff mechanical system. Even for initial values $y_0 = \phi(t_0)$ on the smooth motion, any numerical method inevitably introduces local errors δ_i and consequently, the approximations y_i lie only in some neighborhood of ϕ. This results in small artificial high frequency oscillations. Methods with $|R(\infty)| < 1$, in particular L-stable methods, have a potential to damp these perturbations if the stepsize is in the right range. From this point of view, it makes also no difference if the initial values lie exactly on the smooth motion or only in some neighborhood since methods with numerical damping converge rapidly towards the smooth solution.

Figure 7.8 displays on the left the decay of the stability functions along the imaginary axis for implicit Euler, Radau IIa ($s = 3$) and RODAS. In structural dynamics, such a characteristic curve of a method is called the *spectral radius* [Hug87].

There is a range of small stepsizes where $|R(i\tau\omega)|$ is close to one and information is almost preserved, and another range where $|R(i\tau\omega)|$ tends to zero and strong damping prevails. What we need in practical simulation is a stepsize control mechanism which distinguishes between the two cases damping of perturbations and tracking of excitations with physical significance.

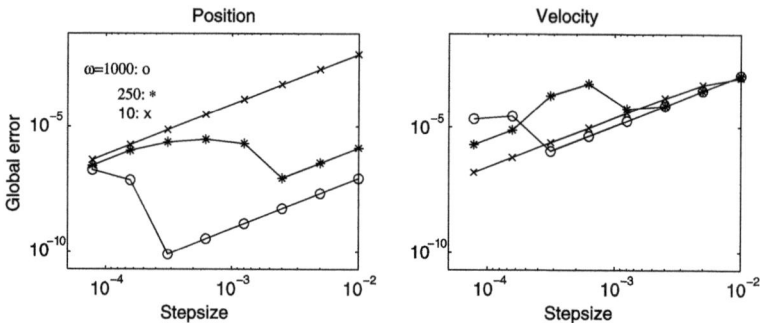

Fig. 7.9 Convergence of BDF-2 for the test problem (7.59)

7.3.5 BDF Methods

For order $k > 2$, the BDF methods are not A-stable. In particular, a certain segment of the imaginary axis is not part of their stability domain in this case. Consequently, only the first order method BDF-1, which is implicit Euler, and the second order method BDF-2 are appropriate for solving stiff mechanical systems.

What is the global convergence behavior of BDF-2 for the test problem (7.59)? A numerical study, based on the data used above for Radau IIa and RODAS, yields the diagrams in Fig. 7.9. With respect to the position variable, the global error decreases for increasing stiffness, and the second order is preserved. However, there is a transition area between the stiff and the non-stiff case $\tau < \epsilon$ where the error curve exhibits a hump. With respect to the velocity, we also observe second order even in the stiff case and a somewhat irregular behavior in the transition area.

In order to analyze the numerical dissipation of the BDF-2 method, we compute the roots ζ_1 and ζ_2 of

$$\frac{3}{2}\zeta^2 - 2\zeta + \frac{1}{2} - i\tau\omega\zeta^2,$$

which is the polynomial (7.23) evaluated along the imaginary axis $\mu = i\tau\omega$. Figure 7.10 displays the resulting root curves depending on the stepsize τ in the complex plane. At $\tau\omega = 0$, both curves start on the real line with $\zeta_1(0) = 1$ and $\zeta_2(0) = 1/3$. In the limit $\tau\omega \to \infty$, they reach the origin. A plot of the corresponding absolute values leads to the spectral radius of the method. In Fig. 7.8 on the right, the decay of these absolute values for increasing stepsize is shown and compared with the spectral radius of the implicit Euler method. The numerical dissipation of BDF-2 is less strong, and though it possesses the equivalent of L-stability in the limit $\tau\omega \to \infty$, the decay to zero is significantly slower than for the one-step methods Radau IIa ($s = 3$) and RODAS.

We remark finally that variable order variable stepsize BDF codes such as DASSL apply a sophisticated order control that detects stability problems and reduces the order to $k = 2$ for stiff mechanical systems. Nevertheless, if one knows

Fig. 7.10 BDF-2, root curves of stability polynomial depending on $\tau\omega$

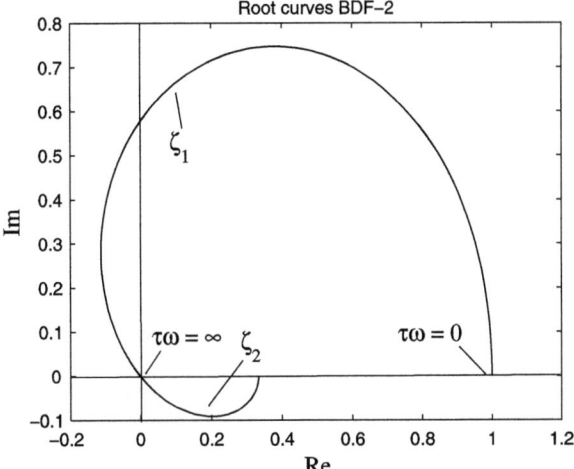

about the presence of high frequencies in the multibody system, it is advisable to reduce the order a priori.

7.3.6 A Nonlinear Test Equation

So far, the discussion concentrated on the linear second order test equation (7.58) and shed new light on the behavior of time integration methods at stiff mechanical systems. Nonlinear stiffness terms, however, involve an additional important effect that we study next. We modify (7.58) by squaring the difference between the state variable d and the smooth motion ϕ,

$$\ddot{d} = -\omega^2(d-\phi)^2 + \ddot{\phi}, \qquad d(t_0) = \phi(t_0), \ \dot{d}(t_0) = \dot{\phi}(t_0). \qquad (7.76)$$

The exact solution of (7.76) is again the smooth motion $d(t) = \phi(t)$.

One step of implicit Euler yields, after eliminating the discretized velocities, the nonlinear equation

$$0 = d_1 + \tau^2\omega^2\Big(d_1 - \phi(t_1)\Big)^2 - d_0 - \tau\dot{d}_0 - \tau^2\ddot{\phi}(t_1). \qquad (7.77)$$

The simplified Newton iteration for solving (7.77) converges if the fixed-point mapping $d_1 = \Phi(d_1)$ is contractive with

$$\Phi(d) := d - \Big(1 + 2\tau^2\omega^2(d^{(0)} - \phi(t_1))\Big)^{-1}$$
$$\cdot \Big(d + \tau^2\omega^2(d-\phi(t_1))^2 - d_0 - \tau\dot{d}_0 - \tau^2\ddot{\phi}(t_1)\Big).$$

We have
$$\Phi'(d) = \frac{2\tau^2\omega^2(d^{(0)} - d)}{1 + 2\tau^2\omega^2(d^{(0)} - \phi(t_1))}.$$

Now assume a starting point $d^{(0)}$ with $d^{(0)} - d = \mathcal{O}(\tau)$, $d^{(0)} - \phi(t_1) = \mathcal{O}(\tau)$. Convergence is then only guaranteed if the stepsize satisfies

$$\tau = \mathcal{O}(\epsilon^{2/3}). \tag{7.78}$$

An analogous restriction holds also in the general nonlinear case (6.7) if the stiff potential U is not a quadratic form [Lub93].

The condition (7.78) represents a severe issue for the time integration and holds for all implicit methods. If the starting values are closer, $d^{(0)} - \phi(t_1) = \mathcal{O}(\epsilon^2)$, the situation improves but it is not clear how to obtain such values or how to modify the iteration such that the convergence is sped up. In practice, stiff nonlinear springs and geometric stiffening terms may lead to nonlinear terms that provoke divergence of the simplified Newton method due to (7.78). As a consequence, the time integration scheme will then reduce the stepsize and resolve the latent high frequency oscillations.

7.3.7 Remarks

Summarizing we have seen that, though at first sight clearly superior, implicit methods suffer at stiff mechanical systems from several drawbacks, which are mainly due to the singular perturbation character of the problem. In variable stepsize codes, the order reduction in the stiff velocity components leads to wrong error estimation and thus to small stepsizes τ. As a remedy, a scaling procedure is suggested in [SS02] that detects the stiff variables and modifies the error estimator such that asymptotically correct data is obtained, cf. also Sect. 8.1.

Another simple—but sometimes questionable—strategy is to specify loose tolerances for the stiff velocities. In this way, these components are filtered out in the error estimator, and the code may proceed with large stepsizes in order to track the smooth motion. In situations where the source of stiffness is easily identified, it sometimes also makes sense to apply different integration schemes in the fashion of multi-rate and co-simulation methods such that the stiff component is treated separately. The bottom line, however, is that such problems are still hard to solve and prone for numerical difficulties.

7.4 Special Integration Methods

In this last section, we finally discuss two second order methods that have their merits for specific applications where higher order is not that important.

7.4 Special Integration Methods

7.4.1 Generalized-α Method

The generalized-α method [CH93] represents the method of choice for time-dependent solid mechanics applications with deformable bodies discretized by the finite element method. Since this field of *structural dynamics* and the field of multibody dynamics are more and more growing together, extensions of the α-method for constrained mechanical systems have recently found much interest.

We begin with the α-method, also called *HHT-method* due to Hilber, Hughes, and Taylor [HHT77], in the unconstrained case with equations of motion

$$\dot{p} = v, \qquad M\dot{v} = f(p, v, t).$$

The mass matrix M has been assumed constant here, which reflects the typical structure of the equations of structural dynamics. The α-method then reads

$$\frac{p_{n+1} - p_n}{\tau} = v_n + \tau(\tfrac{1}{2} - \beta)w_n + \tau\beta w_{n+1}, \tag{7.79a}$$

$$\frac{v_{n+1} - v_n}{\tau} = (1 - \gamma)w_n + \gamma w_{n+1}, \tag{7.79b}$$

$$Mw_{n+1} = \alpha f_n + (1 - \alpha)f_{n+1}, \tag{7.79c}$$

with $f_n := f(p_n, v_n, t_n)$ and discrete accelerations w_n. The method reduces to Newmark's method [New59] for the parameter value $\alpha = 0$. If $\gamma = 1/2 + \alpha$, the α-method is of second order. Moreover, *unconditional stability*, which is equivalent to A-stability, is achieved for $\beta = (1 + \alpha)^2/4$ and $0 \leq \alpha \leq 1/3$. In this way, the parameter α can be used to control the numerical dissipation.

An extension to the α-method named the generalized-α method is introduced in [CH93]. In this method, the relation (7.79c) for the acceleration is replaced by the convex combination

$$\alpha_m M w_n + (1 - \alpha_m) M w_{n+1} = \alpha_f f_n + (1 - \alpha_f) f_{n+1}, \tag{7.80}$$

with parameters α_m and α_f. The coefficients should satisfy the relations

$$\gamma = 1/2 - \alpha_m + \alpha_f \quad \text{second order}, \quad -1 \leq \alpha_m \leq \frac{1}{2} \quad \text{zero-stability}. \tag{7.81}$$

Thus, the remaining parameters α_m, α_f and β can be used to adapt the method to special requirements.

Of particular interest is the behavior of the generalized-α method (7.79a), (7.79b), (7.80) for stiff mechanical systems. As discussed above, numerical dissipation is a desirable property whenever the high frequencies need not be resolved. An attractive feature of the α-method in this context is controllable numerical dissipation, which is mostly expressed in terms of the spectral radius ρ_∞ at infinity. More specifically, we have $\rho_\infty \in [0, 1]$ where $\rho_\infty = 0$ represents asymptotic annihilation of the high frequency response, i.e., the equivalent of L-stability. On the other hand,

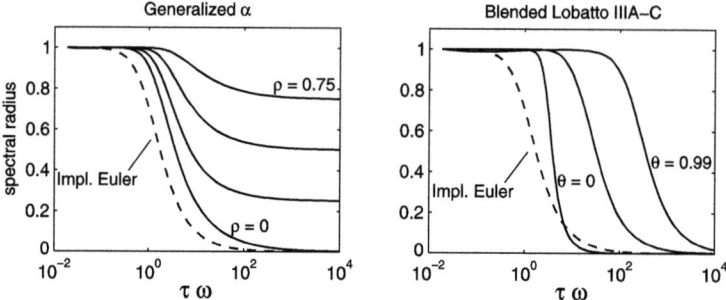

Fig. 7.11 Spectral radii of the generalized α-method for different values of ρ_∞ (*left*) and of the class of blended Lobatto IIIa-c methods, $s = 2$ (*right*)

$\rho_\infty = 1$ stands for the case of no algorithmic dissipation. Unconditional stability is achieved for the parameters

$$\alpha_f = \frac{\rho_\infty}{1+\rho_\infty}, \qquad \alpha_m = \frac{2\rho_\infty - 1}{1+\rho_\infty}, \qquad \beta = \tfrac{1}{4}(1 - \alpha_m + \alpha_f)^2. \qquad (7.82)$$

Notice that $\gamma = 1/2 - \alpha_m + \alpha_f$ is determined by the order condition (7.81).

Figure 7.11 displays on the left the spectral radius of the generalized-α method for the choice (7.82) and different values of ρ_∞. The spectral radius is here defined as

$$s(\tau\omega) := \max_{i=1,2,3} |\sigma_i(\tau\omega)|$$

where the σ_i are the eigenvalues of the 3×3 stability matrix that is obtained when applying the method to the test equation $\ddot{d} = -\omega^2 d$. One observes nicely that the spectral radius indeed converges to ρ_∞ for $\tau\omega \to \infty$.

Extensions to Constrained Mechanical Systems

The first extensions of the generalized-α scheme to constrained mechanical systems were proposed in [CG89, YPR98]. The more recent α-RATTLE method [LS07], which is equivalent to the method of [JN07], is based on the GGL formulation (7.4) and enforces both the constraints at position and velocity level simultaneously. Assuming again a constant mass matrix and using the notation $g_{n+1} := g(p_{n+1})$, $G_{n+1} := G(p_{n+1})$, we write one step of this method as

$$M\frac{p_{n+1} - p_n}{\tau} = M\left(v_n + \tau(\tfrac{1}{2} - \beta)w_n + \tau\beta w_{n+1}\right) - \frac{\tau}{2}G_{n+1}^T \lambda_{n+1},$$

$$M\frac{v_{n+1} - v_n}{\tau} = M((1-\gamma)w_n + \gamma w_{n+1}) - \tfrac{1}{2}G_n^T \lambda_{n+1} - \tfrac{1}{2}G_{n+1}^T \mu_{n+1},$$

$$(1 - \alpha_m)Mw_{n+1} = \alpha_f f_n + (1 - \alpha_f)f_{n+1} - \alpha_m M w_n, \qquad (7.83)$$

7.4 Special Integration Methods

$$0 = G_{n+1} v_{n+1},$$
$$0 = g_{n+1}.$$

As above, the method coefficients should satisfy the conditions (7.81) for second order and zero stability while (7.82) specifies the numerical dissipation. Note that the position and velocity variables in this scheme are of second order but the Lagrange multipliers λ_n and μ_n, which both are approximations of $\lambda(t_n)$, lose one power of τ and are only first order accurate. The name α-RATTLE scheme is inspired by the special case $\alpha_m = \alpha_f = \beta = 0$ and $\gamma = 1/2$ where (7.83) reduces to the RATTLE method of molecular dynamics [HLW02]. Though the definition (7.83) has the form of a one-step method, the convergence analysis turns out to be rather subtle due to the additional acceleration variables. We do not go further into the details here but remark that the special case $\alpha_m = 0$, i.e., the extension of the HHT-method, is easier to analyze in the framework of general linear methods once the local errors have been determined [Hai11]. A convergence proof in the most general case, including state-dependent mass matrices, is found in [JN07].

As alternative to the α-RATTLE method, the approach in [AB07] discretizes the equations of motion (7.1a)–(7.1c) directly in the second order formulation with position constraints only. In brief, the latter algorithm uses discrete values $p_{n+1}, \dot{p}_{n+1}, \ddot{p}_{n+1}, \lambda_{n+1}$ that satisfy the dynamic equations (7.1a)–(7.1c) and auxiliary variables for the accelerations

$$(1 - \alpha_m) w_{n+1} + \alpha_m w_n = (1 - \alpha_f) \ddot{p}_{n+1} + \alpha_f \ddot{p}_n. \tag{7.84}$$

These are then integrated via

$$p_{n+1} = p_n + h \dot{p}_n + h^2 (\tfrac{1}{2} - \beta) w_n + h^2 \beta w_{n+1}, \tag{7.85a}$$
$$\dot{p}_{n+1} = \dot{p}_n + h(1 - \gamma) w_n + h \gamma w_{n+1}. \tag{7.85b}$$

Again, the free coefficients $\alpha_f, \alpha_m, \beta, \gamma$ determine the method.

A remarkable feature of (7.85a), (7.85b) is that second order convergence can be shown for position and velocity variables as well as the Lagrange multipliers. This means that no order reduction occurs, and the main reason for this exceptional behavior is that the algebraic variable is only used to evaluate the acceleration from (7.1a)–(7.1c) but is otherwise not discretized. It thus inherits the order from the differential variables. The size of the nonlinear system to be solved in each time step is somewhat smaller when compared to (7.83), and the iteration matrix has a similar structure as in the BDF case (7.55). In this context, we point out that all variants of the generalized-α method for constrained mechanical systems require careful scaling of the iteration matrix to avoid ill-conditioning, and the approach proposed by [BBC07] can be carried over.

Implicit Runge–Kutta Methods with Adjustable Dissipation

While the idea of adjustable numerical dissipation is widespread in structural dynamics, it has only found little interest in the numerical analysis of ODEs and DAEs.

Table 7.2 Coefficients of Lobatto IIIa, IIIc, and blended method, $s = 2$

0	0	0	0	1/2	−1/2	0	$(1-\theta)/2$	$-(1-\theta)/2$
1	1/2	1/2	1	1/2	1/2	1	1/2	1/2
IIIa	1/2	1/2	IIIc	1/2	1/2		1/2	1/2

A closer look at the class of implicit Runge–Kutta methods reveals that it is indeed very easy to construct such methods, and the class of *SPARK-methods* (Super Partitioned Additive Runge–Kutta) [Jay99] seems particular attractive in this context. We sketch here the method derivation and refer to [Sch04] for further details.

Starting with Dahlquist's equation $\dot{y} = \lambda y$, we split the right hand side into

$$\dot{y} = \theta \lambda y + (1-\theta)\lambda y$$

with parameter $0 \leq \theta \leq 1$. When discretizing the equation, we make use of two different schemes with the same weight vector b but specific coefficient matrices A_1 and A_2. The stability function of the method reads then

$$R(z,\theta) = 1 + zb^T \left(I_s - z(\theta A_1 + (1-\theta)A_2)\right)^{-1} \mathbf{1}.$$

Obviously, the method given by the coefficient matrix $\theta A_1 + (1-\theta)A_2$ requires further analysis with respect to order and stability. In case of the collocation methods of Lobatto type with variants IIIa, IIIb and IIIc [HW96, §IV.5], one can show that these methods can be combined in an almost arbitrary fashion while still maintaining the classical order $2s - 2$ for s stages [Jay99]. Moreover, the combinations or *blended methods* [SS03]

$$\theta A_{IIIa} + (1-\theta)A_{IIIc}, \qquad \theta A_{IIIb} + (1-\theta)A_{IIIc} \tag{7.86}$$

inherit the L-stability from the IIIc-method for $\theta < 1$. Table 7.2 contains as an example the combination IIIa–IIIc for $s = 2$, and Fig. 7.11 presents on the right a plot of the resulting spectral radii for different values of θ.

7.4.2 A Variant of the Implicit Midpoint Rule

In applications such as aerospace engineering, the conservation of energy and momentum is sometimes of importance. Using L-stable methods becomes then prohibitive, and instead A-stable schemes with specific properties are required. Starting with the work of [ST92], such approaches have originally found widespread interest in the field of nonlinear elastodynamics where elastic solids are discretized by the finite element method and then undergo large deformation. Modifications of the midpoint rule are the crucial ingredient for such *energy-momentum schemes* [HLW02, BU07].

7.4 Special Integration Methods

We start with the unconstrained case where the implicit midpoint rule reads

$$\frac{p_{n+1} - p_n}{\tau} = v_{n+1/2}, \tag{7.87a}$$

$$M(p_{n+1/2})\frac{v_{n+1} - v_n}{\tau} = f(p_{n+1/2}, v_{n+1/2}, t_{n+1/2}). \tag{7.87b}$$

Here, the midpoint is defined by $p_{n+1/2} = (p_n + p_{n+1})/2$ and $t_{n+1/2} = (t_n + t_{n+1})/2 = t_n + \tau/2$.

The implicit midpoint rule is a collocation method of Gauss type with $s = 1$ stage. It is A-stable, i.e., unconditionally stable, and symmetric in time, which is favorable if conservation of physical invariants is desired. We skip over the details for modifying the right hand side evaluation of (7.87a), (7.87b) in such a way that, e.g., momentum and angular momentum are preserved. See, e.g., [ST92, BU07] for a discussion of these aspects.

Instead, we focus on extending (7.87a), (7.87b) to the constrained case. We learnt above that stiffly accurate methods are well-suited for solving DAEs, but the Gauss methods do not possess this property. In particular, a straightforward discretization of the constraint $0 = g(p)$ at the midpoint yields $0 = g(p_{n+1/2})$, which implies that the constraint will not be satisfied at the new solution point p_{n+1}. A better discretization, here applied to the GGL formulation (7.4), is given by

$$\frac{p_{n+1} - p_n}{\tau} = v_{n+1/2} - G^T_{n+1/2}\mu_{n+1},$$

$$M_{n+1/2}\frac{v_{n+1} - v_n}{\tau} = f(p_{n+1/2}, v_{n+1/2}, t_{n+1/2}) - G^T_{n+1/2}\lambda_{n+1}, \tag{7.88}$$

$$0 = G(q_{n+1})v_{n+1},$$

$$0 = g(q_{n+1}).$$

While the constraint Jacobian in the right hand side of the dynamics equation is evaluated at the midpoint $G_{n+1/2} := G(p_{n+1/2})$, the constraints are directly enforced at the endpoint. With this modification of the standard midpoint rule, the method (7.88) becomes symmetric and can be interpreted as a certain symmetric projection method as introduced by [Hai00]. Notably, (7.88) has also been suggested as time integration scheme in dynamic contact problems [Sol00].

The method (7.88) is of second order in position and velocity and of first order in the Lagrange multipliers. Due to its structure, it is reversible in time and thus the basis for constructing methods with specific conservation properties, in analogy to the unconstrained case. The computational effort is comparable to that of a BDF method applied to the GGL formulation of the equations of motion.

Application in Dynamic Contact

Finally, we sketch how the midpoint rule is applied to friction-free dynamic contact problems, cf. [Sol00] and Sect. 3.4.2. Contact problems are characterized by inequality constraints that implement the non-penetration condition (3.79c). Discretized by finite elements, the equations of motion take the form [Wri02]

$$M\ddot{p} = f(p) - G(p)^T \lambda,$$
$$0 \le g(p) \quad \textit{componentwise},$$
(7.89)

with discretized contact pressure $\lambda \ge 0$ and *complementarity conditions*

$$(\lambda_i \cdot g_i)_{i=1,\dots,n_c} = 0 \tag{7.90}$$

for all n_c possible contact nodes. In (7.89), we assume for simplicity a constant mass matrix and an autonomous, dissipation-free force vector.

The unilateral constraints $0 \le g(p)$ do not imply that $0 \le G(p)\dot{p}$. However, we have

$$0 = g_j(p) \quad \Rightarrow \quad 0 = \frac{\partial g_j(p)}{\partial p} \cdot \dot{p}$$

for the set of *active constraints*. Based on this observation, a stabilized formulation can be established that reads [AS00]

$$\dot{p} = v - G(p)^T \mu,$$
$$M\dot{v} = f(p) - G(p)^T \lambda,$$
$$0 \le g(p), \tag{7.91}$$
$$0 = (G_j(p)v)_{j \in \{\text{active constraints}\}},$$
$$0 = (\lambda_i \cdot g_i)_{i=1,\dots,n_c} = (\mu_i \cdot g_i)_{i=1,\dots,n_c}$$

where $(G_j(p)v)$ denotes the product of row j in G and v. A particular feature of this formulation is that the velocities in the active contact nodes are constrained, which means that in these nodes only a motion tangential to the contact surface is allowed. Optionally, one can also scale the first equation in (7.91) by the mass matrix M as in (2.83). The extra multipliers μ satisfy furthermore a complementarity condition of the form (7.90).

The integration scheme (7.88) is now extended to the stabilized contact formulation (7.91), which yields

$$\frac{p_{n+1} - p_n}{h} = v_{n+1/2} - G(p_{n+1/2})^T \mu_{n+1},$$
$$M \frac{v_{n+1} - v_n}{h} = f(p_{n+1/2}) - G(p_{n+1/2})^T \lambda_{n+1},$$
$$0 \le g(p_{n+1}), \tag{7.92}$$
$$0 = (G_j(p_{n+1})v_{n+1})_{j \in \{\text{active constraints}\}},$$
$$0 = (\lambda_{n+1,i} \cdot g_i(p_{n+1}))_{i=1,\dots,n_c},$$
$$0 = (\mu_{n+1,i} \cdot g_i(p_{n+1}))_{i=1,\dots,n_c}.$$

An *active set strategy* [Mur88] must be used in each time step do determine those components where the contact condition is active. As the set of active constraints may change even within Newton's method, the solution of the nonlinear system becomes more involved here and requires special attention. More on such specific algorithmic details is found, e.g., in [PGW13, SP05, Wri02, WK03].

The constraint at the velocity level in (7.92) is non-standard in dynamic contact and has a particular side effect in impact situations. Since the velocity in an active node is constrained to motions tangential to the contact surface, impacts are automatically handled as *plastic impacts* [Löt82] with a corresponding instantaneous dissipation of energy.

Chapter 8
Numerical Case Studies

Flexible multibody dynamics has become a key methodology for various engineering fields, and state-of-the-art simulation software offers corresponding modules with powerful libraries for all kinds of applications. Since it would be beyond the scope of this monograph to cover such specific topics to a large extent, we focus instead in this final chapter on case studies that illustrate some of the numerical issues discussed before.

As examples for challenging engineering tasks that can be modelled by means of a flexible multibody system, we mention, among many others, the rotorcraft dynamics in helicopter design [BBN01], the landing of an aircraft [BCG09], the treatment of dynamic contact [Ebe00] and large deformation problems [BS09] in a multibody framework, and the analysis of ride comfort and handling capabilities in vehicle dynamics [AG01]. Interactive applications in the virtual assembly planning of cables and hoses based on a fast Cosserat-type beam model are introduced in [LLA11], and further examples can be found in the proceedings [AS09, PA94, Sch90].

8.1 Slider Crank I

The slider crank mechanism of Fig. 8.1 has been analyzed by various authors, see, e.g., Jahnke et al. [JPD93]. This highly non-linear but small-scale example is well understood and therefore appropriate for a detailed numerical analysis.

In contrast to [JPD93], we use here the DAE model of Sect. 5.3.1 where the angles α_1, α_2 and the translation r_3 describe the rigid body motion. The elastic connecting rod, on the other hand is modelled by an Euler–Bernoulli beam and discretized by two sinus shape functions for the lateral displacement (variables d_1, d_2) and a quadratic polynomial for the longitudinal displacement (variables d_3, d_4). Non-linear or geometric stiffening terms are for the moment excluded but will be taken into account later on. Constraints arise from the joint between connecting rod and sliding block and from the prescribed crank motion, see below.

Summing up, the model features $n_q = 3$ rigid body motion coordinates $q = (\alpha_1, \alpha_2, r_3)$, $n_d = 4$ elastic variables d, and $n_\lambda = 3$ Lagrange multipliers λ. The tech-

Fig. 8.1 Slider crank mechanism with flexible connecting rod. The crank motion $\alpha_1 = \Omega t$ is prescribed

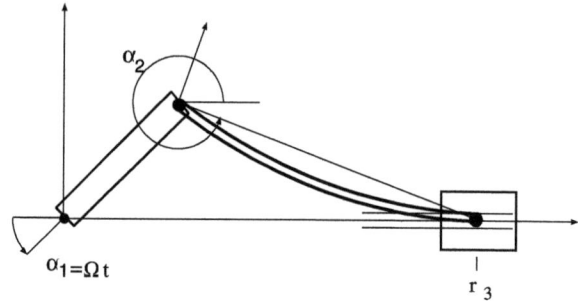

nical data are given by the length $l = 0.3$ m, the cross section area $A = 6.4 \cdot 10^{-5}$ m^2, and the elasticity modulus $E = 2 \cdot 10^{11}$ N/m^2. For these data, the eigenvalues of the elastic body are (in Hz)

$$\omega_1 = 203.2, \quad \omega_2 = 812.8, \quad \omega_3 = 4216.7, \quad \omega_4 = 15171.3.$$

Note that the high frequencies ω_3 and ω_4 belong to the longitudinal displacement of the beam. They make the system stiff.

It is well-known that a separate modal analysis of the beam may not be sufficient for determining its behavior as part of a multibody system. E.g., the eigenvalues may change under rotation. Nevertheless, the above frequencies already indicate where we can expect problems during time integration. Besides this undamped model of the connecting rod, we also consider a version with Rayleigh damping (3.21) and parameters $\zeta_2 = 100$ and $\zeta_1 = (2\hat{\omega} - \zeta_2)/\hat{\omega}^2$ where $\hat{\omega} = 9.5 \cdot 10^4$. In this way, the highest frequency ω_4 is shifted from the imaginary to the negative real line while the lower frequencies are damped less strongly.

Figure 8.2 shows the smooth solution of the mechanism in the undamped case, which is in very good agreement with [JPD93]. The solution was computed by means of the RADAU5 code and tight tolerances ATOL $= 10^{-13}$, RTOL $= 10^{-11}$. The equations of motion were evaluated as system of index 2 with extra acceleration variables (7.5). Moreover, the angular velocity of the crank was prescribed as additional constraint $\alpha_1 = \Omega t$ where $\Omega = 150/(2\pi)$ Hz.

One main difficulty in this context is the computation of consistent initial values which are on or at least close to the smooth motion. In Chap. 6 we saw that the initial values play a crucial role in stiff mechanical systems. If they lie far off the smooth motion, high frequency perturbations will show up instantaneously. Figure 8.3 illustrates that for the very bad choice $d(0) = \dot{d}(0) = 0$ (all deformation variables and velocities set to zero). As the crank velocity is constant, certain inertia forces act already in the beginning of the simulation, but these initial values do not take them into account and the elastic body is not in (quasistatic) equilibrium. Without damping in the model, the displacement variables oscillate around the smooth motion. In this case, any numerical method requires tiny steps in order to track the solution. It should be stressed that these oscillations do not reflect any physical or numerical instability but are merely due to the bad choice of the initial data. They cannot be observed in the physical model.

8.1 Slider Crank I

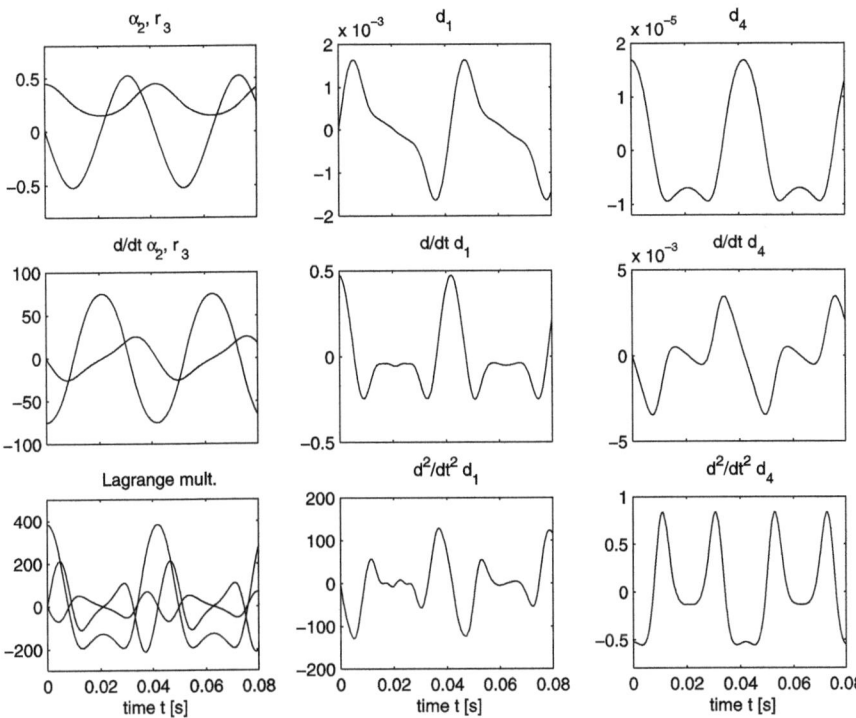

Fig. 8.2 Solution of slider crank. Translations in m, rotations in rad

The selected very strong Rayleigh damping leads to fast decay and is one alternative, Fig. 8.3. However, in multibody dynamics an undamped model is often preferred in order to better detect possible instabilities. Then, reasonable initial values have to be provided in another way. In Fig. 8.2, we used the asymptotic technique for DAEs outlined in Sect. 6.4 to compute initial data very close to the smooth motion.

Let us now turn to the time integration itself. Section 7.3 analyzed the order reduction phenomenon for implicit Runge–Kutta methods like RADAU5. This effect actually shows up when simulating the slider crank mechanism. Moreover, it may slow down the integration since the stepsize control reacts on order reduction with very large error estimates and consequently reduces the stepsize. A closer look at the error estimator reveals that the velocities of the stiff components dominate due to a wrong asymptotic behavior. Thus, in order to apply large steps in a variable stepsize code, the influence of the stiff velocities, here $\dot{d}_4$ and $\dot{d}_3$, must be reduced [Lub93, SS02]. This can be achieved by scaling these components by τ in the error estimator or by specifying loose tolerances for these components. Table 8.1 demonstrates the effect of τ-scaling in case of the slider crank example with stiff velocities $\dot{d}_4$ and $\dot{d}_3$.

In this comparison, a number of different cases was tested—with or without Rayleigh damping and with initial values on the smooth motion or only close by. The gain due to the τ-scaling is significant, in particular for perturbed initial val-

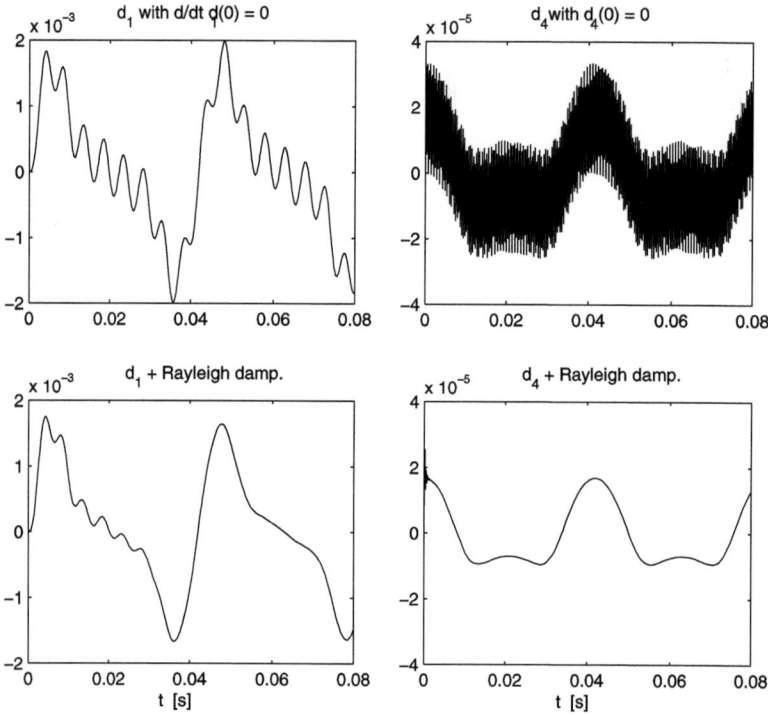

Fig. 8.3 Initial values far off the smooth motion. Solution without (*above*) and with damping in the model (*below*)

Table 8.1 Effect of τ-scaling of $\dot{d}_3$, $\dot{d}_4$ on variable stepsize integration with RADAU5. Tolerance ATOL $= 10^{-8}$, RTOL $= 10^{-3}$. NST: Number STeps, NAC: Accepted, NRJ: ReJected, NFE: Function, NJE: Jacobian Evaluations

Initial Values	τ-scal.	Rayleigh damp.	NST	NAC	NRJ	NFE	NJE
smooth	no	no	562	445	116	3419	267
"	yes	no	140	111	28	972	96
perturbed	no	no	6921	5499	1419	33970	1416
"	yes	no	165	138	23	1149	118
perturbed	no	yes	245	194	48	1697	150
"	yes	yes	144	114	27	1016	98

ues that lead to small oscillations in the beginning. It also turns out that the results obtained by the τ-scaling are as precise as required by the tolerance specification. The diagram in Fig. 8.4 demonstrates this by comparing the results with and without τ-scaling. In particular for loose tolerances, the scaling yields a gain in efficiency and makes it possible to proceed with stepsizes $\tau \gg 1/\omega_4$.

8.1 Slider Crank I

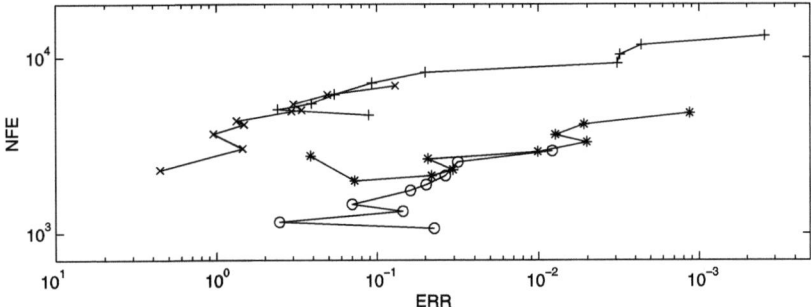

Fig. 8.4 Computational effort in terms of function evaluations (NFE) versus accuracy of the results (ERR: 2-norm of relative error). ∗: index 2, ○: index 2 with τ-scaling, +: GGL, ×: GGL with τ-scaling

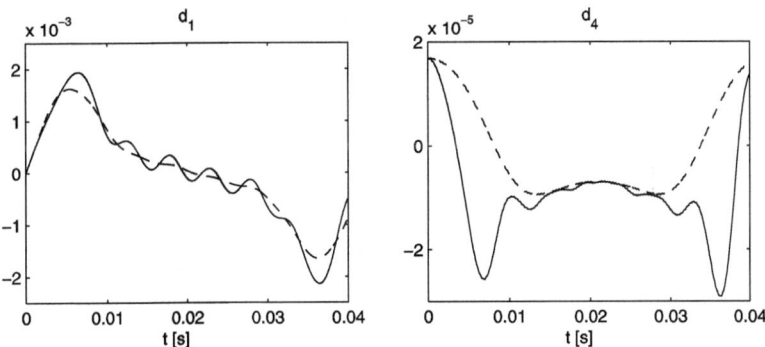

Fig. 8.5 Solution with non-linear (—) and linear beam model (- - -)

Additionally, Fig. 8.4 contains the corresponding results for the GGL formulation (7.4) of the equations of motion. For the slider crank example, this stabilized formulation gives in general much worse results than the formulation of index 2, regardless whether the τ-scaling is applied or not.

Geometric Stiffening We next analyze the numerical solution for the extended beam model (5.63) where an additional non-linear term models coupling effects between lateral and longitudinal displacements. Figure 8.5 shows the lateral displacement d_1 of the midpoint and the longitudinal displacement d_4 of the right pivot for both the linear and the non-linear beam model. One observes a strong difference in the component d_4 and additional oscillations in d_1. The oscillations are again induced by the initial values and, due to the extra coupling term in (5.63), propagate from the bending mode to d_4. Note that at higher angular velocity Ω, the system even turns instable and runs into a singularity as the beam starts to buckle.

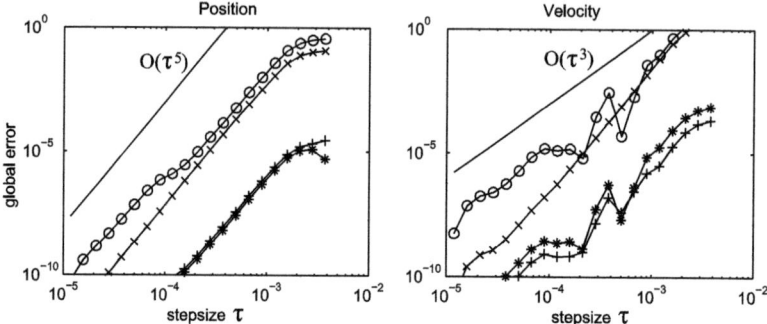

Fig. 8.6 Global error behavior RADAU5 for slider crank example and non-linear beam model, split up into components α_2 (*), r_3 (+), d_1 (x), d_4 (o)

With respect to the performance of the time integrator RADAU5, the non-linearity of the beam model is not strong enough to provoke any additional convergence problems of Newton's method as discussed in Sect. 7.3.6. A simulation run of RADAU5 with linear beam model and tolerances ATOL $= 10^{-8}$, RTOL $= 10^{-3}$ yields the statistics data, cf. the first line of Table 8.1,

NST: 562,   NAC: 445,   NRJ: 116,   NJE: 267,

which indicates a single divergence over all time steps, due to the difference of NST-(NAC+NRJ). In the non-linear case, the data are

NST: 504,   NAC: 414,   NRJ: 87,   NJE: 354.

This means that the code now even needs less steps and suffers from three diverged Newton iterations. In both cases, the number of rejected steps is relatively high.

A source for this large number of rejected steps may lie, once again, in the order reduction that is caused by the stiff components d_3 and d_4. A closer look at the global error behavior, split up into components, is provided by Fig. 8.6. The order in the stiff velocity d_4 drops to about 3 while the non-stiff components perform much better. In the position variables, the error in d_4 also dominates, cf. the estimate of Theorem 7.3.

In summary, the slider crank example shows that stiff mechanical systems require much more than just the application of an implicit solver like RADAU5. Initial values must be chosen carefully, and the order reduction in the stiff velocity components may actually slow down the simulation if no care is taken.

8.2 Slider Crank II

In the next case study, we consider again the slider crank mechanism and refine the model of Sect. 8.1. The connecting rod is now treated as a planar elastic body under

8.2 Slider Crank II

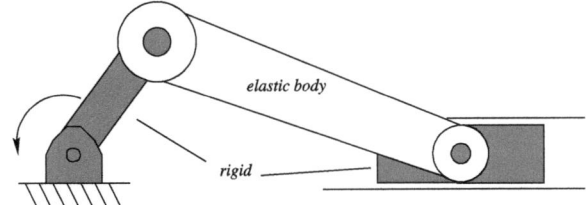

Fig. 8.7 Slider crank with 2D plane stress model of connecting rod

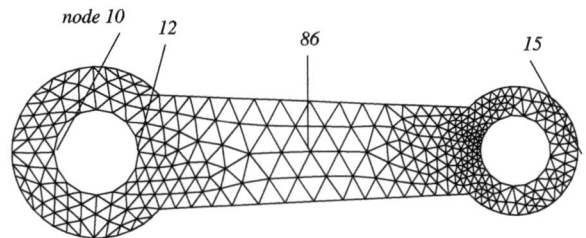

Fig. 8.8 Finite element grid connecting rod, 620 triangles and 376 nodes

the assumption of plane stress, Fig. 8.7. We distinguish two approaches for modeling the revolute joints at both ends of the connecting rod. While the first approach expresses the joints as equality constraints, the second approach makes use of inequality constraints and takes the contact problems in both bearings into account.

For both approaches, a finite element grid with a total of 620 triangles and 376 nodes for the connecting rod is generated, Fig. 8.8. Selecting linear elements (3-node elements) for simplicity as basis functions, we obtain $n_d = 2 \cdot 376 = 752$ elastic displacements d. The $n_q = 3$ gross motion variables q are the same as for the model in Sect. 8.1, and the origin of the floating reference frame is placed again in the center of the left pivot. Both pins in the left and right revolute joint are assumed to be rigid, and the motion should be free of friction. Looking at the left joint in more detail, we define Γ_0 as the inner boundary of the connecting rod and set up the corresponding constraint equations.

Equality Constraints We may postulate that Γ_0 is always in contact with the rigid pin. This leads to the distance constraint

$$(x + u(x,t))^T (x + u(x,t)) = r^2 \quad \text{for all } x \in \Gamma_0$$

where r is the pin's radius. This strong form can be recast in weak form as

$$\int_{\Gamma_0} \vartheta ((x + u(x,t))^T (x + u(x,t)) - r^2) ds = 0 \quad \text{for all } \vartheta$$

with appropriate test functions ϑ. Node-to-surface discretization gives

$$(x_i + u(x_i,t))^T (x_i + u(x_i,t)) = r^2 \quad \text{for } x_i \in \Gamma_0, \ i = 1, \ldots, n_c,$$

with n_c being the number of nodes on the circle Γ_0.

In case of the finite element discretization with nodal basis, we have

$$u(x_i, t) \doteq u_h(x_i, t) = N_u(x_i)d(t) = d_i(t)$$

where $d_i(t) \in \mathbb{R}^2$ is the discrete displacement in node i. The corresponding constraint equations finally read

$$(x_i + d_i(t))^T (x_i + d_i(t)) = r^2 \quad \text{for } x_i \in \Gamma_0, \ i = 1, \ldots, n_c. \tag{8.1}$$

Since we employ a linear elasticity model, we may further simplify these equations by neglecting the quadratic term $d_i^T d$ in (8.1). Due to $x_i^T x_i = r^2$, we then get the linearized constraints

$$x_i^T d_i(t) = 0 \quad \text{for } x_i \in \Gamma_0, \ i = 1, \ldots, n_c. \tag{8.2}$$

The contact with the pin in the right bearing can be treated in the same way. There, however, the gross motion variables need to be taken into account.

Counting all the nodes in both joints and the prescribed crank motion, we obtain $n_\lambda = 39$ constraint equations (8.1) or (8.2), respectively, that define the first approach for the slider crank. Note that (8.2) is a linear constraint and follows the same reasoning as the rigid body elements in (5.35).

Inequality Constraints For the second model, we replace the equality constraints in the inner nodes by inequality constraints

$$(x_i + d_i(t))^T (x_i + d_i(t)) \geq r^2 \quad \text{for } x_i \in \Gamma_0, \ i = 1, \ldots, n_c. \tag{8.3}$$

These constraints express a unilateral contact condition between the pin and Γ_0. Again, a linearization may be applied, leading to

$$x_i^T d_i(t) \geq 0 \quad \text{for } x_i \in \Gamma_0, \ i = 1, \ldots, n_c. \tag{8.4}$$

Depending on the forces acting at the nodes, a gap is allowed to open, and the boundary Γ_0 is here not required to perform a rigid motion.

For the numerical solution of the contact problem, we employ the implicit midpoint rule in the extended form (7.92) with switching between active and inactive nodes and plastic impacts due to the additional constraints for the active velocities. The active set strategy, however, is not iterated within each time step but just updated from step to step. Moreover, the same treatment of the contact problem is combined with the BDF-2 method as basic time integrator to compare both schemes.

8.2 Slider Crank II

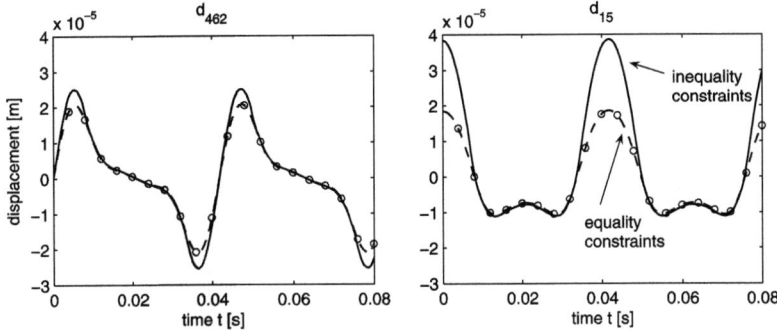

Fig. 8.9 Elastic deformation connecting rod: d_{462} lateral displacement midpoint, d_{15} longitudinal displacement right end, both for approach without and with contact problem in the bearings

Simulation Results As material constants of the connecting rod, standard values $E = 2 \cdot 10^{11}$ N/m^2 for the elasticity modulus and $\nu = 0.3$ for Poisson's number are selected. The numerical solution is then evaluated with respect to the components

d_{462} : lateral displacement in node 86,
d_{15} : longitudinal displacement in node 15,
λ_1 : Lagrange multiplier of constraint in node 10,
λ_9 : Lagrange multiplier of constraint in node 12.

Figure 8.9 displays both d_{462} and d_{15} as obtained from a simulation run with fixed stepsize $\tau = 10^{-4}$. Obviously, we get the same qualitative behavior as for the beam model in the previous section, but the displacements are now in a different order of magnitude. The equality constraints lead to somewhat smaller displacements than the inequality constraints.

The comparison of the implicit midpoint rule and the BDF-2 method reveals tiny oscillations of the former method, Fig. 8.10. Due to the lack of numerical dissipation, these oscillations are not damped out by the midpoint rule but nevertheless, the method is as robust here as the BDF scheme. Whether such oscillations are of physical significance and thus need to be resolved, is in general a delicate issue and depends on the example at hand. E.g., in case of the contact problem, the discontinuities arising from shifting contact points may induce such oscillations, and one might be interested in tracking them. In practice, however, such oscillations die out very soon, which would in turn favor a method with corresponding numerical dissipation.

A look at the Lagrange multipliers reveals stronger differences between the models, Fig. 8.11. Furthermore, we observe the switching between active and inactive nodes in case of the model with inequality constraints.

The differences between both models with respect to the deformed mesh become only visible by scaling the displacements. Figure 8.12 shows a snapshot of the

226 8 Numerical Case Studies

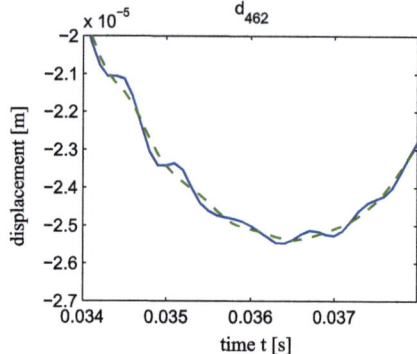

Fig. 8.10 Detail of solution for implicit midpoint (—) and BDF-2 (- - -) methods

deformed mesh where the displacements have been magnified by a factor of 300. While the boundary Γ_0 stays in contact with the left pin in case of the constraints (8.1), a gap opens up for the inequality constraints (8.3), cf. the results in [JPD93].

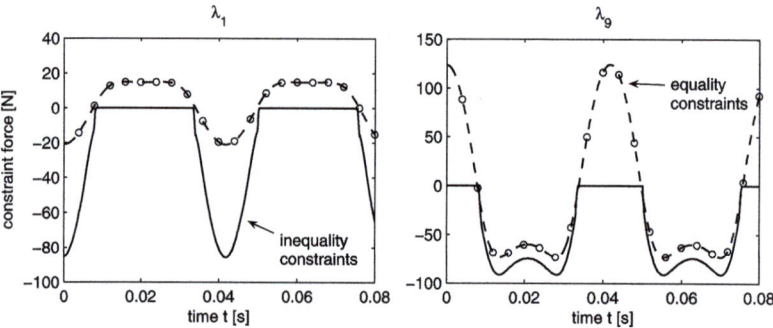

Fig. 8.11 Lagrange multipliers: λ_1 constraint in node 10, λ_9 constraint in node 12, both for approach without and with contact problem in the bearings

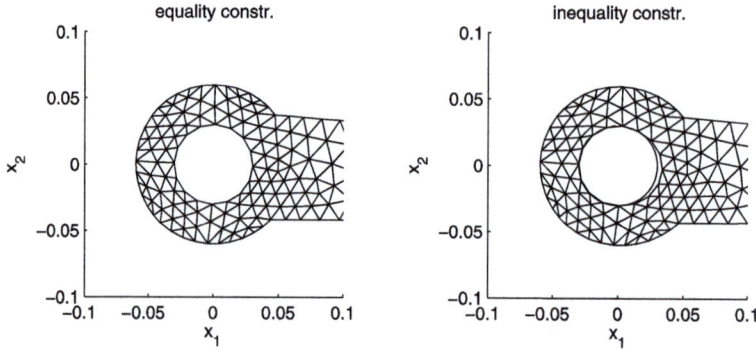

Fig. 8.12 Deformation left joint at $t = 0.08$ [s], displacements scaled by a factor of 300. While the inner boundary remains rigid for the approach with equality constraints (*left*), a gap opens up in case of the contact problem with unilateral constraints (*right*)

8.2 Slider Crank II

Fig. 8.13 Snapshots of slider crank motion with image section of deformed body. The displacements are magnified by a factor of 500, and red color indicates stress peaks

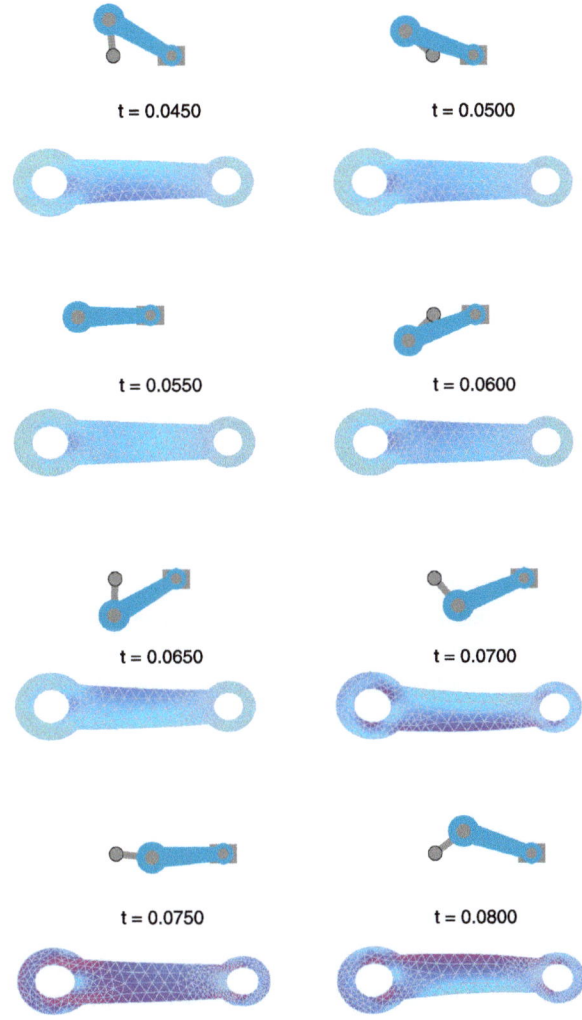

In order to illustrate the dynamics of the system, Fig. 8.13 displays, moreover, eight snapshots at different points of time. The image section below the time counter shows the current state of the flexible connecting rod. For visualization purposes, the displacements are here magnified by a factor of 500, and the von Mises stress [CK06]

$$\sigma_{\text{VM}} = \sqrt{\sigma_{11}^2 + \sigma_{22}^2 - \sigma_{11}\sigma_{22} + 3\sigma_{12}^2}$$

is evaluated, with red color indicating stress peaks (the stress evaluation is also discussed at the end of Sect. 8.5). One observes nicely that whenever the slider moves to its rightmost position, the forces acting on the flexible body reach their maximum.

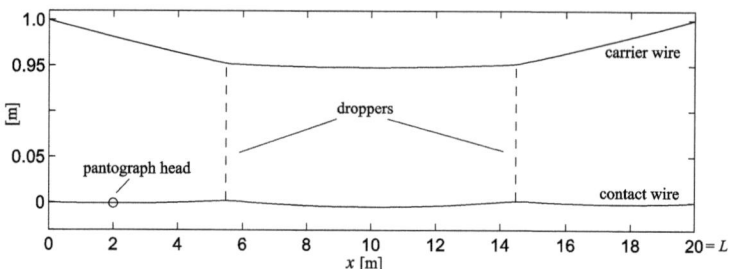

Fig. 8.14 Initial configuration of catenary, range $L = 20$ m

What conclusions can be drawn from this extended model of the slider crank? First, this problem shows the close relation between flexible multibody dynamics and finite element analysis. State-of-the-art finite element codes are able to handle such problems without using the approach of floating reference frames. Instead, each body is modelled under the assumption of large deformation and thus may perform rotations and translations. Secondly, the time integration issues, such as stiffness and the handling of unilateral or bilateral constraints, are common in both worlds, but the rules of the game change as the problem size grows. In large-scale and non-smooth applications such as crashworthiness analysis, implicit methods are often not competitive and explicit low order time integrators that resolve the smallest time scale still prevail [RS03].

8.3 Pantograph and Catenary

We continue with the simplified system of pantograph and catenary as introduced in Sect. 4.5.3. Figure 8.14 displays the chosen setup of the catenary and the initial position of the pantograph head (Fig. 4.8) as specified in [AS00, SA00]. Besides the unilateral contact condition for the interaction of pantograph and catenary, the model features two massless vertical interconnections, the so-called *droppers*. In this way, the carrier wire and the contact wire are coupled by two constraint equations. The droppers may only transmit a tensile force. If compressed, they buckle, which means that the corresponding constraints are also unilateral.

The spatial discretization discussed in Sect. 5.3.3 with linear finite elements for the carrier wire and cubic elements for the contact wire leads to the equations of motion

$$\begin{aligned} M\ddot{p} &= -Ap - D\dot{p} + f - G(t)^T \lambda, \\ 0 &\leq G(t)p \quad \text{(componentwise)}, \\ 0 &= (\lambda_i \cdot G_i(t)p)_{i=1,2,3} \quad \text{(complementarity)}. \end{aligned} \qquad (8.5)$$

8.3 Pantograph and Catenary

Table 8.2 Eigenfrequencies of carrier and contact wire in Hz, 500 elements

	Carrier wire		Contact wire	
	ω_{max}	ω_{min}	ω_{max}	ω_{min}
Frequency	$2.1793 \cdot 10^3$	3.9528	$5.9257 \cdot 10^4$	2.9631

Here, the coordinates are partitioned into $p = (q, d_{ca}, d_{co})$ with

q : vertical displacement pantograph (bodies 1 & 2),
d_{ca} : vertical displacement carrier wire,
d_{co} : vertical displacement & slopes contact wire.

The matrices M, A, D in (8.5) possess a block-diagonal structure that reflects this partitioning. Except for the constraint matrix $G(t)$, which depends explicitly on time t due to the moving contact point, all other matrices are constant. The vector f is also constant and stands for the gravitational force.

In this particular example, we select $N + 1 = 500$ elements both for carrier and contact wire to discretize a range of $L = 20$ m. The corresponding dimensions are $n_q = 2$ for the pantograph, $n_d = 1503$ for all elastic degrees of freedom, and $n_\lambda = 3$ for the unilateral constraints.

Table 8.2 contains the smallest and largest eigenfrequencies of the subsystems for carrier and contact wire in the uncoupled case. The maximal frequency of $5.92 \cdot 10^4$ Hz implies a bound for the stepsize in the time integration, which is also valid for implicit methods if the effect of numerical dissipation is to be avoided. In this respect, the catenary is an interesting example as waves travel back and forth in the system, and the time integration is expected to resolve these physically significant effects in detail. We deal thus not with a stiff mechanical system here.

Similar to the treatment of the slider crank problem in the previous section, the stabilized contact formulation (7.91) is used as basis for the application of time integration methods. More specifically, the work [AS00] employs the BDF-2 method and a half-explicit variant of Störmer's method to solve the equations of motion, in combination with an active set strategy to switch between active and inactive constraints. Moreover, the finite element discretization is cross-checked by means of a finite difference grid for the catenary.

In Fig. 8.15, some components of a simulation run at constant pantograph speed of 48 m/s and stepsize $\tau = 5 \cdot 10^{-5}$ are shown. While the displacements exhibit a relatively smooth behavior, the Lagrange multipliers or constraint forces, respectively, tend to oscillate strongly. Since travelling waves in the catenary are reflected at the right end due to the boundary conditions, an amplification effect can be observed which would not be present in a real technical system. At time $t = 0.2604$ s, the pantograph passes the second dropper. The resulting buckling is reflected by the multiplier λ_2, which vanishes for a short time interval.

We refer to [PEM+97, TSS05] for more elaborate models and refined computational methods for pantograph and catenary dynamics.

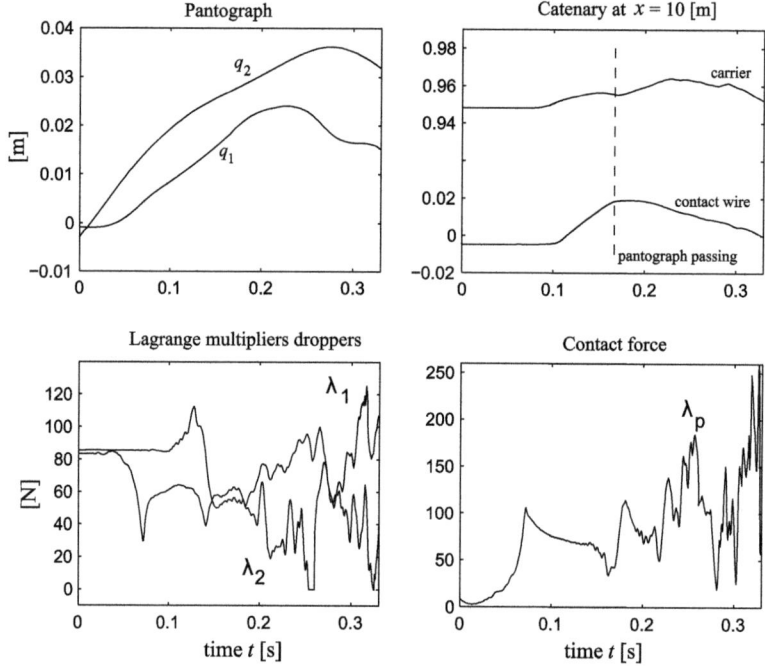

Fig. 8.15 Solution components pantograph and catenary at velocity 48 m/s

8.4 Planar Truck Model

The truck model considered next represents an extension of a benchmark problem proposed in [SGFR94]. As explained in Sect. 5.3.2, the rigid load area of the original model has been replaced by an elastic structure and another tire and a load have been added. Thus, the resulting model consists of one elastic and eight rigid bodies, Fig. 8.16. The gross motion is expressed by $n_q = 16$ coordinates q, and the load and the rotational joint between bodies 3 and 7 imply $n_\lambda = 5$ constraint equations. For the 2D-finite element discretization of the load area, 6-node (quadratic) triangular elements are applied under the assumption of plane stress. In total, this gives $n_d = 644$ nodal variables d for the displacements.

In the following, we study both the full finite element model within the multibody system as well as the reduced models of Sect. 5.3.2. The first reduced model is based on modal condensation of the freely vibrating structure where the rigid body modes have been eliminated a priori by the choice of the body-fixed reference frame and corresponding boundary conditions. Selecting the first five low frequencies and corresponding eigenmodes, we obtain the transformation matrix (5.61) that defines the reduced model.

As more sophisticated alternative, the mixed static and modal condensation or Craig–Bampton approach [CB68] is applied, with the horizontal and vertical displacements in nodes 46 and 51 as master variables and additional five eigenmodes

8.4 Planar Truck Model

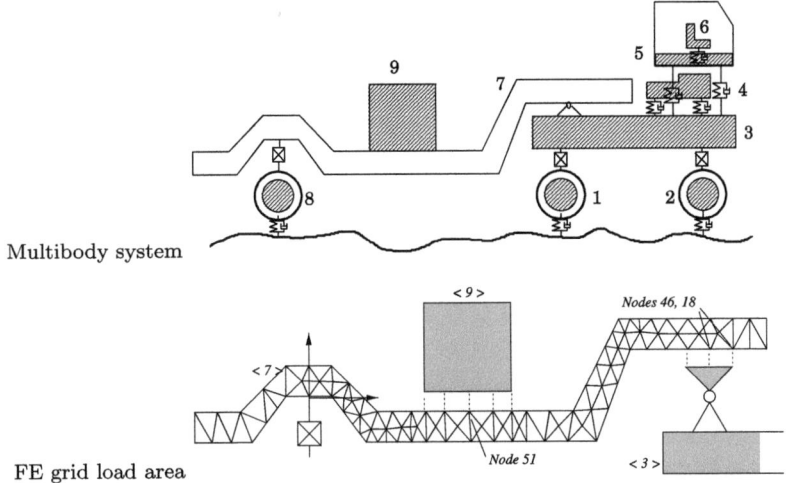

Fig. 8.16 Truck with one elastic (load area 7) and eight rigid bodies

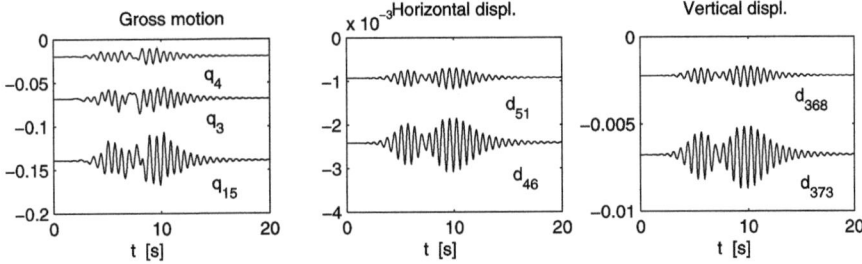

Fig. 8.17 Solution components truck: Translations $q_3, q_{15}, d_{46}, d_{51}, d_{368}, d_{373}$ in m

of the slave subsystem. This defines the transformation matrix (5.62) with a total of nine elastic degrees of freedom. From the results of Tables 5.3 and 5.4 we know already that the mixed condensation technique yields better results for the static equilibrium position while still preserving the low eigenfrequencies. The question now is how the reduced models perform in a dynamic analysis.

Obviously, the reduced models are much less stiff than the full finite element grid, and due to the smaller dimension, they are computationally less expensive. But this is not the point that we are interested in. Before taking a reduced model for simulation purposes, one should verify its approximation properties. For this reason, a comparison run is necessary, and this in turn leads to the challenging task of simulating a DAE which includes the full finite element model.

The simulation result in Fig. 8.17 was again computed by RADAU5, based on the formulation (7.5) of index 2 and the full finite element model. The vehicle travels here with constant speed of 30 m/s over a road profile that has been generated by approximating measurement data via a Fourier series [SGFR94]. By means of a bell-shaped function, the road excitation reaches its maximum at $t = 10$ s. Besides

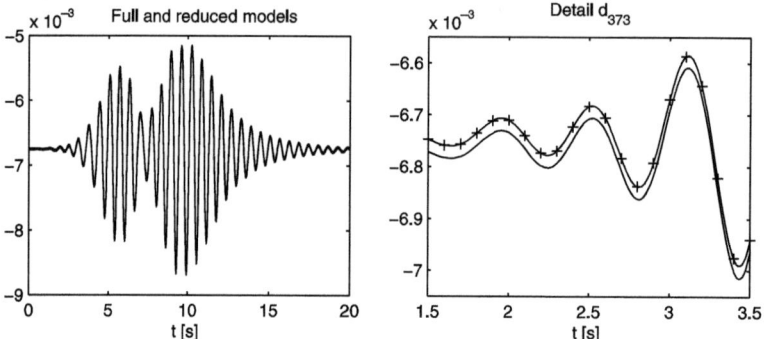

Fig. 8.18 Comparison of reduced models and full finite element structure. The component d_{373} stands for the vertical displacement in node 51. The image section on the right shows that the mixed static and modal approximation coincides with the finite element solution (—) while the reduction to eigenmodes (+) leads to an offset

the rigid body variables q_3 (vertical displacement of chassis), q_4 (rotation of chassis), and q_{15} (vertical displacement load area), Fig. 8.17 displays furthermore the horizontal and vertical displacements in the nodes 46 and 51 of the finite element mesh with respect to the body-fixed reference frame, cf. Fig. 8.16. Initial values for this simulation run are obtained from the equilibrium position for vanishing road excitation.

The road excitation leads to some response in the coordinates and, more important, by comparing both the full and the reduced model, it turns out that the solution curves at first sight almost coincide, Fig. 8.18 on the left.

This is, however, misleading as Fig. 8.18 demonstrates on the right. Actually only the mixed static and modal reduction yields a satisfactory model that comes very close to the full finite element solution. The displacement for the purely modal reduction, in contrast, leads to a certain offset, which is already present in the data for the initial values obtained from the static equilibrium position, cf. Table 5.3.

But what about the simulation times and statistics? Table 8.3 illustrates that, as expected, a reduced model is computationally much less expensive. However, it also shows that the pure system dimension is not all of the game in stiff solvers. A thorough analysis reveals that computing Jacobian matrices by finite differences performs very badly in case of the full finite element grid, while a modified algorithm, which supplies an approximate Jacobian, leads to a more than 25 times faster simulation with equivalent results.

We explain next this approximate Jacobian, which is inspired by [HW96, §VII.7] in the non-stiff case, in more detail. The necessary partial derivatives for Newton's method within RADAU5 are supplied as follows after block Gaussian elimination has been applied to generate the iteration matrix (7.55). Taking the partitioning $p = (q, d)$ into account, only the mass matrix M, the constraint Jacobian G and the

8.5 3D Trailer Frame

Table 8.3 Comparison of different models, integration with RADAU5 for RTOL = 10^{-3}, ATOL adapted to magnitude of components. CPU: Computing times relative to costs of full model

Model	Jacobian	CPU	NST	NAC	NRJ	NFE	NJE
FE, $n_d = 644$	finite diff.	100 %	301	251	50	2160	222
FE, $n_d = 644$	approximate	3.57 %	308	252	56	2245	234
Reduced, $n_d = 9$	finite diff.	0.004 %	445	381	63	3005	323
Reduced, $n_d = 9$	approximate	0.0009 %	469	390	77	3169	344

blocks

$$(M - f + G^T \lambda)_p \doteq \begin{pmatrix} A_r & 0 \\ 0 & A_h \end{pmatrix}, \quad f_v \doteq \begin{pmatrix} D_r & 0 \\ 0 & 0 \end{pmatrix}$$

are provided while all other terms are neglected, Damping and stiffness matrix D_r and A_r of the gross motion are calculated a priori from a linearization procedure, and also the constant stiffness matrix A_h of the elastic body is set up only once. Thus, this approach includes nothing but those terms which are really necessary for stiff solvers, resulting in an enormous speed-up.

Another aspect which is important when solving the large system with full finite element grid concerns the condition number of the iteration matrix. Due to the very large stiffness terms of the FE discretization and comparatively small entries in the constraint Jacobian, the iteration matrix has for large stepsizes—and that is, once again, what we want—a condition number of about 10^{14}. Here we notice an effect of the stiff mechanical system that was predicted in Sect. 7.2.5. However, by scaling the constraints, it is possible to achieve a condition number of 10^6, and based on this remedy and the above approximate Jacobian, the numerical simulation with RADAU5 proves reliable and efficient even for the present full finite element model of the elastic body.

Finally, we mention briefly two additional aspects. In case of a still finer discretization, the linear algebra should additionally be changed to include sparse matrix algorithms because then, the computational effort spent in solving linear systems becomes more and more dominant. Further savings can be achieved by fine-tuning the code RADAU5. In Table 8.3, default integration parameters were specified, resulting in quite frequent Jacobian evaluations. As Jacobians and matrix decompositions are very costly for large systems, the corresponding control parameters (WORK(3), WORK(5), WORK(6)) should be adjusted so that the code tries to use the Jacobian for several consequent steps. At the truck model with full FE grid, this tuning results in additional savings of roughly 50 % computing time.

8.5 3D Trailer Frame

This last example consists of a realistic 3D trailer frame for a commercial road vehicle [DSW08]. The trailer features three axles and is connected to the tractor truck

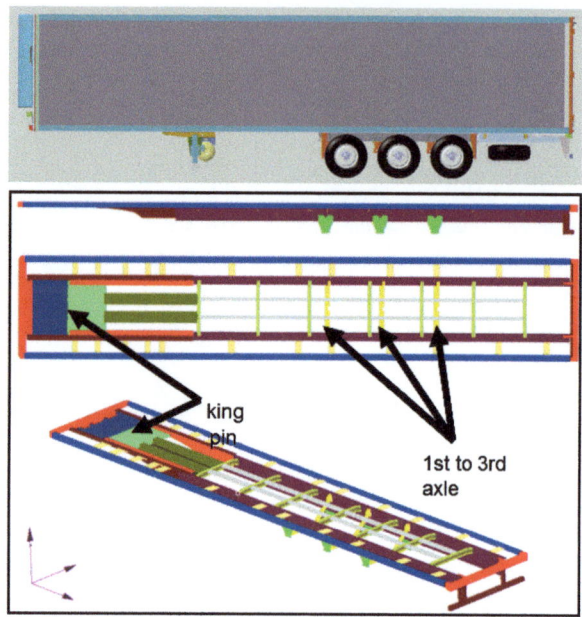

Fig. 8.19 Trailer with platform and three axles (*top*) and details of frame (*bottom*)

at the so-called *king-pin*, Fig. 8.19. We use this complex frame structure to illustrate the huge savings that result from a careful and experienced model reduction. Since engineering judgment is crucial in this context and cannot be replaced by a blackbox procedure, the trailer frame marks at the same time the limitations of this state-of-the-art approach in flexible multibody dynamics.

In Fig. 8.19 at the bottom, the frame is displayed in detail, and it is obvious that this geometry requires a fine finite element mesh to resolve the structure. Due to the thin-walled components of the frame, shell models with quadrilateral and triangular elements can be applied, in combination with additional contact elements and rigid couplings between the components. Overall, this leads to 939,198 elements and 828,980 nodes in the finite element mesh. Most of the elements used feature not only translatory but also rotatory degrees of freedom, which explains the total of 4,167,114 elastic degrees of freedom. The incorporation of such a huge elastic substructure in a multibody dynamics simulation is still beyond the capabilities of today's software.

As standard approach for the model reduction, the Craig–Bampton technique (5.62) is selected. Among the master variables $\boldsymbol{d}_M$ are 5 degrees of freedom at the king pin and six vertical displacements (two pneumatic springs and four dampers) for each axle plus 66 additional nodes where interconnection elements such as bushings between the frame and the platform are attached. These 89 static modes are then augmented by 30 eigenmodes of the slave system, leading to a total of $n_d = 119$ condensated elastic degrees of freedom $\boldsymbol{d}_c$. The first 10 eigenfrequencies of this model for the elastic body are listed in Table 8.4. The largest frequency of the reduced model, on the other hand, is 1.656 Hz, which indicates moderate stiffness. In other words, the model reduction step chops off the higher frequencies and thus alleviates

8.5 3D Trailer Frame

Table 8.4 First 10 eigenfrequencies of reduced frame model (in Hz). (B) indicates a bending mode and (T) a torsional mode

5,8 (T)	11,4 (B)
17,0 (T)	20,0 (B)
22,8 (B)	25,2 (T)
27,1 (B)	32,0 (T)
38,9 (B)	43,4 (B)

the implications of stiffness. A torsional and a bending eigenmode are displayed in Fig. 8.20.

The condensated elastic degrees of freedom d_c and the discretized displacement field u_h are related via (5.54), i.e.,

$$u_h = N_u d \underset{\text{condensation}}{\longmapsto} u_H = N_u T d_c$$

with the transformation matrix T defined by the selection of master nodes and eigenmodes (5.62) as described above. While the nodal variables d in the finite element method are approximations of the displacement in the nodes of the mesh and are thus measured in terms of the physical unit m (or rad in case of rotational degrees of freedom), the variables d_c have no technical interpretation and are dimensionless. Only the product $T d_c$ carries again physical significance. In this context, one calls d_c the *modal participation factors* (MPF) as each component of this vector stands for the contribution of a mode to the total displacement.

Figure 8.21 shows some of the modal participation factors for the reduced trailer model when traveling over rough road conditions. Such plots are of interest for the numerical analyst since they give hints on smoothness and non-linearity of the problem, but they are not really relevant for the practitioners who prefer to look at physical quantities such as the stress in certain selected points of the structure.

Such an evaluation of the stresses is typically performed as a post-processing step after the time integration. Recall from (3.10) that Hooke's law (3.8) can be written as $\underline{\sigma} = C \underline{\varepsilon}$ with the constant 6×6 matrix C containing the material parameters. The strain tensor ε, in turn, is defined by (3.5) where we now replace the displacement

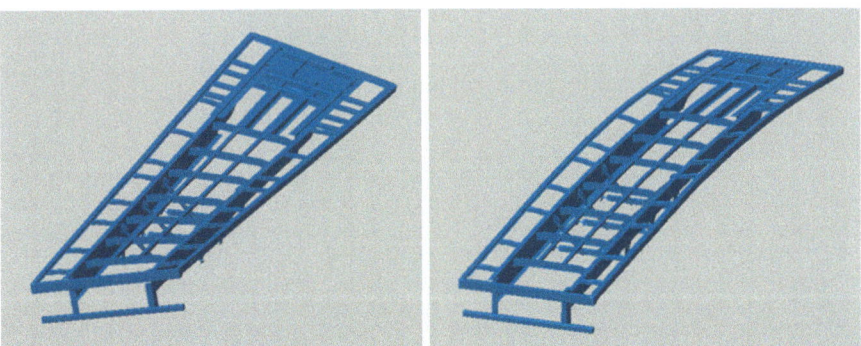

Fig. 8.20 Selected eigenmodes of the structure

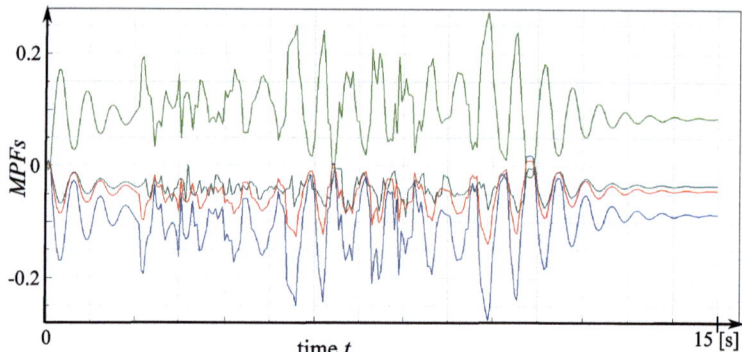

Fig. 8.21 Modal participation factors when traveling over rough road conditions

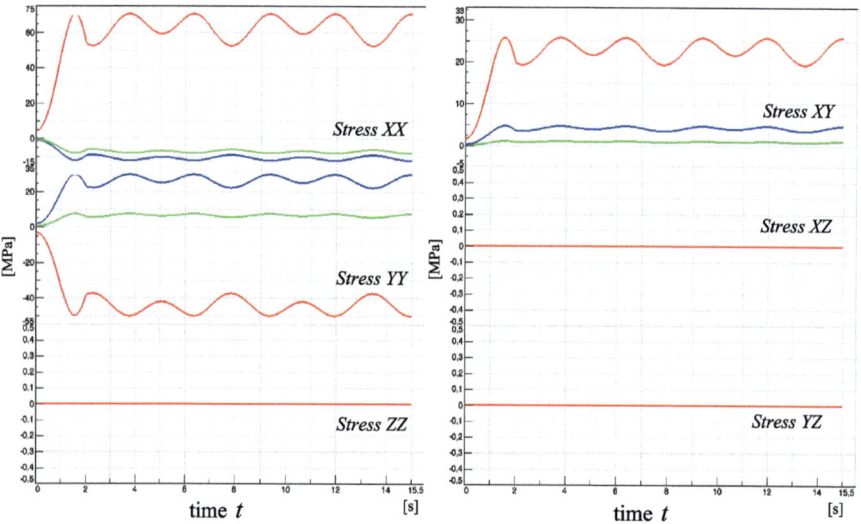

Fig. 8.22 Evaluation of stress in three different points of the frame

field u by its finite element approximation u_h, which leads to

$$\varepsilon(u_h) = \frac{1}{2}(\nabla u_h + \nabla u_h^T). \tag{8.6}$$

The stress evaluation proceeds then as follows. After the modal participation factors d_c have been computed by means of a time integration of the reduced model inside a multibody simulation package, the nodal displacements are determined by $d \doteq T d_c$ and form the discrete displacement field

$$u_h(x, t) \doteq N_u(x) T d_c(t).$$

8.5 3D Trailer Frame

Next, u_h is differentiated according to (8.6) and fed into Hooke's law to calculate the stress values.

Note that the differentiation step contains a subtle detail since the finite element shape functions are usually only continuous across element boundaries, and thus the partial derivatives $\partial/\partial x$ of N_u are well-defined inside each element but exhibit jumps across elements. For this reason, the stress evaluation is performed in the Gauss points of each element, and if the user has requested a different point for the evaluation, interpolation is applied.

State-of-the-art multibody software offers powerful interfaces to finite element packages to perform the above stress evaluation [DHS03], and the results obtained in this way can then be processed further, e.g., in terms of a durability analysis [WLD97]. As example, Fig. 8.22 displays the stress values in three different points of the trailer frame for the above simulation run at rough road conditions. The stresses $\sigma_{33} = \sigma_{13} = \sigma_{23}$ vanish here, which is due to the thin-walled structure, cf. the plane stress model in Sect. 4.4.1.

Though the results for the trailer frame are impressive and demonstrate the power of the available computational tools, we finally point out that the quality of the reduced model rests on engineering judgment and is not supported by an a posteriori estimate that would yield a reliable error bound, cf. [LS08]. This issue remains one of the great challenges ahead in flexible multibody dynamics.

References

[AMR88] R. Abraham, J.E. Marsden, T. Ratiu, *Manifolds, Tensor Analysis, and Applications* (Springer, Berlin, 1988)

[AG01] J. Ambrósio, J. Gonçalves, Complex flexible multibody systems with application to vehicle dynamics. Multibody Syst. Dyn. **6**(2), 163–182 (2001)

[AC03] K.S. Anderson, J.H. Critchley, Improved order-n performance algorithm for the simulation of constrained multi-rigid-body systems. Multibody Syst. Dyn. **9**, 185–212 (2003)

[Ant05] A.C. Antoulas, *Approximation of Large-Scale Dynamical Systems* (SIAM, Philadelphia, 2005)

[Arn81] V.I. Arnold, *Ordinary Differential Equations* (MIT Press, Cambridge, 1981)

[Arn95] M. Arnold, A perturbation analysis for the dynamical simulation of mechanical multibody systems. Appl. Numer. Math. **18**, 37–56 (1995)

[AB07] M. Arnold, O. Brüls, Convergence of the generalized-α scheme for constrained mechanical systems. Multibody Syst. Dyn. **18**, 185–202 (2007)

[AM98] M. Arnold, A. Murua, Non-stiff integrators for differential–algebraic systems of index 2. Numer. Algorithms **19**(1–4), 25–41 (1998)

[AS09] M. Arnold, W. Schiehlen, *Simulation Techniques for Applied Dynamics* (Springer, Berlin, 2009)

[AS00] M. Arnold, B. Simeon, Pantograph and catenary dynamics: a benchmark problem and its numerical solution. Appl. Numer. Math. **34**, 345–362 (2000)

[AL97] U. Ascher, P. Lin, Sequential regularization methods for nonlinear higher index DAEs. SIAM J. Sci. Comput. **18**, 160–181 (1997)

[ACPR95] U. Ascher, H. Chin, L. Petzold, S. Reich, Stabilization of constrained mechanical systems with DAEs and invariant manifolds. J. Mech. Struct. Mach. **23**, 135–158 (1995)

[Bab73] I. Babuška, The finite element method with Lagrange multipliers. Numer. Math. **20**(3), 179–192 (1973)

[Bau10] O. Bauchau, *Flexible Multibody Dynamics* (Springer, Berlin, 2010)

[BBN01] O.A. Bauchau, C.L. Bottasso, Y.G. Nikishkov, Modeling rotorcraft dynamics with finite element multibody procedures. Math. Comput. Model. **33**(10–11), 1113–1137 (2001)

[Bau72] J. Baumgarte, Stabilization of constraints and integrals of motion in dynamical systems. Comput. Methods Appl. Mech. **1**, 1–16 (1972)

[Bet05] P. Betsch, The discrete null space method for the energy consistent integration of constrained mechanical systems. Part I: holonomic constraints. Comput. Methods Appl. Mech. Eng. **194**, 5159–5190 (2005)

[BL06] P. Betsch, S. Leyendecker, The discrete null space method for the energy consistent integration of constrained mechanical systems. Part II: multibody systems. Int. J. Numer. Methods Eng. **67**, 499–552 (2006)

[BS09] P. Betsch, N. Sänger, On the use of geometrically exact shells in a conserving framework for flexible multibody dynamics. Comput. Methods Appl. Mech. Eng. **198**, 1609–1630 (2009)

[BU07] P. Betsch, S. Uhlar, Energy-momentum conserving integration of multibody dynamics. Multibody Syst. Dyn. **17**, 243–289 (2007)

[Bor98] F. Bornemann, *Homogenization in Time of Singularly Perturbed Conservative Mechanical Systems*. Lecture Notes in Mathematics (Springer, Berlin, 1998)

[BBC07] C.L. Bottasso, O.A. Bauchau, A. Cardona, Time-step-size independent conditioning and sensitivity to perturbations in the numerical solution of index three differential-algebraic equations. SIAM J. Sci. Comput. **29**, 397–414 (2007)

[Bra01] D. Braess, *Finite Elements: Theory, Fast Solvers and Applications in Solid Mechanics* (Cambridge University Press, Cambridge, 2001)

[BJO86] H. Brandl, R. Johanni, M. Otter, A very efficient algorithm for the simulation of robots and similar multibody systems without inversion of the mass matrix, in *IFAC/IFIP/IMACS Symposium on Theory of Robots* (1986), pp. 95–100

[Bre08] H. Bremer, *Elastic Multibody Dynamics: A Direct Ritz Approach* (Springer, Berlin, 2008)

[BP92] H. Bremer, F. Pfeiffer, *Elastische Mehrkörpersysteme* (Teubner, Stuttgart, 1992)

[BCP96] K.E. Brenan, S.L. Campbell, L.R. Petzold, *The Numerical Solution of Initial Value Problems in Ordinary Differential-Algebraic Equations* (SIAM, Philadelphia, 1996)

[BS91] S. Brenner, R. Scott, *The Mathematical Theory of Finite Element Methods* (Springer, New York, 1991)

[BF91] F. Brezzi, M. Fortin, *Mixed and Hybrid Finite Element Methods* (Springer, New York, 1991)

[Bri08] A. Brizard, *An Introduction to Lagrangian Mechanics* (World Scientific Publishing, Singapore, 2008)

[BCG09] O. Brüls, A. Cardona, M. Géradin, Modelling, simulation and control of flexible multibody systems, in *Simulation Techniques for Applied Dynamics*, ed. by M. Arnold, W. Schiehlen (2009), pp. 21–74

[But64] J.C. Butcher, Integration processes based on Radau quadrature formulas. Math. Comput. **18**, 233–244 (1964)

[But08] J.C. Butcher, *Numerical Methods for Ordinary Differential Equations* (Wiley, New York, 2008)

[Cam82] S.L. Campbell, *Singular Systems of Differential Equations II*. Research Notes in Mathematics, vol. 61 (Pitman, London, 1982)

[Cam93] S.L. Campbell, Least squares completions for nonlinear differential-algebraic equations. Numer. Math. **65**, 77–94 (1993)

[CG95] S.L. Campbell, C.W. Gear, The index of general nonlinear DAEs. Numer. Math. **72**, 173–196 (1995)

[CG89] A. Cardona, M. Géradin, Time integration of the equations of motion in mechanism analysis. Comput. Struct. **33**(3), 801–820 (1989)

[CH93] J. Chang, G. Hulbert, A time integration algorithm for structural dynamics with improved numerical dissipation. ASME J. Appl. Mech. **60**, 371–375 (1993)

[Cia88] P.G. Ciarlet, *Mathematical Elasticity I* (North Holland, New York, 1988)

[CJM97] F. Collino, P. Joly, F. Millot, Fictitious domain method for unsteady problems: application to electromagnetic scattering. J. Comput. Phys. **138**, 907–938 (1997)

[CHB09] J.A. Cottrell, T.J.R. Hughes, Y. Bazilevs, *Isogeometric Analysis: Toward Integration of CAD and FEA* (Wiley, New York, 2009)

[CH68] R. Courant, D. Hilbert, *Methoden der mathematischen Physik I* (Springer, Heidelberg, 1968)

References

[CB68] R. Craig, M. Bampton, Coupling of substructures for dynamic analysis. AIAA J. **6**, 1313–1319 (1968)

[CK06] R.R. Craig, A.J. Kurdila, *Fundamentals of Structural Dynamics* (Wiley, New York, 2006)

[Dah63] G. Dahlquist, A special stability problem for linear multistep methods. BIT Numer. Math. **3**, 27–43 (1963)

[DK01] W. Dahmen, A. Kunoth, Appending boundary conditions by Lagrange multipliers: analysis of the LBB condition. Numer. Math. **88**, 9–42 (2001)

[DHS03] S. Dietz, G. Hippmann, G. Schupp, Interaction of vehicles and flexible tracks by co-simulation of multibody vehicle systems and finite element track models. Veh. Syst. Dyn. **37**, 372–384 (2003)

[Dor89] M. Dorr, On the discretization of interdomain coupling in elliptic boundary value problems, in *Domain Decomposition Methods*, ed. by T. Chan (SIAM, Philadelphia, 1989)

[DSW08] K. Dreßler, M. Speckert, Th. Weyh, *Mehrkörpersimulation in der Betriebsfestigkeit*. Lecture Notes MDF Seminar Series 2008 (Fraunhofer ITWM, Kaiserslautern, 2008)

[Dup73] T. Dupont, L_2-estimates for Galerkin methods for second order hyperbolic equations. SIAM J. Numer. Anal. **10**, 880–890 (1973)

[DL75] B. Duvaut, J.L. Lions, *Inequalities in Mechanics and Physics* (Springer, New York, 1975)

[Ebe00] P. Eberhard, *Kontaktuntersuchungen durch hybride Mehrkörpersystem/Finite Elemente Simulationen* (Shaker, Aachen, 2000)

[ESF98] E. Eich-Soellner, C. Führer, *Numerical Methods in Multibody Dynamics* (Teubner, Stuttgart, 1998)

[EFS94] A. Eichberger, C. Führer, R. Schwertassek, The benefits of parallel multibody simulation. Int. J. Numer. Methods Eng. **37**(8), 1557–1572 (1994)

[FE11] J. Fehr, P. Eberhard, Simulation process of flexible multibody systems with non-modal model order reduction techniques. Multibody Syst. Dyn. **25**(3), 313–334 (2011)

[Füh88] C. Führer, Differential-algebraische Gleichungssysteme in mechanischen Mehrkörpersystemen. PhD thesis, Mathematisches Institut, Technische Universität München, 1988

[FL91] C. Führer, B. Leimkuhler, Numerical solution of differential-algebraic equations for constrained mechanical motion. Numer. Math. **59**, 55–69 (1991)

[Gan59] F.R. Gantmacher, *Matrizenrechnung*, vol. 2 (VEB Deutscher Verlag der Wissenschaften, Berlin, 1959)

[GB94] J. Garcia de Jalón, E. Bayo, *Kinematic and Dynamic Simulation of Multibody Systems* (Springer, Berlin, 1994)

[GK89] R. Gasch, K. Knothe, *Strukturdynamik* (Springer, Berlin, 1989)

[Gea71] C.W. Gear, *Numerical Initial Value Problems in Ordinary Differential Equations* (Prentice-Hall, Englewood Cliffs, 1971)

[Gea88] C.W. Gear, Differential-algebraic equation index transformation. SIAM J. Sci. Stat. Comput. **9**, 39–47 (1988)

[Gea90] C.W. Gear, Differential-algebraic equations, indices, and integral algebraic equations. SIAM J. Numer. Anal. **27**, 1527–1534 (1990)

[GGL85] C.W. Gear, G.K. Gupta, B.J. Leimkuhler, Automatic integration of the Euler-Lagrange equations with constraints. J. Comput. Appl. Math. **12 & 13**, 77–90 (1985)

[GC00] M. Géradin, A. Cardona, *Flexible Multibody Dynamics* (Wiley, New York, 2000)

[GL96] G.H. Golub, C.F. van Loan, *Matrix Computations* (The John Hopkins University Press, Baltimore, 1996)

[GPHAM10] S. González-Pinto, D. Hernández-Abreu, J.I. Montijano, An efficient family of strongly A-stable Runge–Kutta collocation methods for stiff systems and DAE's.

Part I: stability and order results. J. Comput. Appl. Math. **234**, 1105–1116 (2010)
[GDSV13] A. Goyal, M.R. Dörfel, B. Simeon, A.-V. Vuong, Isogeometric shell discretizations for flexible multibody dynamics. Multibody Syst. Dyn., 1–13 (2013)
[GS69] A. Guillou, J.L. Soulé, La résolution numérique des problemes differentiels aux conditions initiales par des méthodes de collocation. RAIRO. Anal. Numér. **3**, 17–44 (1969)
[Gün01] M. Günther, *Partielle differential-algebraische Systeme in der numerischen Zeitbereichsanalyse elektrischer Schaltungen* (VDI-Verlag, Düsseldorf, 2001)
[GF99] M. Günther, U. Feldmann, CAD based electric circuit modeling in industry I: mathematical structure and index of network equations. Surv. Math. Ind. **8**, 97–129 (1999)
[Guy65] R.J. Guyan, Simultaneous stiffness and nondiagonal mass matrix reduction in structural analysis. AIAA J. **3**(2), 380 (1965)
[Hai00] E. Hairer, Symmetric projection methods for differential equations on manifolds. BIT Numer. Math. **40**, 726–734 (2000)
[Hai11] E. Hairer, Private communication, 2011
[HW96] E. Hairer, G. Wanner, *Solving Ordinary Differential Equations II: Stiff and Differential-Algebraic Problems* (Springer, Berlin, 1996)
[HW99] E. Hairer, G. Wanner, Stiff differential equations solved by Radau methods. J. Comput. Appl. Math. **111**, 93–111 (1999)
[HLR89] E. Hairer, Ch. Lubich, M. Roche, *The Numerical Solution of Differential-Algebraic Equations by Runge–Kutta Methods*. Lecture Notes in Mathematics, vol. 1409 (Springer, Heidelberg, 1989)
[HNW93] E. Hairer, S.P. Nørsett, G. Wanner, *Solving Ordinary Differential Equations I: Non-stiff Problems* (Springer, Berlin, 1993)
[HLW02] E. Hairer, Ch. Lubich, G. Wanner, *Geometric Numerical Integration* (Springer, Berlin, 2002)
[Han90] M. Hanke, On the regularization of index 2 differential-algebraic equations. J. Math. Anal. Appl. **151**(1), 236–253 (1990)
[Hau89] E. Haug, *Computer-Aided Kinematics and Dynamics of Mechanical Systems* (Allyn and Bacon, Boston, 1989)
[Her08] S. Herkt, Model reduction of nonlinear problems in structural mechanics: towards a finite element Tyre model for multibody simulation. PhD thesis, TU Kaiserslautern, FB Mathematik, 2008
[Hig01] I. Higueras, Numerical methods for stiff index 3 DAEs. Math. Comput. Model. Dyn. Syst. **7**(2), 239–262 (2001)
[HHT77] H. Hilber, T. Hughes, R. Taylor, Improved numerical dissipation for time integration algorithms in structural dynamics. Earthq. Eng. Struct. Dyn. **5**, 283–292 (1977)
[HF92] M. Hiller, S. Frik, Road vehicle benchmark 2: five link suspension, in *Application of Multibody Computer Codes to Vehicle System Dynamics, IAVSD Symposium Lyon 1991* (1992), pp. 198–203
[HBG+05] A.C. Hindmarsh, P.N. Brown, K.E. Grant, S.L. Lee, R. Serban, D.E. Shumaker, C.S. Woodward, Sundials: suite of nonlinear and differential/algebraic equation solvers. ACM Trans. Math. Softw. **31**(3), 363–396 (2005)
[Hug87] T.J. Hughes, *The Finite Element Method* (Prentice Hall, Englewood Cliffs, 1987)
[JPD93] M. Jahnke, K. Popp, B. Dirr, Approximate analysis of flexible parts in multibody systems using the finite element method, in *Advanced Multibody System Dynamics* (Kluwer Academic Publishers, Stuttgart, 1993), pp. 237–256
[Jay99] L. Jay, Structure preservation for constrained dynamics with super partitioned additive Runge–Kutta methods. SIAM J. Sci. Comput. **20**, 416–446 (1999)
[JN07] L. Jay, D. Negrut, Extensions of the HHT-method to differential-algebraic equations in mechanics. Electron. Trans. Numer. Anal. **26**, 190–208 (2007)
[KO88] N. Kikuchi, J.T. Oden, *Contact Problems in Elasticity* (SIAM, Philadelphia, 1988)
[KH90] S. Kim, E. Haug, Selection of deformation modes for flexible multibody dynamics. Mech. Struct. Mach. **18**, 565–586 (1990)

References

[Kop89] W. Koppens, The dynamics of systems of deformable bodies. PhD thesis, Technische Universiteit Eindhoven, 1989

[Kro90] L. Kronecker, Algebraische Reduktion der Schaaren bilinearer Formen. Akad. Wiss. Berl. III, 141–155 (1890)

[KM96] P. Kunkel, V. Mehrmann, A new class of discretization methods for the solution of linear differential-algebraic equations with variable coefficients. SIAM J. Numer. Anal. **33**, 1941–1961 (1996)

[KM98] P. Kunkel, V. Mehrmann, Regular solutions of nonlinear differential-algebraic equations and their numerical determination. Numer. Math. **79**, 581–600 (1998)

[KM06] P. Kunkel, V. Mehrmann, *Differential-Algebraic Equations—Analysis and Numerical Solution* (EMS Publishing House, Zurich, 2006)

[Lag88] J.L. Lagrange, *Méchanique analytique* (Libraire chez la Veuve Desaint, Paris, 1788)

[LMT13] R. Lamour, R. März, C. Tischendorf, *Differential-Algebraic Equations: A Projector Based Analysis*. Differential-Algebraic Equations Forum (Springer, Berlin, 2013)

[LLA11] H. Lang, J. Linn, M. Arnold, Multi-body dynamics simulation of geometrically exact Cosserat rods. Multibody Syst. Dyn. **25**(3), 285–312 (2011)

[Löt82] P. Lötstedt, Mechanical systems of rigid bodies subject to unilateral constraints. SIAM J. Appl. Math. **42**, 281–296 (1982)

[LP86] P. Lötstedt, L.R. Petzold, Numerical solution of nonlinear differential equations with algebraic constraints I: convergence results for BDF. Math. Comput. **46**, 491–516 (1986)

[Lub89] Ch. Lubich, h^2 extrapolation methods for differential-algebraic equations of index-2. Impact Comput. Sci. Eng. **1**, 260–268 (1989)

[Lub93] Ch. Lubich, Integration of stiff mechanical systems by Runge–Kutta methods. Z. Angew. Math. Phys. **44**, 1022–1053 (1993)

[LS07] Ch. Lunk, B. Simeon, Solving constrained mechanical systems by the family of Newmark and α-methods. Z. Angew. Math. Mech. **86**, 772–784 (2007)

[LS08] Ch. Lunk, B. Simeon, The reverse method of lines in flexible multibody dynamics, in *Multibody Dynamics: Computational Methods and Applications*, ed. by C.L. Bottasso (Springer, Berlin, 2008), pp. 95–118

[Mag78] K. Magnus, *Dynamics of Multibody Systems* (Springer, Berlin, 1978)

[MH83] J. Marsden, T.J. Hughes, *Mathematical Foundations of Elasticity* (Prentice Hall, Englewood Cliffs, 1983)

[Mär96] R. März, Canonical projectors for linear differential-algebraic equations. Comput. Math. Appl. **31**, 121–135 (1996)

[MT97] R. März, C. Tischendorf, Recent results in solving index-2 differential-algebraic equations in circuit simulation. SIAM J. Sci. Comput. **18**, 139–159 (1997)

[Mur88] K.G. Murty, *Linear Complementarity, Linear and Nonlinear Programming* (Heldermann Verlag, Berlin, 1988)

[New59] N. Newmark, A method of computation for structural dynamics. J. Eng. Mech. Div. **85**, 67–94 (1959)

[O'M74] R.E. O'Malley, *Introduction to Singular Perturbations* (Academic Press, New York, 1974)

[O'M88] R.E. O'Malley, On nonlinear singularly perturbed initial value problems. SIAM Rev. **30**, 193–212 (1988)

[PA94] M. Pereia, J. Ambrosio, *Computer Aided Analysis of Rigid and Flexible Mechanical Systems* (Kluwer Academic Publishers, Dordrecht, 1994)

[Pet86] L.R. Petzold, Order results for implicit Runge–Kutta methods applied to differential/algebraic systems. SIAM J. Numer. Anal. **23**, 837–852 (1986)

[Pit79] J. Pitkäranta, Boundary subspaces for the finite element method with Lagrange multipliers. Numer. Math. **33**, 273–289 (1979)

[PEM+97] G. Poetsch, J. Evans, R. Meisinger, W. Kortüm, W. Baldauf, A. Veitl, J. Wallaschek, Pantograph/catenary dynamics and control. Veh. Syst. Dyn. **28**, 159–195 (1997)

[PGW13] A. Popp, M.W. Gee, W.A. Wall, A primal-dual active set strategy for finite deformation dual mortar contact, in *Recent Advances in Contact Mechanics*, ed. by G.E. Stavroulakis. Lecture Notes in Applied and Computational Mechanics, vol. 56 (Springer, Berlin, 2013), pp. 151–171

[PR91] F. Potra, W.C. Rheinboldt, On the numerical integration for Euler-Lagrange equations via tangent space parametrization. Mech. Struct. Mach. **19**(1), 1–18 (1991)

[PR74] A. Prothero, A. Robinson, On the stability and accuracy of one-step methods for solving stiff systems of ordinary differential equations. Math. Comput. **28**, 145–162 (1974)

[RR00] P. Rabier, W.C. Rheinboldt, *Nonholonomic Motion of Rigid Mechanical Systems from a DAE Point of View* (SIAM, Philadelphia, 2000)

[RR02] P. Rabier, W.C. Rheinboldt, Theoretical and numerical analysis of differential-algebraic equations, in *Handbook of Numerical Analysis*, vol. 8, ed. by P.G. Ciarlet, J.L. Lions (Elsevier, Amsterdam, 2002)

[Rav74] P. Raviart, Hybrid finite element methods for solving 2nd order elliptic equations, in *Topics in Numerical Analysis II*, ed. by J. Miller (Academic Press, New York, 1974)

[Rei00] S. Reich, Smoothed Langevin dynamics of highly oscillatory systems. Physica D **138**(3–4), 210–224 (2000)

[RS08] T. Reis, T. Stykel, Balanced truncation model reduction of second-order systems. Math. Comput. Model. Dyn. Syst. **14**(5), 391–406 (2008)

[RS97] P. Rentrop, G. Steinebach, Model and numerical techniques for the alarm system of river Rhine. Surv. Math. Ind. **6**, 245–265 (1997)

[Rhe84] W.C. Rheinboldt, Differential—algebraic systems as differential equations on manifolds. Math. Comput. **43**(168), 2473–2482 (1984)

[Rhe96] W.C. Rheinboldt, Manpak: A set of algorithms for computations on implicitly defined manifolds. Comput. Math. Appl. **32**, 15–28 (1996)

[RS99] W.C. Rheinboldt, B. Simeon, On computing smooth solutions of DAE's for elastic multibody systems. Comput. Math. Appl. **37**, 69–83 (1999)

[RS10] G. Rill, T. Schaeffer, *Grundlagen und Methodik der Mehrkörpersimulation mit Anwendungsbeispielen* (Vieweg, Wiesbaden, 2010)

[RS88] R.E. Roberson, R. Schwertassek, *Dynamics of Multibody Systems* (Springer, Heidelberg, 1988)

[RS03] W. Rust, K. Schweizerhof, Finite element limit load analysis of thin-walled structures by ANSYS (implicit) LS-DYNA (explicit) and in combination. Thin-Walled Struct. **41**(2), 227–244 (2003)

[Sac96] D. Sachau, Berücksichtigung von flexiblen Körpern und Fügestellen in Mehrkörpersystemen zur Simulation aktiver Raumfahrtstrukturen. PhD thesis, Universität Stuttgart, 1996

[Sch04] M. Schaub, Numerische Integration steifer mechanischer Systeme mit impliziten Runge-Kutta-Verfahren. PhD thesis, TU München, Zentrum Mathematik, 2004

[SS02] M. Schaub, B. Simeon, Automatic h-scaling for the efficient time integration of stiff mechanical systems. Multibody Syst. Dyn. **8**, 329–345 (2002)

[SS03] M. Schaub, B. Simeon, Blended Lobatto methods in multibody dynamics. Z. Angew. Math. Mech. **83**(10), 720–728 (2003)

[Sch90] W. Schiehlen (ed.), *Multibody System Handbook* (Springer, Heidelberg, 1990)

[Sch89] S. Scholz, Order barriers for the B-convergence of ROW methods. Computing **41**, 219–235 (1989)

[ST00] D.E. Schwarz, C. Tischendorf, Structural analysis of electric circuits and consequences for MNA. J. Circuit Theory Appl. **28**, 131–162 (2000)

[SW99] R. Schwertassek, O. Wallrapp, *Dynamik flexibler Mehrkörpersysteme* (Vieweg Verlag, Braunschweig, 1999)

[Sha98] A. Shabana, *Dynamics of Multibody Systems* (Cambridge University Press, Cambridge, 1998)

References

[Sim95] B. Simeon, MBSPACK—numerical integration software for constrained mechanical motion. Surv. Math. Ind. **5**, 169–202 (1995)

[Sim96] B. Simeon, Modelling a flexible slider crank mechanism by a mixed system of DAEs and PDEs. Math. Model. Syst. **2**, 1–18 (1996)

[Sim98a] B. Simeon, Flexible slider crank. Technical report, University of Bari, 1998. pitagora.dm.uniba.it/~testset/

[Sim98b] B. Simeon, Order reduction of stiff solvers at elastic multibody systems. Appl. Numer. Math. **28**, 459–475 (1998)

[Sim00] B. Simeon, *Numerische Simulation gekoppelter Systeme von partiellen und differential-algebraischen Gleichungen in der Mehrkörperdynamik* (VDI-Verlag, Düsseldorf, 2000)

[SA00] B. Simeon, M. Arnold, Coupling DAE's and PDE's for simulating the interaction of pantograph and catenary. Math. Comput. Model. Dyn. Syst. **6**, 129–144 (2000)

[SFR91] B. Simeon, C. Führer, P. Rentrop, Differential-algebraic equations in vehicle system dynamics. Surv. Math. Ind. **1**, 1–37 (1991)

[SGFR94] B. Simeon, F. Grupp, C. Führer, P. Rentrop, A nonlinear truck model and its treatment as a multibody system. J. Comput. Appl. Math. **50**, 523–532 (1994)

[SSP09] B. Simeon, R. Serban, L.R. Petzold, A model of macroscale deformation and microvibration in skeletal muscle tissue. Modél. Math. Anal. Numér. **43**, 805–823 (2009)

[ST92] J.C. Simo, N. Tarnow, The discrete energy-momentum method. Conserving algorithms for nonlinear elastodynamics. Z. Angew. Math. Phys. **43**, 757–792 (1992)

[Sol00] J. Solberg, Finite element methods for frictionless dynamic contact between elastic materials. PhD thesis, UC Berkeley, Dept. of Mechanical Engineering, 2000

[SP05] J.M. Solberg, P. Papadopoulos, An analysis of dual formulations for the finite element solution of two-body contact problems. Comput. Methods Appl. Mech. Eng. **194**, 2734–2780 (2005)

[Ste70] E.M. Stein, *Singular Integrals and Differentiability Properties of Functions* (Princeton University Press, Princeton, 1970)

[SB02] J. Stoer, R. Bulirsch, *Introduction to Numerical Analysis* (Springer, Berlin, 2002)

[SF73] G. Strang, G. Fix, *An Analysis of the Finite Element Method* (Prentice Hall, Englewood Cliffs, 1973)

[SW95] K. Strehmel, R. Weiner, *Numerik gewöhnlicher Differentialgleichungen* (Teubner-Verlag, Stuttgart, 1995)

[Stu08] Th. Stumpp, Asymptotic expansions and attractive invariant manifolds of strongly damped mechanical systems. Z. Angew. Math. Mech. **88**(8), 630–643 (2008)

[TSS05] G. Teichelmann, M. Schaub, B. Simeon, Modelling and simulation of railway cable systems. Z. Angew. Math. Mech. **85**(12), 864–877 (2005)

[TW05] A. Toselli, O. Widlund, *Domain Decomposition Methods—Algorithms and Theory* (Springer, New York, 2005)

[vdHS87] P.J. van der Houwen, B.P. Sommeijer, Explicit Runge–Kutta–Nyström methods with reduced phase errors for computing oscillating solutions. SIAM J. Numer. Anal. **24**, 595–617 (1987)

[Sch99] R. von Schwerin, *Multibody System Simulation* (Springer, Berlin, 1999)

[Wal91] O. Wallrapp, Linearized flexible multibody dynamics including geometric stiffening effects. Mech. Struct. Mach. **19**, 385–409 (1991)

[Wal70] W. Walter, *Differential and Integral Inequalities* (Springer, Berlin, 1970)

[WLD97] P. Weal, C. Liefooghe, K. Dreßler, Product durability engineering: improving the process. J. Sound Vib. **31**(1), 68–79 (1997)

[WAV12] S. Weber, M. Arnold, M. Valášek, Quasistatic approximations for stiff second order differential equations. Appl. Numer. Math. **62**(10), 1579–1590 (2012)

[WH82] R.A. Wehage, E.J. Haug, Generalized coordinate partitioning for dimension reduction in analysis of constrained dynamic systems. J. Mech. Des. **134**, 247–255 (1982)

[Wit77] J. Wittenburg, *Dynamics of Systems of Rigid Bodies* (Teubner, Stuttgart, 1977)

[Woh01] B. Wohlmuth, *Discretization Methods and Iterative Solvers Based on Domain Decomposition* (Springer, New York, 2001)

[WK03] B. Wohlmuth, R. Krause, Monotone methods on non-matching grids for non-linear contact problems. SIAM J. Sci. Comput. **25**(1), 324–347 (2003)

[Wri02] P. Wriggers, *Computational Contact Mechanics* (Wiley, Chichester, 2002)

[Wri70] K. Wright, Some relationships between implicit Runge–Kutta collocation and Lanczos methods. BIT Numer. Math. **10**, 217–227 (1970)

[YPR98] J. Yen, L. Petzold, S. Raha, A time integration algorithm for flexible mechanism dynamics: the Dae-alpha method. Comput. Methods Appl. Mech. Eng. **158**, 341–355 (1998)

[ZV98] J. Znamenáček, M. Valášek, An efficient implementation of the recursive approach to flexible multibody dynamics. Multibody Syst. Dyn. **2**(3), 227–251 (1998)

Index

Symbols
α-method, *see* HHT-method
α-RATTLE method, 210

A
A-stable, 181, 188
Active set strategy, 215, 224
Admissible displacement, 57, 90, 92, 97, 127, 139
Algebraic variable, 29, 33, 182, 191, 211
Angular velocity, 24, 91
Asymptotic expansion, 160, 163

B
Baumgarte stabilization, 45
BDF method, 180, 182, 184, 190, 191, 206, 224, 229
Boundary condition
  Dirichlet, 54, 64, 67, 91, 93, 114, 135
  Neumann, 58, 91

C
Calculus of variations, 6, 58, 68
Cardan angles, 25
Centroid, 15, 23, 93
Closed kinematic loop, 19
Coercive, 63
Collocation method, 181, 186, 190, 212
Component-mode synthesis (CMS), 149
Conservative system, 17, 20, 138
Consistent initial value, 32–34, 41, 175
Constraint
  at acceleration level, 40, 41, 46, 174
  at position level, 14, 44, 174, 178
  at velocity level, 40, 41, 174
  force, 119
  hidden, 40, 41, 44, 169
  in linear elasticity, 67, 68
  in weak form, 69, 71, 72, 75, 77, 97, 101, 106, 135
  incompressibility, 84
  inequality, 119, 214, 224, 226
  Jacobian, 14, 16, 40, 43, 44, 146, 213, 233
  joint, 99
  manifold, 35, 48, 178
  matrix, 130, 142
  minimax characterization, 42
  point, 98, 111, 142, 153
  unilateral, 214
Coordinates, 14
  absolute, 21
  Cartesian, 15, 18
  generalized, 94, 145
  minimal, 15, 17, 19
  natural, 21
  relative, 21, 26
Coriolis force, 17
Craig–Bampton method, *see* mixed static and modal condensation

D
DASSL, 185, 191
Defect, 197
Deformation, 54, 88, 92
  gradient, 54
Differential equation on manifold, 35
Differential variable, 29, 33, 182, 190, 191, 211
Differential-algebraic equation (DAE), 3, 9, 17–21, 26–28, 30, 34, 39, 40, 44, 51, 70, 129, 130, 142, 145, 159, 160, 169, 170, 174, 181, 188, 189, 217, 219, 231
  fully implicit, 28, 32, 34
  linear constant coefficient, 30, 33

Differential-algebraic equation (DAE) (*cont.*)
  linear-implicit, 28, 34, 181, 186, 188, 191
  saddle point form, 38
  semi-explicit, 29, 33, 35, 37, 182, 184, 186, 189
  stiff, 169, 173
Direction cosine matrix, 25
Dispersive wave equation, 6
Domain decomposition, 80
Durability analysis, 237
Dynamic contact, 82, 213
Dynamics, 16

E
Eigenfrequency, 7, 148, 218, 234
Eigenmode, 148, 230, 235
Eigenvalue problem, 148
Elliptic, 63, 73
Energy momentum method, 212
Energy norm, 128, 129
Equations of motion
  of flexible multibody system, 106, 145
  of rigid multibody system, 16, 18
Euler angles, 25, 88
Euler parameters, 25, 37, 88
Euler–Bernoulli beam, 110, 112, 150

F
Finite element method (FEM), 125, 128, 131, 223, 228, 230, 234
First integral, 37
Floating reference frame, 87, 96, 99, 139, 143, 151
Force vector, 8, 16, 18, 146

G
Galerkin projection, 126, 129, 139, 143, 147
Gauss method, 187
Generalized-α method, 209
Geometric stiffening, 112, 138, 151, 221
GGL formulation, 46, 174, 221
Global error, 176, 197, 199
Green–Lagrangian strain tensor, *see* strain tensor
Guyan reduction, *see* static condensation

H
Hamiltonian system, 20
Hamilton's principle, 5, 16, 56, 59, 68
HHT-method, 209
Hooke's law, 55, 237

I
IDA, 185, 191
Incompressible material, 84

Index of a DAE, 30, 74, 130, 131, 142, 146, 164, 199, 203
  differential, 35
  differentiation, 32, 41
  nilpotency, 32
  perturbation, 34, 43, 79
  reduction, 44
Inertia tensor, 25
Inf-sup condition, 71

J
Joint modeling, 99

K
Kinematics, 14, 88
Kinetic energy, 5, 17, 20, 24, 89, 105, 109
Korn's inequality, 63
Kronecker canonical form, 30

L
L-stable, 188, 198
Lagrange, 3, 13
  element, 128, 135
  equations of the first kind, 16, 17, 20, 26
  equations of the second kind, 17, 19, 20
  function, 20
  multiplier, 9, 15, 17, 18, 20, 37–41, 43, 44, 46, 50, 129, 130, 132, 133, 135, 145, 157, 158, 160, 174–176, 179, 191, 193, 201, 211, 213, 217, 225, 226, 229
  polynomial, 187
Lamé constant, 56
Leading coefficient of BDF method, 181
Linear oscillator, 7
Lobatto method, 212
Local error, 177, 197, 199, 202
Local parametrization, 36
Local state space form, 35

M
Manifold, *see* constraint manifold
Mass density, 56
Mass matrix, 8, 16, 18, 130, 136, 146, 152, 161, 163, 174
Matrix pencil, 30
Midpoint rule, 212, 224
Mixed static and modal condensation, 149, 155, 230
Modal condensation, 148, 155
Modulus of elasticity, 56
Moment of inertia, 7, 19
Multibody system, 13, 19, 21
  flexible, 87, 104, 139, 143, 145

Index

N
Newton–Euler equations, 23, 95
Node-to-surface, 223
Null space
 matrix, 15, 16, 18, 49
 method, 51
Numerical dissipation, 8, 209

O
Order reduction, 184, 190, 191, 193, 211
Ordinary differential equation (ODE), 3, 17, 159, 170

P
Pantograph and catenary, 117, 156, 228
Partial differential-algebraic equation (PDAE), 80
Pendulum
 elastic, 4, 160, 167
 rigid, 15, 18, 41, 43, 160
Plane strain, 109
Plane stress, 109, 116, 223, 230, 237
Poisson's equation, 24
Poisson's number, 56
Potential energy, 5, 20, 89, 106
Principle of D'Alembert, 21
Principle of Jourdain, 21
Principle of least action, *see* Hamilton's principle
Principle of virtual work, 59, 92

Q
Quaternions, *see* Euler parameters

R
RADAU, 190, 191
Radau IIa method, 188, 190, 191, 194, 201
RADAU5, 190, 191, 218, 231
Rayleigh damping, 59, 108, 218
Rigid body element, 137
RODAS, 203
RODASP, 203
Rodrigues parameters, 25
Rosenbrock–Wanner method (ROW), 202
Runge–Kutta method
 blended implicit, 211
 half-explicit, 179
 implicit, 186, 187, 189, 194, 196, 201

S
Saddle point problem
 matrix, 38, 40, 178
 stationary, 71, 79, 129, 131
 transient, 69, 70, 73, 76, 79, 80, 82, 97, 101, 121, 129, 131
SAFERK method, 190
Segment condition, 65
Singular value decomposition (SVD), 28, 42
Singularly perturbed system, 29, 160, 163, 169
Slider crank, 21, 112, 150, 217, 222
Sobolev space, 59
Solution invariant, 36
SPARK method, 212
Spectral radius, 205, 210
Stabilized index-2 system, *see* GGL formulation
State space form, *see* Lagrange equations of the second kind
Static condensation, 147
Stiff mechanical system, 30, 159, 163, 193, 207, 218
Stiffness matrix, 130
Strain, 54, 237
 energy, 56
 tensor, 55
Stress, 55, 237
 tensor, 55
 von Mises, 227
Strong form, 58, 69, 91
SUNDIALS, 185
Surface traction, 56, 73

T
Tangent bundle, 48
Tangent space, 35
Trace space, 64, 70
Trailer frame, 234
Truck model, 115, 153, 230

U
Unconditional stability, 209
Underlying ODE, 33

V
Vector product, 24
Virtual displacement, 59

W
Weak derivative, 60
Weak form, 58, 69, 96, 97, 104
Wheel suspension, 26

Z
Zero-stable, 181

MIX
Papier aus verantwortungsvollen Quellen
Paper from responsible sources
FSC® C105338

If you have any concerns about our products,
you can contact us on
ProductSafety@springernature.com

In case Publisher is established outside the EU,
the EU authorized representative is:
Springer Nature Customer Service Center GmbH
Europaplatz 3, 69115 Heidelberg, Germany

Printed by Libri Plureos GmbH
in Hamburg, Germany